# Text Mining and Analysis
## Practical Methods, Examples, and Case Studies Using SAS®

Goutam Chakraborty, Murali Pagolu,
Satish Garla

support.sas.com/bookstore

The correct bibliographic citation for this manual is as follows: Chakraborty, Goutam, Murali Pagolu, and Satish Garla. 2013. *Text Mining and Analysis: Practical Methods, Examples, and Case Studies Using SAS®,*. Cary, NC: SAS Institute Inc.

**Text Mining and Analysis: Practical Methods, Examples, and Case Studies Using SAS®**

SAS Institute Inc., SAS Campus Drive, Cary, North Carolina 27513-2414.

November 2013

SAS provides a complete selection of books and electronic products to help customers use SAS® software to its fullest potential. For more information about our offerings, visit **support.sas.com/bookstore** or call 1-800-727-3228.

SAS® and all other SAS Institute Inc. product or service names are registered trademarks or trademarks of SAS Institute Inc. in the USA and other countries. ® indicates USA registration.

Other brand and product names are trademarks of their respective companies.

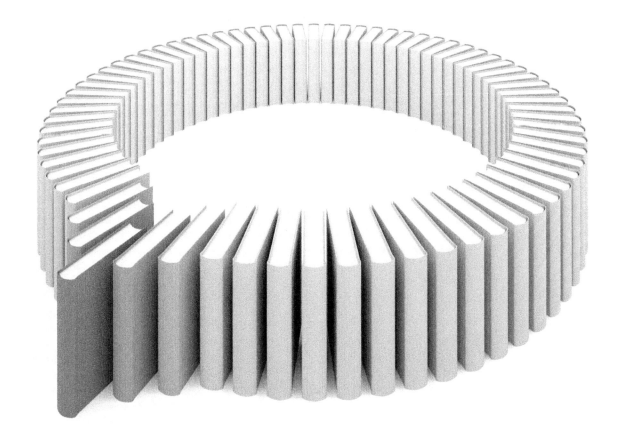

# Gain Greater Insight into Your SAS® Software with SAS Books.

Discover all that you need on your journey to knowledge and empowerment.

support.sas.com/bookstore
*for additional books and resources.*

THE POWER TO KNOW®

# Contents

x

# About This Book

## Purpose

Analytics is the key driver of how organizations make business decisions to gain competitive advantage. While the popular press has been abuzz with Big Data, we believe in "it is the analysis, stupid." Having Big Data means little if that data is not leveraged via analytics to create better value for all stakeholders. One of the primary drivers of Big Data is the advent of social media that has exponentially increased the rate at which textual data is generated on the Internet and the World Wide Web. In addition to data generated via the Internet and the web, organizations have large repositories of textual data collected via forms, reports, customer surveys, voice-of-customers, call-center records and so on. There are numerous organizations that simply collect and store large volumes of unstructured text data, which are yet to be explored to uncover hidden nuggets of useful information that can benefit their business. However, there are not a lot of resources available that can efficiently handle text data for the business analyst community. This book is designed to help industries leverage their textual data and SAS tools to perform comprehensive text analytics.

## Is This Book for You?

Typical readers are business analysts, data analysts, customer intelligence analysts, customer insights analysts, web analysts, social media analysts, students in professional courses related to analytics and SAS programmers. Anyone who wants to retrieve, organize, categorize, analyze, and interpret textual data for generating insights about customer and prospects' behaviors, their sentiments and want to use such insights for making better decisions will find this book useful.

## Prerequisites

While some familiarity with SAS products will be beneficial, this book is intended for anyone who is willing to learn how to apply text analytics using primarily the point-and-click interfaces of SAS Enterprise Miner, SAS Text Miner, SAS Content Categorization Studio, SAS Information Retrieval Studio and SAS Sentiment Analysis studio.

## Software Used to Develop the Book's Content

Below is a list of software used in this book. Be sure to check out the SAS website for updates and changes to the software. The SAS support website contains the latest Online Help documents that have enhancements and changes in new releases of the software.

- SAS® Enterprise Miner (Release 7.1 and Release 12.1)
- SAS® Text Miner (Release 4.1 and 5.1)
- SAS® Crawler, SAS® Search and Indexing (Release 12.1)
- SAS® Enterprise Content Categorization Studio (Release 12.1)
- SAS® Sentiment Analysis Studio (Release 12.1)

**Note:** SAS® Information Retrieval Studio is a graphical user interface (GUI) based framework using SAS® Crawler, SAS® Search and Indexing components that can be configured and maintained.

## Example Code and Data

You can access the example code and data for this book at http://support.sas.com/publishing/authors. From this website select Goutam Chakraborty or Murali Pagolu or Satish Garla. Then look for the cover image of this book, and select "Example Code and Data" to download the SAS programs and data that are included in this book. The data and programs are organized by chapter and case study.

The case studies in this book contain step-by-step instructions for performing a specific type of analysis with the given data. A lot of text mining tasks are subjective and iterative. It is difficult to list each and every task performed in the analysis. The results that you see in your analysis when you follow the exact steps as listed in the case study might differ slightly from the screenshots in the case study. Hence, we also provide you with the SAS® Enterprise Miner projects that the case study authors created in their analysis. These projects can be accessed from the authors' website.

For an alphabetical listing of all books for which example code and data is available, see http://support.sas.com/bookcode. Select a title to display the book's example code.

If you are unable to access the code through the Web site, send e-mail to saspress@sas.com.

## Additional Resources

SAS offers you a rich variety of resources to help build your SAS skills and explore and apply the full power of SAS software. Whether you are in a professional or academic setting, we have learning products that can help you maximize your investment in SAS.

| Bookstore | http://support.sas.com/bookstore/ |
|---|---|
| Training | http://support.sas.com/training/ |
| Certification | http://support.sas.com/certify/ |
| SAS Global Academic Program | http://support.sas.com/learn/ap/ |
| SAS OnDemand | http://support.sas.com/learn/ondemand/ |
| Knowledge Base | http://support.sas.com/resources/ |
| Support | http://support.sas.com/techsup/ |
| Training and Bookstore | http://support.sas.com/learn/ |
| Community | http://support.sas.com/community/ |
| | |

## Keep in Touch

We look forward to hearing from you. We invite questions, comments, and concerns. If you want to contact us about a specific book, please include the book title in your correspondence.

## To Contact the Authors through SAS Press

By e-mail: saspress@sas.com

Via the Web: http://support.sas.com/author_feedback

## SAS Books

For a complete list of books available through SAS, visit http://support.sas.com/bookstore.

Phone: 1-800-727-3228

Fax: 1-919-677-8166

E-mail: sasbook@sas.com

## SAS Book Report

Receive up-to-date information about all new SAS publications via e-mail by subscribing to the SAS Book Report monthly eNewsletter. Visit http://support.sas.com/sbr.

# About The Authors

Dr. Goutam Chakraborty has a B. Tech (Honors) in mechanical engineering from the Indian Institute of Technology, Kharagpur; a PGCGM from the Indian Institute of Management, Calcutta; and an MS in statistics and a PhD in marketing from the University of Iowa. He has held managerial positions with a subsidiary of Union Carbide, USA, and with a subsidiary of British American Tobacco, UK. He is a professor of marketing at Oklahoma State University, where he has taught business analytics, marketing analytics, data mining, advanced data mining, database marketing, new product development, advanced marketing research, web-business strategy, interactive marketing, and product management for more than 20 years.

Murali Pagolu is a Business Analytics Consultant at SAS and has four years of experience using SAS software in both academic research and business applications. His focus areas include database marketing, marketing research, data mining and customer relationship management (CRM) applications, customer segmentation, and text analytics. Murali is responsible for implementing analytical solutions and developing proofs of concept for SAS customers. He has presented innovative applications of text analytics, such as mining text comments from YouTube videos and patent portfolio analysis, at past SAS Analytics conferences. He currently holds six SAS certification credentials.

Satish Garla is an Analytical Consultant in Risk Practice at SAS. He has extensive experience in risk modeling for healthcare, predictive modeling, text analytics, and SAS programming. He has a distinguished academic background in analytics, databases, and business administration. Satish holds a master's degree in Management Information Systems at Oklahoma State University and has completed the SAS and OSU Data Mining Certificate program. He is a SAS Certified Advanced Programmer for SAS 9 and a Certified Predictive Modeler using SAS Enterprise Miner 6.1.

Learn more about these authors by visiting their author pages, where you can download free chapters, access example code and data, read the latest reviews, get updates, and more:

http://support.sas.com/chakraborty

http://support.sas.com/pagolu

http://support.sas.com/garla

# Acknowledgments

The authors would like to extend their gratitude to Radhika Kulkarni and Saratendu Sethi who have been consistently extending their support, encouragement and guidance throughout the development of this book. A special mention of the technical experts James Cox and Terry Woodfield for their invaluable input and suggestions that have greatly helped us in shaping the book. We also would like to thank Lise Cragen for her valuable input and suggestions.

We would like to express our appreciation and thanks to all of the technical reviewers of the book: Barry deVille, Fiona McNeill, Meilan Ji, Penny (Ping) Ye, Praveen Lakkaraju, Vivek Ajmani, Youqin Pan (Salem State University), and Zhongyi Liu for spending their precious time in reviewing the content for the book and providing constructive feedback.

We are also thankful to Arila Barnes, Dan Zaratsian, Gary Gaeth, Jared Peterson, Jiawen Liu, Maheshwar Nareddy, Mantosh Sarkar, Mary Osborne, Saratendu Sethi, and Zubair Shaikh for their valuable contributions to the case studies in this book.

We would also like to express our deepest gratitude to the SAS Publications Production team: Aimee Rodriguez, Amy Wolfe, Brenna Leath, Denise T. Jones, John West and Shelley Sessoms. Without their patience, help, advice and support through the thick and thin, this book would have never seen the light of day.

# Chapter 1 Introduction to Text Analytics

## Overview of Text Analytics

Text analytics helps analysts extract meanings, patterns, and structure hidden in unstructured textual data. The information age has led to the development of a wide variety of tools and infrastructure to capture and store massive amounts of textual data. In a 2009 report, the International Data Corporation (IDC) estimated that approximately 80% percent of the data in an organization is text based. It is not practical for any individual (or group of individuals) to process huge textual data and extract meanings, sentiments, or patterns out of the data. A paper written by Hans Peter Luhn, titled "The Automatic Creation of Literature Abstracts," is perhaps one of the earliest research projects conducted on text analytics. Luhn writes about applying machine methods to automatically generate an abstract for a document. In a traditional sense, the term "text mining" is used for automated machine learning and statistical methods that encompass a bag-of-words approach. This approach is typically used to examine content collections versus assessing individual documents. Over time, the term "text analytics" has evolved to encompass a loosely integrated framework by borrowing techniques from data mining, machine learning, natural language processing (NLP), information retrieval (IR), and knowledge management.

Text analytics applications are popular in the business environment. These applications produce some of the most innovative and deeply insightful results. Text analytics is being implemented in many industries. There are new types of applications every day. In recent years, text analytics has been heavily used for discovering trends in textual data. Using social media data, text analytics has been used for crime prevention and fraud detection. Hospitals are using text analytics to improve patient outcomes and provide better care. Scientists in the pharmaceutical industry are using this technology to mine biomedical literature to discover new drugs.

Text analytics incorporates tools and techniques that are used to derive insights from unstructured data. These techniques can be broadly classified as the following:

- information retrieval
- exploratory analysis
- concept extraction
- summarization
- categorization
- sentiment analysis
- content management
- ontology management

In these techniques, exploratory analysis, summarization, and categorization are in the domain of text mining. Exploratory analysis includes techniques such as topic extraction, cluster analysis, etc. The term "text analytics" is somewhat synonymous with "text mining" (or "text data mining"). Text mining can be best conceptualized as a subset of text analytics that is focused on applying data mining techniques in the domain of textual information using NLP and machine learning. Text mining considers only syntax (the study of structural relationships between words). It does not deal with phonetics, pragmatics, and discourse.

Sentiment analysis can be treated as classification analysis. Therefore, it is considered predictive text mining. At a high level, the application areas of these techniques divide the text analytics market into two areas: search and descriptive and predictive analytics. (See Display 1.1.) Search includes numerous information retrieval techniques, whereas descriptive and predictive analytics include text mining and sentiment analysis.

**Display 1.1: High-Level Classification of Text Analytics Market and Corresponding SAS Tools**

| Text Analytics | | | | |
|---|---|---|---|---|
| Search (Information Organization and Access) | | | Descriptive and Predictive Analysis (Discovering Trends, Patterns, and Modeling) | |
| Information Retrieval | Content Categorization | Ontology Management | Text Mining | Sentiment Analysis |

| SAS Text Analytics Suite | | | | |
|---|---|---|---|---|
| SAS Crawler, SAS Search and Indexing | SAS Enterprise Content Categorization | SAS Ontology Management | SAS Text Miner | SAS Sentiment Analysis Studio |
| Chapter 2 | Chapter 7 | Not covered in this book | Chapters 3, 4, 5, and 6 | Chapter 8 |

SAS has multiple tools to address a variety of text analytics techniques for a range of business applications. Display 1.1 shows the SAS tools that address different areas of text analytics. In a typical situation, you might need to use more than one tool for solving a text analytics problem. However, there is some overlap in the underlying features that some of these tools have to offer. Display 1.2 provides an integrated view of SAS Text Analytics tools. It shows, at a high level, how they are organized in terms of functionality and scope. SAS Crawler can extract content from the web, file systems, or feeds, and then send it as input to SAS Text Miner, SAS Sentiment Analysis Studio, or SAS Content Categorization. These tools are capable of sending content to the indexing server where information is indexed. The query server enables you to enter search queries and retrieve relevant information from the indexed content.

SAS Text Miner, SAS Sentiment Analysis Studio, and SAS Content Categorization form the core of the SAS Text Analytics tools arsenal for analyzing text data. NLP features such as tokenization, parts-of-speech recognition, stemming, noun group detection, and entity extraction are common among these tools. However, each of these tools has unique capabilities that differentiate them individually from the others. In the following section, the functionality and usefulness of these tools are explained in detail.

**Display 1.2: SAS Text Analytics Tools: An Integrated Overview**

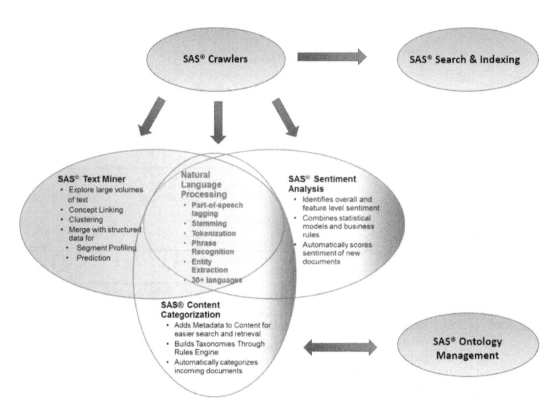

The following paragraphs briefly describe each tool from the SAS Text Analytics suite as presented in Display 1.2:

- **SAS Crawler, SAS Search and Indexing** – Useful for extracting textual content from the web or from documents stored locally in an organized way. For example, you can download news articles from websites and use SAS Text Miner to conduct an exploratory analysis, such as extracting key topics or themes from the news articles. You can build indexes and submit queries on indexed documents through a dedicated query interface.

- **SAS Ontology Management** – Useful for integrating existing document repositories in enterprises and identifying relationships between them. This tool can help subject matter experts in a knowledge domain create ontologies and establish hierarchical relationships of semantic terms to enhance the process of search and retrieval on the document repositories.

  *Note:* SAS Ontology Management is not discussed in this book because we primarily focus on areas where the majority of current business applications are relevant for textual data.

- **SAS Content Categorization** – Useful for classifying a document collection into a structured hierarchy of categories and subcategories called taxonomy. In addition to categorizing documents, it can be used to extract facts from them. For example, news articles can be classified into a predefined set of categories such as politics, sports, business, financial, etc. Factual information such as events, places, names of people, dates, monetary values, etc., can be easily retrieved using this tool.

- **SAS Text Miner** – Useful for extracting the underlying key topics or themes in textual documents. This tool offers the capability to group similar documents—called clusters—based on terms and their frequency of occurrence in the corpus of documents and within each document. It provides a feature called "concept linking" to explore the relationships between terms and their strength of association.

  For example, textual transcripts from a customer call center can be fed into this tool to automatically cluster the transcripts. Each cluster has a higher likelihood of having similar problems reported by customers. The specifics of the problems can be understood by reviewing the descriptive terms explaining each of the clusters. A pictorial representation of these problems and the associated terms, events, or people can be viewed through concept linking, which shows how strongly an event can be related to a problem.

SAS Text Miner enables the user to define custom topics or themes. Documents can be scored based on the presence of the custom topics. In the presence of a target variable, supervised classification or prediction models can be built using SAS Text Miner. The predictions of a prediction model with numerical inputs can be improved using topics, clusters, or rules that can be extracted from textual comments using SAS Text Miner.

- **SAS Sentiment Analysis** – Useful for identifying the sentiment toward an entity in a document or the overall sentiment toward the entire document. An entity can be anything, such as a product, an attribute of a product, brand, person, group, or even an organization. The sentiment evaluated is classified as positive or negative or neutral or unclassified. If there are no terms associated with an entity or the entire document that reflect the sentiment, it is tagged "unclassified."

  Sentiment analysis is generally applied to a class of textual information such as customers' reviews on products, brands, organizations, etc., or to responses to public events such as presidential elections. This type of information is largely available on social media sites such as Facebook, Twitter, YouTube, etc.

## Text Mining Using SAS Text Miner

A typical predictive data mining problem deals with data in numerical form. However, textual data is typically available only in a readable document form. Forms could be e-mails, user comments, corporate reports, news articles, web pages, etc. Text mining attempts to first derive a quantitative representation of documents. Once the text is transformed into a set of numbers that adequately capture the patterns in the textual data, any traditional statistical or forecasting model or data mining algorithm can be used on the numbers for generating insights or for predictive modeling.

A typical text mining project involves the following tasks:

1. **Data Collection:** The first step in any text mining research project is to collect the textual data required for analysis.
2. **Text Parsing and Transformation:** The next step is to extract, clean, and create a dictionary of words from the documents using NLP. This includes identifying sentences, determining parts of speech, and stemming words. This step involves parsing the extracted words to identify entities, removing stop words, and spell-checking. In addition to extracting words from documents, variables associated with the text such as date, author, gender, category, etc., are retrieved.

   The most important task after parsing is text transformation. This step deals with the numerical representation of the text using linear algebra-based methods, such as latent semantic analysis (LSA), latent semantic indexing (LSI), and vector space model. This exercise results in the creation of a term-by-document matrix (a spreadsheet or flat-like numeric representation of textual data as shown in Table 1.1). The dimensions of the matrix are determined by the number of documents and the number of terms in the collection. This step might involve dimension reduction of the term-by-document matrix using singular value decomposition (SVD).

   Consider a collection of three reviews (documents) of a book as provided below: Document 1: I am an avid fan of this sport book. I love this book.

   Document 2: This book is a must for athletes and sportsmen. Document 3: This book tells how to command the sport.

   Parsing this document collection generates the following term-by-document matrix in Table 1.1:

**Table 1.1: Term-By-Document Matrix**

| Term/Document | Document 1 | Document 2 | Document 3 |
|---|---|---|---|
| the | 0 | 0 | 1 |
| I | 2 | 0 | 0 |
| am | 1 | 0 | 0 |
| avid | 1 | 0 | 0 |
| fan | 1 | 0 | 0 |
| this | 2 | 1 | 1 |
| book | 2 | 1 | 1 |
| athletes | 0 | 1 | 0 |
| sportsmen | 0 | 1 | 0 |
| sport | 1 | 0 | 1 |
| command | 0 | 0 | 1 |
| tells | 0 | 0 | 1 |
| for | 0 | 1 | 0 |
| how | 0 | 0 | 1 |
| love | 1 | 0 | 0 |
| an | 1 | 0 | 0 |
| of | 1 | 0 | 0 |
| is | 0 | 1 | 0 |
| a | 0 | 1 | 0 |
| must | 0 | 1 | 0 |
| and | 0 | 1 | 0 |
| to | 0 | 0 | 1 |

3. **Text Filtering:** In a corpus of several thousands of documents, you will likely have many terms that are irrelevant to either differentiating documents from each other or to summarizing the documents. You will have to manually browse through the terms to eliminate irrelevant terms. This is often one of the most time-consuming and subjective tasks in all of the text mining steps. It requires a fair amount of subject matter knowledge (or domain expertise). In addition to term filtering, documents irrelevant to the analysis are searched using keywords. Documents are filtered if they do not contain some of the terms or filtered based on one of the other document variables such as date, category, etc. Term filtering or document filtering alters the term-by-document matrix. As shown in Table 1.1, the term-by-document matrix contains the frequency of the occurrence of the term in the document as the value of each cell. Instead, you could have a log of the frequency or just a 1 or 0 value indicating the presence of the term in a document as the value for each cell. From this frequency matrix, a weighted term-by-document matrix is generated using various term-weighting techniques.

4. **Text Mining:** This step involves applying traditional data mining algorithms such as clustering, classification, association analysis, and link analysis. As shown in Display 1.3, text mining is an iterative process, which involves repeating the analysis using different settings and including or excluding terms for better results. The outcome of this step can be clusters of documents, lists of single-term or multi-term topics, or rules that answer a classification problem. Each of these steps is discussed in detail in Chapter 3 to Chapter 7.

**Display 1.3: Text Mining Process Flow**

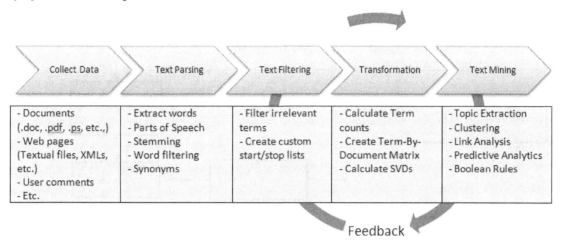

## Information Retrieval

Information retrieval, commonly known as IR, is the study of searching and retrieving a subset of documents from a universe of document collections in response to a search query. The documents are often unstructured in nature and contain vast amounts of textual data. The documents retrieved should be relevant to the information needs of the user who performed the search query. Several applications of the IR process have evolved in the past decade. One of the most ubiquitously known is searching for information on the World Wide Web. There are many search engines such as Google, Bing, and Yahoo facilitating this process using a variety of advanced methods.

Most of the online digital libraries enable its users to search through their catalogs based on IR techniques. Many organizations enhance their websites with search capabilities to find documents, articles, and files of interest using keywords in the search queries. For example, the United States Patent and Trademark Office provides several ways of searching its database of patents and trademarks that it has made available to the public. In general, an IR system's efficiency lies in its ability to match a user's query with the most relevant documents in a corpus. To make the IR process more efficient, documents are required to be organized, indexed, and tagged with metadata based on the original content of the documents. SAS Crawler is capable of pulling information from a wide variety of data sources. Documents are then processed by parsers to create various fields such as title, ID, URL, etc., which form the metadata of the documents. (See Display 1.4.) SAS Search and Indexing enables you to build indexes from these documents. Users can submit search queries on the indexes to retrieve information most relevant to the query terms. The metadata fields generated by the parsers can be used in the indexes to enable various types of functionality for querying.

**Display 1.4: Overview of the IR Process with SAS Search and Indexing**

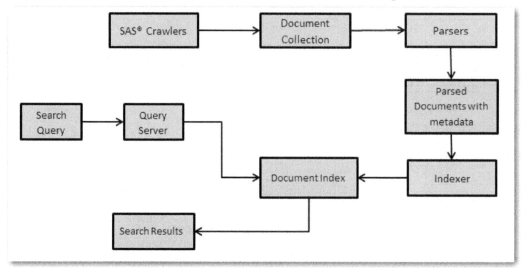

## Document Classification

Document classification is the process of finding commonalities in the documents in a corpus and grouping them into predetermined labels (supervised learning) based on the topical themes exhibited by the documents. Similar to the IR process, document classification (or text categorization) is an important aspect of text analytics and has numerous applications.

Some of the common applications of document classification are e-mail forwarding and spam detection, call center routing, and news articles categorization. It is not necessary that documents be assigned to mutually exclusive categories. Any restrictive approach to do so might prove to be an inefficient way of representing the information. In reality, a document can exhibit multiple themes, and it might not be possible to restrict them to only one category. SAS Text Miner contains the text topic feature, which is capable of handling these situations. It assigns a document to more than one category if needed. (See Display 1.5.) Restricting documents to only one category might be difficult for large documents, which have a greater chance of containing multiple topics or features. Topics or categories can be either automatically generated by SAS Text Miner or predefined manually based on the knowledge of the document content.

In cases where a document should be restricted to only one category, text clustering is usually a better approach instead of extracting text topics. For example, an analyst could gain an understanding of a collection of classified ads when the clustering algorithm reveals the collection actually consists of categories such as Car Sales, Real Estate, and Employment Opportunities.

**Display 1.5: Text Categorization Involving Multiple Categories per Document**

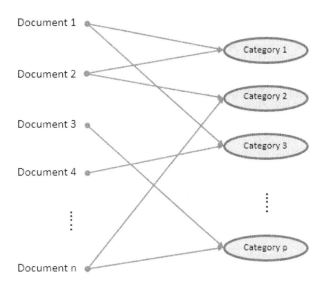

SAS Content Categorization helps automatically categorize multilingual content available in huge volumes that is acquired or generated or that exists in an information repository. It has the capability to parse, analyze, and extract content such as entities, facts, and events in a classification hierarchy. Document classification can be achieved using either SAS Content Categorization or SAS Text Miner. However, there are some fundamental differences between these two tools. The text topic extraction feature in SAS Text Miner completely relies on the quantification of terms (frequency of occurrences) and the derived weights of the terms for each document using advanced statistical methods such as SVD.

On the other hand, SAS Content Categorization is broadly based on statistical and rule-based models. The statistical categorizer works similar to the text topic feature in SAS Text Miner. The statistical categorizer is used as a first step to automatically classify documents. Because you cannot really see the rules behind the classification methodology, it is called a black box model. In rule-based models, you can choose to use linguistic rules by listing the commonly occurring terms most relevant for a category. You can assign weights to these terms based on their importance. Boolean rule-based models use Boolean operators such as AND, OR, NOT, etc., to specify the conditions with which terms should occur within documents. This tool has additional custom-built operators to assess positional characteristics such as whether the distance between the two terms is within a distance of $n$ terms, whether specific terms are found in a given sequence, etc. There is no limit on how complex these rules can be (for example, you can use nested Boolean rules).

## Ontology Management

Ontology is a study about how entities can be grouped and related within a hierarchy. Entities can be subdivided based on distinctive and commonly occurring features. SAS Ontology Management enables you to create relationships between pre-existing taxonomies built for various silos or departments. The subject matter knowledge about the purpose and meaning can be used to create rules for building information search and retrieval systems. By identifying relationships in an evolutionary method and making the related content available, queries return relevant, comprehensive, and accurate answers. SAS Ontology Management offers the ability to build semantic repositories and manage company-wide thesauri and vocabularies and to build relationships between them.

To explain its application, consider the simple use case of an online media house named ABC. (The name was changed to maintain anonymity.) ABC uses SAS Ontology Management. ABC collects a lot of topics over a period of time. It stores each of these topics, along with metadata (properties), including links to images and textual descriptions. SAS Ontology Management helps ABC store relationships between the related topics. ABC regularly queries its ontology to generate a web page for each topic, showing the description, images, related topics, and other metadata that it might have selected to show. (See Display 1.6.) ABC uploads the information from SAS Ontology Management to SAS Content Categorization, and then tags news articles with topics that appear in the articles using rules that it's created. All tagged articles are included in a list on the topic pages.

**Display 1.6: Example Application of SAS Ontology Management from an Online Media Website**

---

# Information Extraction

In a relational database, data is stored in tables within rows and columns. A structured query on the database can help you retrieve the information required if the names of tables and columns are known. However, in the case of unstructured data, it is not easy to extract specific portions of information from the text because there is no fixed reference to identify the location of the data. Unstructured data can contain small fragments of information that might be of specific interest, based on the context of information and the purpose of analysis. Information extraction can be considered the process of extracting those fragments of data such as the names of people, organizations, places, addresses, dates, times, etc., from documents.

Information extraction might yield different results depending on the purpose of the process and the elements of the textual data. Elements of the textual data within the documents play a key role in defining the scope of information extraction. These elements are tokens, terms, and separators. A document consists of a set of tokens. A token can be considered a series of characters without any separators. A separator can be a special character, such as a blank space or a punctuation mark. A term can be a defined as a token with specific semantic purpose in a given language.

There are several types of information extraction that can be performed on textual data.

- Token extraction
- Term extraction or term parsing
- Concept extraction
- Entity extraction

- Atomic fact extraction
- Complex fact extraction

Concept extraction involves identifying nouns and noun phrases. Entity extraction can be defined as the process of associating nouns with entities. For example, although the word "white" is a noun in English and represents a color, the occurrence of "Mr. White" in a document can be identified as a person, not a color. Similarly, the phrase "White House" can be attributed to a specific location (the official residence and principal workplace of the president of the United States), rather than as a description of the color of paint used for the exterior of a house. Atomic fact extraction is the process of retrieving fact-based information based on the association of nouns with verbs in the content (i.e., subjects with actions).

## Clustering

Cluster analysis is a popular technique used by data analysts in numerous business applications. Clustering partitions records in a data set into groups so that the subjects within a group are similar and the subjects between the groups are dissimilar. The goal of cluster analysis is to derive clusters that have value with respect to the problem being addressed, but this goal is not always achieved. As a result, there are many competing clustering algorithms. The analyst often compares the quality of derived clusters, and then selects the method that produces the most useful groups. The clustering process arranges documents into nonoverlapping groups. (See Display 1.7.) Each document can fall into more than one topic area after classification. This is the key difference between clustering and the general text classification processes, although clustering provides a solution to text classification when groups must be mutually exclusive, as in the classified ads example.

In the context of text mining, clustering divides the document collection into mutually exclusive groups based on the presence of similar themes. In most business applications involving large amounts of textual data, it is often difficult to profile each cluster by manually reading and considering all of the text in a cluster. Instead, the theme of a cluster is identified using a set of descriptive terms that each cluster contains. This vector of terms represents the weights measuring how the document fits into each cluster. Themes help in better understanding the customer, concepts, or events. The number of clusters that are identified can be controlled by the analyst.

The algorithm can generate clusters based on the relative positioning of documents in the vector space. The cluster configuration is altered by a start and stop list.

**Display 1.7: Text Clustering Process Assigning Each Document to Only One Cluster**

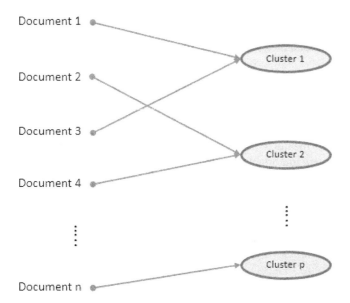

For example, consider the comments made by different patients about the best thing that they liked about the hospital that they visited.

1. Friendliness of the doctor and staff.
2. Service at the eye clinic was fast.
3. The doctor and other people were very, very friendly.
4. Waiting time has been excellent and staff has been very helpful.
5. The way the treatment was done.
6. No hassles in scheduling an appointment.
7. Speed of the service.
8. The way I was treated and my results.
9. No waiting time, results were returned fast, and great treatment.

The clustering results from text mining the comments come out similar to the ones shown in Table 1.2. Each cluster can be described by a set of terms, which reveal, to a certain extent, the theme of the cluster. This type of analysis helps businesses understand the collection as a whole, and it can assist in correctly classifying customers based on common topics in customer complaints or responses.

**Table 1.2: Clustering Results from Text Mining**

| Cluster No. | Comment | Key Words |
|---|---|---|
| 1 | 1, 3, 4 | doctor, staff, friendly, helpful |
| 2 | 5, 6, 8 | treatment, results, time, schedule |
| 3 | 2, 7 | service, clinic, fast |

The derivation of key words is accomplished using a weighting strategy, where words are assigned a weight using features of LSI. Text mining software products can differ in how the keywords are identified, resulting from different choices for competing weighting schemes.

SAS Text Miner uses two types of clustering algorithms: expectation maximization and hierarchical clustering. The result of cluster analysis is identifying cluster membership for each document in the collection. The exact nature of the two algorithms is discussed in detail in "Chapter 6 Clustering and Topic Extraction."

## Trend Analysis

In recent years, text mining has been used to discover trends in textual data. Given a set of documents with a time stamp, text mining can be used to identify trends of different topics that exist in the text. Trend analysis has been widely applied in tracking the trends in research from scientific literature. It has also been widely applied in summarizing events from news articles. In this type of analysis, a topic or theme is first defined using a set of words and phrases. Presence of the words across the documents over a period of time represents the trend for this topic. To effectively track the trends, it is very important to include all related terms to (or synonyms of) these words.

For example, text mining is used to predict the movements of stock prices based on news articles and corporate reports. Evangelopoulos and Woodfield (2009) show how movie themes trend over time, with male movies dominating the World War II years and female movies dominating the Age of Aquarius. As another example, mining social networks to identify trends is currently a very hot application area. Google Trends, a publicly available website, provides a facility to identify the trends in your favorite topics over a period of time. Social networking sites such as Twitter and blogs are great sources to identify trends. Here is a screenshot of the trend for the topic "text mining" from Google Trends. It is clearly evident that the growth in search traffic and online posts for the term "text mining" peaked after 2007. This is when the popularity of text mining applications in the business world jump-started.

**Display 1.8: Trend for the Term "text mining" from Google Trends**

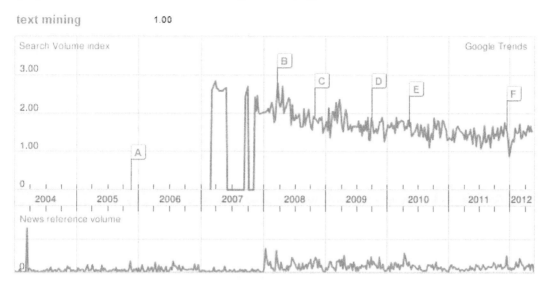

The concept linking functionality in SAS Text Miner helps in identifying co-occurring terms (themes), and it reveals the strength of association between terms. With temporal data, the occurrence of terms from concept links can be used to understand the trend (or pattern) of the theme across the time frame. Case Study 1 explains how this technique was applied to reveal the trend of different topics that have been presented at SAS Global Forum since 1976.

## Enhancing Predictive Models Using Exploratory Text Mining

Although text mining customer responses can reveal valuable insights about a customer, plugging the results from text mining into a typical data mining model can often significantly improve the predictive power of the model. Organizations often want to use customer responses captured in the form of text via e-mails, customer survey questionnaires, and feedback on websites for building better predictive models. One way of doing this is to first apply text mining to reveal groups (or clusters) of customers with similar responses or feedback. This cluster membership information about each customer can then be used as an input variable to augment the data mining model. With this additional information, the accuracy of a predictive model can improve significantly.

For example, a large hospital conducted a post-treatment survey to identify the factors that influence a patient's likelihood to recommend the hospital. By using the text mining results from the survey, the hospital was able to identify factors that showed an impact on patient satisfaction, which was not measured directly through the survey questions. Researchers observed a strong correlation between the theme of the cluster and the ratings given by the patient for the likelihood for the patient to recommend the hospital.

In a similar exercise, a large travel stop company observed significant improvement in predicting models by using customers' textual responses and numerical responses from a survey. Display 1.9 shows an example receiver operating characteristic (ROC) curve of the models with and without textual comments. The ROC curve shows the performance of a binary classification model. The larger the area under the curve, the better the model performance. The square-dashed curve (green), which is an effect of including results from textual responses, has a larger area under the curve compared to the long-dashed-dotted curve (red), which represents the model with numerical inputs alone.

**Display 1.9: ROC Chart of Models With and Without Textual Comments**

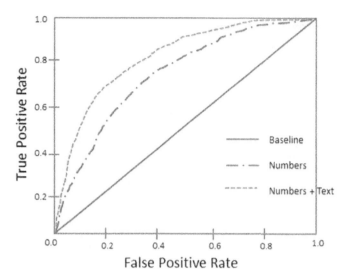

With the widespread adoption by consumers of social media, a lot of data about any prospect or customer is often available on the web. If businesses can cull and use this information, they can often generate better predictions of consumer behavior. For example, credit card companies can track customers' posts on Twitter and other social media sites, and then use that information in credit scoring models. However, there are challenges to using text mining models and predictive models together because it can be difficult to get textual data for every member in the data mining model for the same time period.

## Sentiment Analysis

The field of sentiment analysis deals with categorization (or classification) of opinions expressed in textual documents. Often these text units are classified into multiple categories such as positive, negative, or neutral, based on the valence of the opinion expressed in the units. Organizations frequently conduct surveys and focus group studies to track a customer's perception of their products and services. However, these methods are time-consuming and expensive and cannot work in real time because the process of analyzing text is done manually by experts. Using sentiment analysis, an organization can identify and extract a customers' attitude, sentiment, or emotions toward a product or service. This is a more advanced application of text analytics that uses NLP to capture the polarity of the text: positive, negative, neutral, or mixed. With the advent of social networking sites, organizations can capture enormous amounts of customers' responses instantly. This gives real-time awareness to customer feedback and enables organizations to react fast. Sentiment analysis works on opinionated text while text mining is good for factual text. Sentiment analysis, in combination with other text analytics and data mining techniques, can reveal very valuable insights.

Sentiment analysis tools available from SAS offer a very comprehensive solution to capture, analyze, and report customer sentiments. The polarity of the document is measured at the overall document level and at the specific feature level.

Here is an example showing the results of a sentiment analysis on a customer's review of a new TV brand:

The TV is wonderful. Great **size**, great **picture**, easy **interface**. It makes a cute little song when you **boot** it up and when you shut it off. I just want to point out that the **43"** does not in fact play videos from the USB. This is really annoying because that was one of the major perks I wanted from a new TV. Looking at the product description now, I realize that the feature list applies to the X758 series as a whole, and that each model's capabilities are listed below. Kind of a dumb oversight on my part, but it's equally stupid to put a description that does not apply on the listing for a very specific model.

In the previous text, green color represents positive tone, red color represents negative tone, and product features and model names are highlighted in blue and brown, respectively. In addition to extracting positive and negative sentiments, names of product models and their features are identified. This level of identification helps identify the sentiment of the overall document and tracks the sentiment at a product-feature level, including the characteristics and sub-attributes of features.

"Chapter 8 Sentiment Analysis" discusses sentiment analysis using SAS Sentiment Analysis Studio through an example of tracking sentiment in feedback comments from customers of a leading travel stop company.

## Emerging Directions

Although the number of applications in text analytics has grown in recent years, there continues to be a high level of excitement about text analytics applications and research. For example, many of the papers presented at the Analytics 2011 Conference and SAS Global Forum 2013 were based on different areas of text analytics. In a way, the excitement about text analytics reminds us of the time when data mining and predictive modeling was taking off at business and academic conferences in the late 90s and early 2000s. The text analytics domain is constantly evolving with new techniques and new applications. Text analytics solutions are being adopted at the enterprise level and are being used to operationalize and integrate the voice of the customer into business processes and strategies. Many enterprise solution vendors are integrating some form of text analytics technology into their product line. This is evident from the rate of acquisitions in this industry. One of the key reasons that is fueling the growth of the field of text analytics is the increasing amount of unstructured data that is being generated on the web. It is expected that 90% of the digital content in the next 10 years will be unstructured data.

Companies across all industries are looking for solutions to handle the massive amounts of data, also popularly known as big data. Data is generated constantly from various sources such as transaction systems, social media interactions, clickstream data from the web, real-time data captured from sensors, geospatial information, and so on. As we have already pointed out, by some estimates, 80% of an organization's current data is not numeric!

This means that the variety of data that constitutes big data is unstructured. This unstructured data comes in various formats: text, audio, video, images, and more. The constant streaming of data on social media outlets and websites means the velocity at which data is being generated is very high. The variety and the velocity of the data, together with the volume (the massive amounts) of the data organizations need to collect, manage, and process in real time, creates a challenging task. As a result, the three emerging applications for text analytics will likely address the following:

1. Handling big (text) data
2. Voice mining
3. Real-time text analytics

## Handling Big (Text) Data

Based on the industry's current estimations, unstructured data will occupy 90% of the data by volume in the entire digital space over the next decade. This prediction certainly adds a lot of pressure to IT departments, which already face challenges in terms of handling text data for analytical processes. With innovative hardware architecture, analytics application architecture, and data processing methodologies, high-performance computing technology can handle the complexity of big data. SAS High-Performance Text Mining helps you decrease the computational time required for processing and analyzing bulk volumes of text data significantly. It uses the combined power of multithreading, a distributed grid of computing resources, and in-memory processing. Using sophisticated implementation methodologies such as symmetric multiprocessing (SMP) and massively parallel processing (MPP), data is distributed across computing nodes. Instructions are allowed to execute separately on each node. The results from each node are combined to produce meaningful results. This is a cost-effective and highly scalable technology that addresses the challenges posed by the three **V**s. (variety, velocity, and volume) of big data.

SAS High-Performance Text Mining consists of three components for processing very large unstructured data. These components are document parsing, term handling, and text processing control. In the document parsing component, several NLP techniques (such as parts-of-speech tagging, stemming, etc.) are applied to the input text to derive meaningful information. The term handling component accumulates (corrects misspelled terms using a synonyms list), filters (removes terms based on a start or stop list and term frequency), and assigns weights to terms. The text processing control component manages the intermediate results and the inputs and outputs generated by the document parsing and term handling components. It helps generate the term-by-document matrix in a condensed form. The term-by-document matrix is then summarized using the SVD method, which produces statistical representations of text documents. These SVD scores can be later included as numeric inputs to different types of models such as cluster or predictive models.

## Voice Mining

Customer feedback is collected in many forms—text, audio, and video—and through various sources—surveys, e-mail, call center, social media, etc. Although the technology for analyzing videos is still under research and development, analyzing audio (also called voice mining) is gaining momentum. Call centers (or contact centers) predominantly use speech analytics to analyze the audio signal for information that can help improve call center effectiveness and efficiency. Speech analytics software is used to review, monitor, and categorize audio content. Some tools use phonetic index search techniques that automatically transform the audio signal into a sequence of phonemes (or sounds) for interpreting the audio signal and segmenting the feedback using trigger terms such as "cancel," "renew," "open account," etc. Each segment is then analyzed by listening to each audio file manually, which is daunting, time-intensive, and nonpredictive. As a result, analytical systems that combine data mining methods and linguistics techniques are being developed to quickly determine what is most likely to happen next (such as a customer's likelihood to cancel or close the account). In this type of analysis, metadata from each voice call, such as call length, emotion, stress detection, number of transfers, etc., that is captured by these systems can reveal valuable insights.

## Real-Time Text Analytics

Another key emerging focus area that is being observed in text analytics technology development is real-time text analytics. Most of the applications of real-time text analytics are addressing data that is streaming continuously on social media. Monitoring public activity on social media is now a business necessity. For example, companies want to track topics about their brands that are trending on Twitter for real-time ad placement. They want to be informed instantly when their customers post something negative about their brand on the Internet. Less companies want to track news feeds and blog posts for financial reasons. Government agencies are relying on real-time text analytics that collect data from innumerate sources on the web to learn about and predict medical epidemics, terrorist attacks, and other criminal actions. However, real time can mean different things in different contexts. For companies involved in financial trading by tracking current events and news feeds, real time could mean milliseconds. For companies tracking customer satisfaction or monitoring brand reputation by collecting customer feedback, real time could mean hourly. For every business, it is of the utmost importance to react instantly before something undesirable occurs.

The future of text analytics will surely include the next generation of tools and techniques with increased usefulness for textual data collection, summarization, visualization, and modeling. Chances are these tools will become staples of the business intelligence (BI) suite of products in the future. Just as SAS Rapid Predictive Modeler today can be used by business analysts without any help from trained statisticians and modelers, so will be some of the future text analytics tools. Other futuristic trends and applications of text analytics are discussed by Berry and Kogan (2010).

## Summary

Including textual data in data analysis has changed the analytics landscape over the last few decades. You have witnessed how traditional machine learning and statistical methods to learn unknown patterns in text data are now replaced with much more advanced methods combining NLP and linguistics. Text mining (based on a traditional bag-of-words approach) has evolved into a much broader area (called text analytics). Text analytics is regarded as a loosely integrated set of tools and methods developed to retrieve, cleanse, extract, organize, analyze, and interpret information from a wide range of data sources. Several techniques have evolved, with each focused to answer a specific business problem based on textual data. Feature extraction, opinion mining, document classification, information extraction, indexing, searching, etc., are some of the techniques that we have dealt with in great detail in this chapter. Tools such as SAS Text Miner, SAS Sentiment Analysis Studio, SAS Content Categorization, SAS Crawler, and SAS Search and Indexing are mapped to various analysis methods. This information helps you distinguish and differentiate the specific functionalities and features that each of these tools has to offer while appreciating the fact that some of them share common features.

In the following chapters, we use SAS Text Analytics tools (except SAS Ontology Management, which is not discussed in this book) to address each methodology discussed in this chapter. Chapters are organized in a logical sequence to help you understand the end-to-end processes involved in a typical text analysis exercise. In Chapter 2, we introduce methods to extract information from various document sources using SAS Crawler. We show you how to deal with the painstaking tasks of cleansing, collecting, transforming, and organizing the unstructured text into a semi-structured format to feed that information into other SAS Text Analytics tools. As you progress through the chapters, you will get acquainted with SAS Text Analytics tools and methodologies that will help you adapt them at your organization.

## References

Albright, R., Bieringer, A., Cox, J., and Zhao, Z. 2013. "Text Mine Your Big Data: What High Performance Really Means". Cary, NC: SAS Institute Inc. Available at: http://www.sas.com/content/dam/SAS/en_us/doc/whitepaper1/text-mine-your-big-data-106554.pdf

Berry, M.W., and Kogan, J. Eds. 2010. *Text Mining: Applications and Theory.* Chichester, United Kingdom: John Wiley & Sons.

Dale, R., Moisl, H. and Somers, H. 2000. Handbook of Natural Language Processing. New York: Marcel Dekker.

Dorre, J. Gerstl, P., and Seiffert, R. 1999. "Text Mining: Finding Nuggets in Mountains of Textual Data".

*KDD-99: Proceedings of the Fifth ACM SIGKDD International Conference on Knowledge Discovery and Data Mining.* San Diego, New York: Association for Computing Machinery, 398-401.

Evangelopoulos, N., and Woodfield, T. 2009. "Understanding Latent Semantics in Textual Data". M2009 12th Annual Data Mining Conference, Las Vegas, NV.

Feldman, R. 2004. "Text Analytics: Theory and Practice". *ACM Thirteenth Conference on Information and Knowledge Management (CIKM) CIKM and Workshops 2004.* Available at: http://web.archive.org/web/20041204224205/http://ir.iit.edu/cikm2004/tutorials.html

Grimes, S. 2007. "What's Next for Text. Text Analytics Today and Tomorrow: Market, Technology, and Trends". Text Analytics Summit 2007.

Halper, F.,Kaufman, M., and Kirsh, D. 2013. "Text Analytics: The Hurwitz Victory Index Report". Hurwitz & Associates 2013. Available at: http://www.sas.com/news/analysts/Hurwitz_Victory_Index-TextAnalytics_SAS.PDF

H.P.Luhn. 1958. "The Automatic Creation of Literature Abstracts*". IBM Journal of Research and Development*, 2(2):159-165.

Manning, C. D., and Schutze, H. 1999. Foundations of Statistical Natural Language Processing. Cambridge, Massachusetts: The MIT Press.

McNeill, F. and Pappas, L. 2011. "Text Analytics Goes Mobile". *Analytics Magazine*, September/October 2011. Available at: http://www.analytics-magazine.org/septemberoctober-2011/403-text-analytics- goes-mobile

Mei, Q. and Zhai, C. 2005. "Discovering Evolutionary Theme Patterns from Text: An Exploration of Temporal Text Mining". KDD 05: *Proceedings of the Eleventh ACM SIGKDD International Conference on Knowledge Discovery in Data Mining,* 198 – 207.

Miller, T. W, 2005. Data and Text Mining: A Business Applications Approach. Upper Saddle River, New Jersey: Pearson Prentice Hall.

Radovanovic, M. and Ivanovic, M. 2008. "Text Mining: Approaches and Applications". *Novi Sad Journal of Mathematics.* Vol. 38: No. 3, 227-234.

Salton, G., Allan, J., Buckley C., and Singhal, A. 1994. "Automatic Analysis, Theme Generation, and Summarization of Machine-Readable Texts". *Science*, 264.5164 (June 3): 1421-1426.

SAS® Ontology Management, Release 12.1. Cary, NC: SAS® Institute Inc.

Shaik, Z., Garla, S., Chakraborty, G. 2012. "SAS® since 1976: an Application of Text Mining to Reveal Trends". *Proceedings of the SAS Global Forum 2012 Conference.* SAS Institute Inc., Cary, NC.

Text Analytics Using SAS® Text Miner. Course Notes. Cary, NC, SAS Institute. Inc. Course information: https://support.sas.com/edu/schedules.html?ctry=us&id=1224

Text Analytics Market Perspective. White Paper, Cary, NC: SAS Institute Inc. Available at: http://smteam.sas.com/xchanges/psx/platform%20sales%20xchange%20on%20demand%20%202007%20sessions/text%20analytics%20market%20perspective.doc

Wakefield, T. 2004. "A Perfect Storm is Brewing: Better Answers are Possible by Incorporating Unstructured Data Analysis Techniques." DM Direct Newsletter, August 2004.

Weiss S, Indurkhya N, Zhang T, and Damerau F. 2005. *Text Mining: Predictive Methods for Analyzing Unstructured Information.* Springer-Verlag.

# Chapter 2  Information Extraction Using SAS Crawler

## Introduction to Information Extraction and Organization

In a little less than two decades, we have witnessed how the Internet has evolved and become an integral part of a common man's life. The emergence of many e-commerce websites and social media channels has quickly turned the web into a breeding powerhouse of massive data, mostly unstructured in nature. In a scientific study conducted in 2011, the IDC estimated that the digital world produced 1.8 zettabytes (equivalent to 1.8 trillion gigabytes) of information in that year alone. It is approximated that the amount of data will multiply at least 50 times by the year 2020. This figure applies not just to content on the web, but also to data in documents and files stored locally on PCs and servers across the globe. This irrepressible explosion of data poses quite a challenge for organizations that like to tap into even a portion of this data and transform it into a format suitable for text analytics. Advanced scientific methods such as NLP greatly help organizations in the process of discovering useful nuggets of information hidden within a gazillion bytes of data.

As a first step, organizations should be able to separate useful information (signal) from vast piles of unstructured data (noise). This process can be termed "information extraction." It should not be confused with the science of information retrieval. Information retrieval is an umbrella term that covers many advanced concepts and techniques such as page ranking, document clustering, etc., which are much broader than information extraction. In this chapter, we discuss how SAS proprietary software can be used to extract unstructured data stored on the web or in file storage systems in a systematic and efficient manner. The extracted data can be parsed, transformed, and even exported to other SAS tools, which can consume this information, analyze, and then produce useful results.

## SAS Crawler

Businesses can have multiple needs with textual data. Although text mining serves as a generic process to explore textual data, it is not necessarily the only need. For example, a product firm might require extracting its user reviews and understanding the sentiment of its products or brands in the market. An online media company might choose to categorize its document base into various predefined themes to help its users easily navigate and locate relevant articles or files based on their topics of interest. Some organizations might like to gather information, build an index, and facilitate a search interface for their office staff to access the content in an efficient manner. With such diverse needs, almost all organizations face challenges in managing such large volumes of text data.

To gather and manage such massive volumes of data in a systematic fashion, you need sophisticated tools capable of extracting text data from almost any data source or data format. SAS offers exclusive web crawlers capable of performing directed crawls on the web, files, and even feeds to grab relevant content and store it in raw files. These raw files can be used by other SAS Text Analytics products such as SAS Text Miner for analysis.

## SAS Search and Indexing

SAS Search and Indexing helps you choose and then create either a simple or advanced index to efficiently handle user-submitted search queries to retrieve relevant information. Queries are submitted against these prebuilt indexes to return the most relevant matching documents. SAS provides additional components, which are used to parse, categorize, and export the crawled documents to other SAS applications. These components form the heart of the entire IR framework.

## SAS Information Retrieval Studio Interface

SAS web, file, and feed crawlers, SAS Search and Indexing, and other supporting components are all integrated on a single framework—SAS Information Retrieval Studio. SAS Information Retrieval Studio is a graphical user interface in which you can configure, control, and monitor the end-to-end information extraction process. The high-level architecture diagram of SAS Information Retrieval Studio provides a general overview of all of the components and how they interact with each other. (See Display 2.1.) Throughout this chapter, we discuss various components of the SAS Information Retrieval Studio interface based on version 12.1. Screenshots referring to these components are subject to slight changes in later versions.

**Display 2.1: Simple Architectural Diagram of the SAS Information Retrieval Studio Graphical Framework**

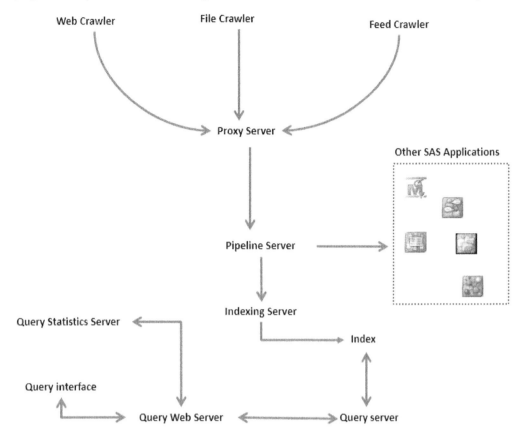

## Components of SAS Crawler

**Web Crawler** – Extracts text from web pages on the Internet or on intranet sites.

**File Crawler** – Extracts text from files or documents stored locally on PCs, shared network drives, etc.

**Feed Crawler** – Extracts text from web feed formats such as RSS or Atom.

## Core Components of the SAS Information Retrieval Studio Interface

**Proxy Server** – Controls the flow of documents pushed by the web or file and feed crawlers to the pipeline server. Functions like a pressure valve to maintain a steady movement of files fed by crawlers into the pipeline server.

**Pipeline Server** – Documents the flow into the pipeline server from the proxy server. The pipeline server performs actions such as parsing, modifying fields, analyzing content, etc., on the documents. It then passes them to either the indexing server or another SAS application. This is arguably the most important component of the SAS IR framework. It works like the heart for a human body. The main function of the heart is to purify and pump oxygenated blood to various parts of the body. Similarly, a pipeline server helps cleanse, parse, convert, extract, and pass meaningful content to various SAS applications for further processing.

## Components of SAS Search and Indexing

**Indexing Server** – Builds an index of documents to help with searching them using the query server.

**Query Server** – Helps with searching the documents based on the index built using the indexing server.

**Query Web Server** – Hosts the query interface to help users search documents interactively.

**Query Statistics Server** – Monitor user search queries through the query web server.

**Virtual Indexing Server*** – Distributes documents across a set of multiple index segments and maintains them as if they are all under one single large index.

**Virtual Query Server[i]** – Distributes the influx of user-submitted queries to several back-end query servers.

## Other Important Components

**SAS Markup Matcher Server** – Helps you create custom templates to target specific sections of HTML web pages or XML files and to extract only the wanted content. For example, you can do a simple web crawl to extract the contents of articles posted on a blog site. However, web pages on blogs generally contain links to other sites, articles, advertisements, banners, and comments sections. As a result, you might end up retrieving additional unwanted content during the web crawl. The SAS Markup Matcher helps you control what you need to extract.

**SAS Document Conversion Server** – Document Conversion is an essential component of the IR process. It provides a means to convert documents in various file formats into individual TXT files. Documents in formats such as HTML, PDF, PPT, MS Word, etc., need to be normalized and converted to a common format that is understood by other SAS applications and the indexing server. This step is crucial in situations where a file crawl or a web crawl can potentially retrieve documents in one or more of these formats.

# Web Crawler

A web crawler is a software program that traverses pages on the World Wide Web and leverages the embedded linkage structure through which web pages are interconnected. Web crawlers require a web page link to begin crawling (also known as a seed page or entry point).

A typical web crawler starts parsing a seed page and scans for links in the seed page that point to other web pages. The newly found pages are again searched, parsed, and scanned for links that point to other web pages. This process continues recursively until all links referenced in this way are searched and all links are collected. While parsing these web pages for links, the web crawler grabs the actual content from them and stores it as document files. These files are often passed to other programs that work in conjunction with web crawlers to build indexes based on the document files. These indexes are helpful in retrieving the information based on search queries entered by users. Parsing generally involves cleaning up HTML and other tags scattered across the source code of the crawled web pages. Display 2.2 shows how a typical web crawler works. In this scenario, the seed page references two hypertext links (Link 1 and Link 2). Each link points to web pages that refer to other links (for example, Page 1 refers to Links 3, 4, and 5, and Page 2 refers to Links 6 and 7).

**Display 2.2: A Typical Web Crawler Traversing through Web Pages**

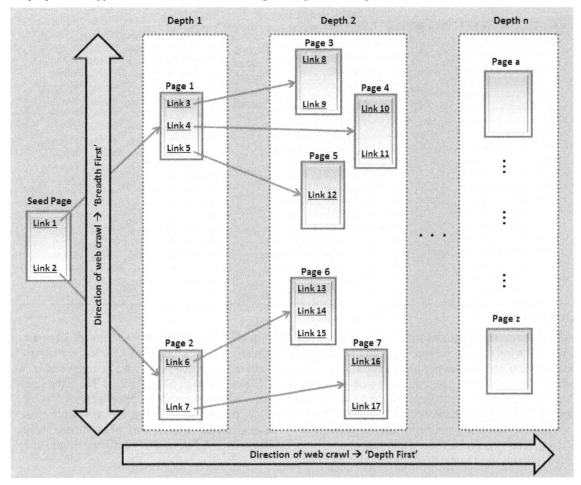

**Link and Page:** A link is nothing but a URL that acts as a pointer to a web page. When you click on a link, it retrieves the associated content from the web and displays it as a web page in a browser. Throughout this chapter, link *n* represents a reference to the web page *n*.

**Child Page:** All web pages that are linked to the seed page directly or indirectly through the linkage structure are considered child pages. In this example, any of the links shown in Display 2.2 can be reached by traversing through the links starting from the seed page. They are all child pages in this example.

**Modes of Crawl:** Primarily, there are two modes that represent the two different directions in which a web crawler can be guided to move through the linked list of web pages and to retrieve content.

## Breadth First

The breadth first direction guides the web crawler to search through all of the child pages at Depth 1 referenced by the seed page first. Then, it searches through all of the child pages at Depth 2 referenced by the pages in Depth 1. This process repeats until all of the child pages related to the seed page are completely crawled. In Display 2.2, in the breadth first mode, pages are searched starting with the Seed Page, Depth 1 (Pages 1 and 2), Depth 2 (Pages 3 through 7), and so on, crawling through all links at each depth before proceeding to the links at the next depth. Display 2.3 shows the web crawl path in breadth first mode for this current example.

**Display 2.3: Traversing Path of a Web Crawler in Breadth First Mode**

Seed Page → Depth 1 → Depth 2 → Depth 3 ............................ → Depth n

| | | | |
|---|---|---|---|
| Page 1 | Page 3 | Page 8 | Page a |
| Page 2 | Page 4 | Page 9 | . |
| | Page 5 | Page 10 | . |
| | Page 6 | Page 11 | . |
| | Page 7 | Page 11 | . |
| | | Page 12 | . |
| | | Page 13 | . |
| | | Page 14 | . |
| | | Page 15 | . |
| | | Page 16 | . |
| | | Page 17 | . |
| | | | . |
| | | | . |
| | | | Page z |

## Depth First

The depth first option guides the crawler to select one link first, and then completely go through all of the child links following it, until the end. Once all of child links within this link are completely crawled, the control goes back to the next link, and follows all of its child links. In the current example, the depth first mode of crawl follows the path starting from Seed Page, Link 1 (Depth 1), Links 3 through 5 (Depth 2), Links 8 through 12 (Depth 3), and so on, until all of the child links for Link 1 are completely crawled. (See Display 2.4.) Once done, the control goes back to Link 2, and then recursively goes through its child links. As the depth of the crawling increases, the number of links to crawl exponentially increases regardless of the mode of crawling. Starting with version 12.1, the depth first option is not available in SAS Information Retrieval Studio and breadth first is used as the default method.

**Display 2.4: Traversing Path of a Web Crawler in Depth First Mode**

(Depth 1) Page 1→ (Depth 2) Pages 3 through 5→ (Depth 3) Pages 8 through 12→ .......till the end
(Depth 1) Page 2→ (Depth 2) Pages 6 through 7→ (Depth 3) Pages 13 through 17→ ........till the end

Though web crawlers are also known by several other names such as spiders, robots, worms, etc., implying they move physically, they neither execute nor migrate from one physical location to another. Unlike other categories of software programs such as viruses, crawlers do not spread across networks, but reside and run on one machine. A web crawler does nothing but automatically follow links, send page requests, and fetch required information hosted on machines over the Internet through a standard HTTP protocol.

SAS Web Crawler has many configurable settings and features that are covered in great detail in *SAS Information Retrieval Studio: Administration Guide*. We discuss some of the key features when we walk you through a step-by-step demonstration on how to conduct web crawling toward the end of this chapter.

## Web Crawling: Real-World Applications and Examples

There are many real-world applications of using a web crawler. Almost all search engines such as Google, Bing, etc., use web crawlers to gather documents for building large indexes. Whenever a user submits a query to these search engines, the indexes are used to locate web pages with related information, and they are returned as a result of the search query. McCallum *et al.* used crawlers and built portals that were repositories of indexed research papers related to computer science. Site maps in a graphical layout can be developed by initiating a web crawl from the seed page going in the breadth first mode, and then connecting the related pages visually.

Lieberman *et al.* developed a browsing agent, which monitors the real-time browsing patterns of users, understands which links the user is spending time with, and tries to suggest more links that might be suitable for the user's interests and needs.

The most prominent application of web crawling adapted by many companies in the contemporary business world is to crawl the competitors' websites to collect information about their new products, future plans, investment strategies, etc. In the field of biomedical studies, researchers are often required to access huge volumes of literary content about genes. They deploy web crawling to gather gene-related information from various sources on the web. On a downside, a popular but rather litigious application of using web crawling is to extract the e-mail addresses of individuals from various websites and spam their inboxes with phishing e-mails. Through phishing, hackers can gain access to individual passwords, credit card information, or personally identifiable data.

## File Crawler

A file crawler searches through the files or documents located physically on local disk storage or on a shared network. It then downloads these files to send them to the pipeline server for processing. The file crawler can grab content contained in document types such as PDF, HTML, Microsoft Word, XML, text files, etc. A document conversion process can be used to convert the data from these formats into raw text files. Remember to select **Yes** for the **Encapsulate XML Files** option if you are crawling HTML or XML files and to include a markup_matcher document processor in the pipeline server for normalizing the text in the files.

## Feed Crawler

A feed crawler gathers information like news updates, weather updates, press releases, blog entries, etc., that are distributed on web feeds. It can gather either full text or a summary of the information. A feed crawler always pulls data from the web links found within the feeds, whereas a web crawler typically grabs content off standard web pages. Feeds are generally embedded as hyperlinks within web pages and sometimes users can confuse them with web pages. In the following subsections, we attempt to explain a feed, commonly used feed types in the industry, and a way to identify or distinguish them from web pages.

## What Is a Feed?

A feed can be technically described as a mechanism to publish frequently updated information about websites. Feeds (also termed web feeds) are simply formatted documents on the web (often built in an XML file format) containing web links redirecting to the information sources that are added frequently at regular time periods. A software program called a feed reader (also called an aggregator) is required to read and pull information from these web feeds whenever they are updated with new information. A user can subscribe to a feed and monitor its updates from a web browser. Many browsers such as Internet Explorer and Google Chrome have built-in feed readers, which check the user-registered feeds for new posts and avoids the task of manually visiting all of the interested sites for information. SAS Feed Crawler in SAS Information Retrieval Studio performs the job of a feed reader and can be configured based on the user needs.

## Types of Feeds

Primarily, there are two types of feeds based on their format and structure—RSS and Atom. RSS stands for RDF Site Summary and is nicknamed Really Simple Syndication. The SAS Information Retrieval Studio feed reader is capable of reading content from both RSS and Atom feeds. Feeds can be classified depending on the content. Full content and summary only are the two major categories of feeds. Full content (as the name implies) refers to feeds that contain all of the information that is needed. Summary-only feeds contain only a short summary of the actual content. You need to click on the link within the feed and navigate to the original web page where full information is available.

## How to Identify Feeds?

Generally, feeds are found in the websites with an icon (⬛) next to the URL pointing to them. Remember, feeds are not web pages. Exercise caution and check whether the URL that you provide in the tab is, indeed, a feed URL or a web page. For example, the web page http://www.sas.com/rss provides a comprehensive list of all SAS RSS feeds and their links. (See Display 2.5.)

**Display 2.5: Partial Screen Capture of SAS RSS Feeds Web Page**

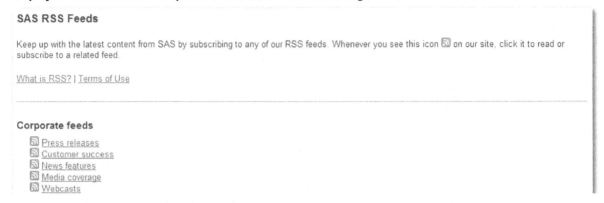

Once you click on a link (for example, **Media coverage**), you can see the actual feed. This is a summary-only feed, and the links found here actually refer to the web pages of external media sites. You can click on a link to navigate to the web page that contains the full content of the news article. (See Display 2.6.)

**Display 2.6: Partial Screen Capture of the SAS Media Coverage Feed**

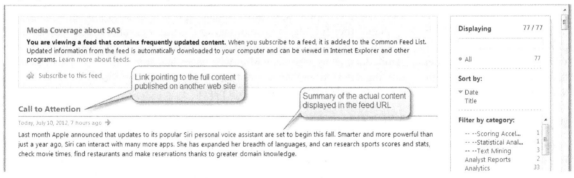

# Understanding Core Component Servers

Now that you understand how the web, file, and feed crawlers can work independently to fetch data from various sources, look at the core component servers and their functionalities.

## Proxy Server

The proxy server acts as an intermediary layer between the crawlers and the pipeline server. It receives documents from the crawlers and sends them to the pipeline server. The component proxy server that we are referencing should not be confused with the more traditional web proxy server. A web proxy server serves as a cache of web pages to help quickly retrieve the web content when a client residing locally requests a web page. The proxy server in SAS Information Retrieval Studio fundamentally serves two purposes:

a.  Enables you to pause the flow of documents to the pipeline server. When paused, the incoming documents from the crawlers are queued until the proxy server is resumed. This is useful in situations where the pipeline server or other downstream servers require ad hoc changes (or maintenance work) without having to pause or stop the crawlers.

b.  Facilitates the creation of multiple pipeline servers to which copies of documents coming in from the crawlers can be sent. Each of these servers runs simultaneously on different machines and provides support as a backup mechanism in case of emergencies such as hardware failure.

## Pipeline Server

The pipeline server is a critical component in SAS Information Retrieval Studio used to perform actions on the incoming documents from the proxy server based on your needs. These actions can be defined by document processors. Document processors form the core components of the pipeline server and solve one or both of the following purposes:

    a.  Process documents into an appropriate form that is ready to be used by other SAS applications such as SAS Text Miner (to analyze trends) or SAS Sentiment Analysis Workbench (to analyze sentiment).

    b.  Send processed documents to another component of SAS Information Retrieval Studio such as an indexer to build indexes for these documents for information search and retrieval.

## Document Processors

A document processor can be defined as a functional module designed to operate on documents to achieve the right result. The document processor can be used to execute simple text normalization tasks such as parsing documents, removing unwanted tags (for example, markup tags in HTML documents), modifying fields, exporting data in a different format (for example, exporting TXT file as CSV file), converting documents (for example, PDF to TXT), etc. Document processors can be used to perform tasks involving advanced semantic analysis such as content categorization, concept or factual extraction, etc. All of these tasks and a few more can be accomplished using the available set of prebuilt document processors provided with SAS Information Retrieval Studio. If the tasks involved are more complex, users can build their own custom processors.

Documents fed into the pipeline server go through at least one document processor before they are sent to other SAS applications or pushed to the indexing server. When more than one processor is required to act on the documents, you should exercise caution to ensure that the processors are in the proper order to obtain the right results. For example, a web crawler might gather HTML pages and send them to the pipeline server via the proxy server. First, these pages should be parsed using an appropriate document processor such as parse_html or heuristic_parse_html before exporting them to another SAS application. Adding processors in the wrong order might yield wrong results or stop the pipeline server in some cases. Each document passes through the pipeline server in a field-value pair format. A document can have many fields such as filename, body, ID, URL, etc. These fields form the metadata of the document, and actual content is assigned to the values for each of these fields. These fields can be configured for a prebuilt document processor when added to the pipeline server based on the requirements of users. Many prebuilt document processors are readily available with SAS Information Retrieval Studio to use in the pipeline server. Some of these processors are described in this section.

**document_converter:** Converts documents in sophisticated file formats such as PDF and MS Word into plain text documents using the SAS Document Conversion server.

**extract_abstract:** Creates an abstract extracting the first 25 to 50 words from the body of the incoming document. Useful to create summaries of scientific journals, technical papers, etc., where generally the abstract or introduction is placed at the beginning of the document.

**heuristic_parse_html:** Heuristically determines which sections of an HTML page can be ignored and which sections should be used to parse and extract the main content.

**export_to_files:** Creates a copy of the incoming documents into separate files. Files are named using a hash function, which converts the character strings in a document into values that are useful in indexing and searching.

## Component Servers of SAS Search and Indexing

SAS Search and Indexing components fit into the overall architecture of the SAS Information Retrieval Studio framework.

### Indexing Server

Index (as the term suggests) is similar to the index provided at the end of a book. The index in a book acts as a quick reference of all of the terms and points to the specific pages of the book in which they are found. Similarly, an index built in SAS Information Retrieval Studio is based on the data in the incoming documents from the pipeline server and the index helps quickly retrieve documents in which terms entered in the search query are found. This component in SAS Information Retrieval Studio helps build an index of the incoming documents to help in the information search and retrieval process. You might opt out of indexing if the documents collected using the crawlers are immediately sent to other SAS applications.

A document in the indexing server is in the field-value pair format. When documents in the pipeline server start flowing into the indexing server, fields are populated by matching similarly named fields. Though all the three crawlers might run simultaneously, passing documents to the indexing server via the pipeline server, it is possible to build only one index at a time. However, you can build indexes on various fields of the document. You might assign various predesigned functionalities to these fields to ensure that they are included in the search and query process. Choosing a specific functionality for a field controls the limitations for the types of queries that you can perform.

Another component in SAS Information Retrieval Studio, the query server, assists in returning the relevant information by executing user queries on indexed documents. You might perform customized searches via a user search interface that runs on a query web server. You might perform searches using either a term or a combination of terms. An intuitive search is possible if you assign the required fields with the functionality label. By default, English is the language used in building indexes. Regardless of the language selection, documents in any language are indexed; though, indexes are optimized for the language selected.

You should pay careful attention when building indexes because they are impacted by other components. Here are some of the caveats to consider whenever the indexing server is functional:

- The same documents are gathered by the crawlers when they are started, stopped, and restarted. Thus, duplicate documents might flow into the indexing server via the pipeline server. Each document indexed by the indexing server must have a unique identification field (ID), such as the URL of a web page. If two documents with the same ID are pushed to the indexing server, the document that comes in last overwrites the document that came in first, ensuring that only one copy of a unique document resides in the index.
- Whenever document processors in the pipeline server are modified when the indexing server is running, be sure to click **Apply Changes** located in the top pane.
- When configuration properties of the indexing server are modified, it does not change the existing index, but it does affect only the next index built. You can perform either of the two following options to remedy this situation:
  - Delete the current index to allow the indexing server to build a new index with the modified properties whenever the crawlers are restarted.
  - Click **Apply Changes** while the indexing server is still running. This deletes the current index and restarts the indexing server to build a new index.

### Query Server

A query server takes the search query issued by the user, runs it against the document index previously built using the indexing server, and returns the matching documents as search results back to the user. You can execute these search queries using a query API in a custom-built program and pass the search queries to the query server. You also have the option to perform these searches and review returned results in an interactive query interface. This web application runs on another component of SAS Information Retrieval Studio called the query web server. Another component of SAS Information Retrieval Studio called the query statistics server can be used to monitor, measure, and analyze query metrics periodically such as the frequency of searches and hit rate. By default, the query server is running behind the scenes all the time. Unlike other components of SAS Information Retrieval Studio, there are no configurable settings on this server that a user can modify.

## Query Web Server

The query web server lets you configure how the search results returned to the user should be displayed in the query interface. It helps you specify how documents should be matched against the search queries and sort results. It enables you to format labels, matched documents, and the theme of the search interface.

There are primarily two types of searches possible on the indexed documents in SAS Information Retrieval Studio:

**Simple Search:** In this format, you can specify the terms in a search query. They are used to look up in the index of documents and to return results in which they occur. Operators such as a plus (+) sign and a minus (-) sign can be used to either include or exclude a word or phrase in the search query. For example, a query like +sas -analytics submits a search on the index to return documents containing the term 'sas' but not when the term 'analytics' appears in the document.

**Advanced Search:** You can use the fields and search terms together in the query expression. It enables you to use Boolean, positional, and counting operators in combination with the search words or phrases. This method improves the quality of search by narrowing down to return the most relevant results. In this case, only those fields marked with a functionality of Search when building the index can be used. For example, a query like (AND, _body:"SAS", _title:"analytics") searches for the term 'SAS' in the body and 'analytics' in the title of the document. 'AND' is a Boolean operator whereas _body and _title are the markers for body and title fields of the document, respectively.

## Query Statistics Server

This component can be used to analyze the patterns of user queries sent to the query server. The query statistics server helps you monitor the topmost submitted user queries by frequency. It also monitors which matches have been found in the indexed document repository. It lets you monitor monthly, daily, and hourly query rates.

## SAS Markup Matcher Server

SAS Markup Matcher is a component provided within the SAS Information Retrieval Studio framework to help you create custom templates for accessing and extracting specific portions of an XML or HTML document. SAS Markup Matcher enables you to specify rules so that whenever an XML or HTML file is sent to it, it can create a new document and fields to add to that document. Once you define a matcher, you can upload it to the SAS Markup Matcher Server and add it to the pipeline server as a document processor. It applies the matcher rules to incoming documents. The output document generated by the SAS Markup Matcher can be either indexed or exported in XML or CSV file format. These output documents can be fed into other SAS applications such as SAS Text Miner, SAS Content Categorization, and SAS Sentiment Analysis to conduct further analysis as required. It is important to remember that the fields that are newly created by SAS Matchup Marker should be listed in the indexing server if you need to build indexes on those fields. In the later sections of this chapter, we demonstrate a simple scenario explaining how to use SAS Markup Matcher. We use this tool to define the rules required to perform guided search and information extraction from HTML web pages.

In the following sections, we demonstrate how you can leverage SAS Information Retrieval Studio to extract, parse, index, and search content in typical use case scenarios.

### Scenario 1: Use SAS Web Crawler to extract, parse, and index content from a website and use simple searches to retrieve information.

1.  Launch SAS Information Retrieval Studio on your web browser.
2.  Click **Projects** in the top pane to open the Manage Projects window. Click **New**, and enter a name for the project, for example, **SASCrawl**. Click **OK** to create a new project. The newly created project appears in the list. Click **Close**.
3.  Select the newly created project, and click **Web Crawler**. Populate the fields for **General Settings** as shown. (See Display 2.7.) As highlighted in the display, you need to specify the HTTP proxy server to use for crawling. You can click **Auto-detect** to automatically show you the proxy server available in your environment.

**Display 2.7: General Settings for SAS Web Crawler**

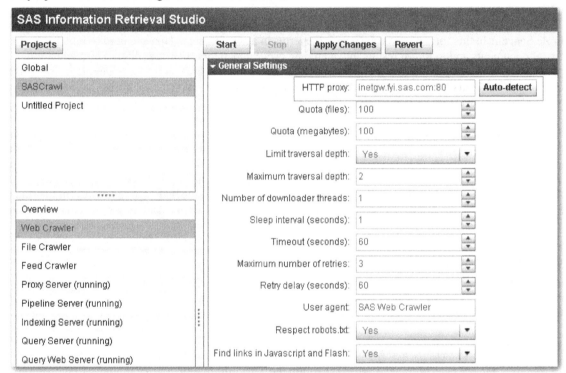

4.  In the Entry Points pane, click **Add** to open the Add Entry Point window. Enter the URL http://www.sas.com, and change the default value for **Quota** (files) to **100**, and click **OK**. You see the entry point for crawling defined as shown in Display 2.8.

**Display 2.8: Entry Points in SAS Web Crawler**

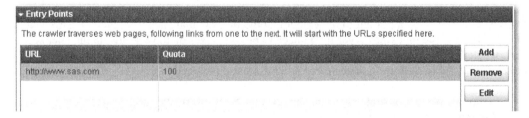

5.  By default, the same URL is added to the scope. However, you can modify the scope of the web crawl to exclude certain links in the website. For example, you can choose to exclude support.sas.com, the SAS job portal, and RSS feeds from crawling. Click **Add**, and enter the URL patterns, match types, and action as shown in Display 2.9. Leave all other settings with their defaults in SAS Web Crawler.

Click **Apply Changes** to make sure that all of the changes that you made are applied to the SAS Web Crawler.

**Display 2.9: Configuration for Managing the Scope in SAS Web Crawler**

▼ **Scope**

By default the crawler will follow links as far as they lead, across the entire web. You can control what sites it goes to by specifying what should be allowed and what should be excluded.

| URL Pattern | Match Type | Action | |
|---|---|---|---|
| http://www.sas.com | Prefix | Allow | Add |
| http://support.sas.com | Prefix | Exclude | Remove |
| http://www.sas.com/jobs | Prefix | Exclude | Edit |
| .+rss.* | Regular expression | Exclude | |
| .+rssfeed[s]*.* | Regular expression | Exclude | |
| .+rss-feed[s]*.* | Regular expression | Exclude | |
| .+feed[s]*.* | Regular expression | Exclude | |

6. Click **Pipeline Server** from the left pane. Add some document processors to process the documents coming from SAS Web Crawler through the proxy server. Click **Add**, and select **heuristic_parse_html** from the list of available document processors. Click **Next**. Accept the default settings, and click **Finish** to add it to the queue.

7. Click **Add**, and select **export-to-files** from the list, and then click **Next**. Change the **Export format** to **Plain Text**, and the click **Finish**, leaving all others with default settings. You should have the two document processors added to the queue. (See Display 2.10.) Click **Apply Changes** to ensure that the changes that you made are applied to the server.

**Display 2.10: Document Processors Added to the Pipeline Server**

▼ **Document Processors**

heuristic_parse_html

export_to_files

[Add] [Remove] [Edit] [Move Up] [Move Down]

8. Click **Indexing Server**, and change the language to **English** under **General Settings** in Configuration. Expand the Advanced Settings, and select Limited by disk usage for Index size. Enter 1000 for Maximum index size (megabytes) to limit the size of the index and potentially avoid any disk space or memory errors. (See Display 2.11.) Click **Apply Changes** to restart the indexing server and allow the changes that you made to take effect.

**Display 2.11: Language and Index Size Options in the Indexing Server**

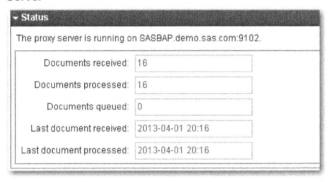

9.  Leave all other options with default settings. Go back to the SAS Web Crawler. Click **Start** to start the SAS Web Crawler. You should notice the word "running" displayed next to the SAS Web Crawler indicating that web crawling is in progress. Click **Proxy Server**, and you should see that the documents are being pumped into the proxy server by the SAS Web Crawler. Simultaneously, documents are transferred to the next stage (i.e., to the pipeline server). (See Display 2.12.)

**Display 2.12: Status Windows Showing Number of Documents Received and Processed by the Proxy Server**

> **▾ Status**
>
> The proxy server is running on SASBAP.demo.sas.com:9102.
>
> | | |
> |---|---|
> | Documents received: | 16 |
> | Documents processed: | 16 |
> | Documents queued: | 0 |
> | Last document received: | 2013-04-01 20:16 |
> | Last document processed: | 2013-04-01 20:16 |

10. Go to the **Pipeline Server**, and click on **heuristic_parse_html** in the document processors. Click **Take Snapshot** under **Document Inspector** to check how the document content is modified at any given stage. Select **heuristic_parse_html** in **Processing Stage**, **1** in **Document**, and **body** in **Field** to view a snapshot of the parsed content from a web page at that point in time. (See Display 2.13.)

**Display 2.13: Status Window Showing Number of Documents Received and Processed by the Proxy Server**

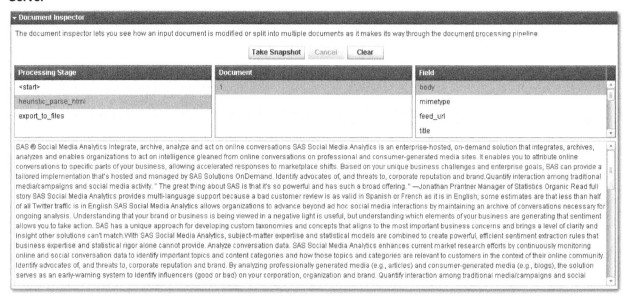

11. Because you have also used the export_to_files document processor in the pipeline server, each of the extracted and parsed documents are exported to the default location of \work\export-to-files in the SAS Information Retrieval Studio installation path. You can navigate to that location on your machine and find those documents as raw text files. (See Display 2.14.) These files contain the body of web pages crawled after the heuristic_parse_html document processor has removed the unwanted HTML tags. These files can be used by other SAS applications such as SAS Text Miner for further analysis as required.

**Display 2.14: Status Window Showing Number of Documents Received and Processed by the Proxy Server**

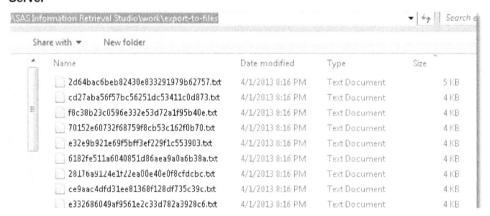

12. Launch the query web server on your browser. Go to the query web server and click on the URL provided in the Status window. You can also launch it by navigating to the SAS Information Retrieval Studio folder in your Start menu. Select **Start ▶ All Programs ▶ SAS Information Retrieval Studio ▶ Query Interface**. By default, the query web server enables you to do simple searches. If you search using the keyword "data mining" and using the inclusion operator (+), it returns all documents that have the term "data mining" in them. (See Display 2.15.)

**Display 2.15: Partial Screen Capture of the Query Interface Displaying Search Results**

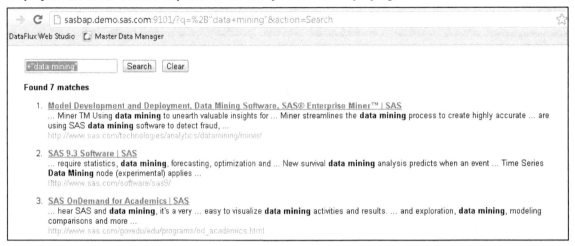

13. In this scenario, you learned how content from web pages can be crawled, parsed, indexed, and searched within the SAS Information Retrieval Studio framework. This scenario is very simple and easy to follow. Often, you might require only specific contents of a website. In the next scenario, you will see a demonstration of how to accomplish this using custom templates.

## Scenario 2: Use SAS Markup Matcher to extract, parse, and index content from specific fields of an HTML web page and export the content to CSV or XML file(s).

1. Click **Global** under **Projects**, and then click **Markup Matcher Server**. You should see the SAS Markup Matcher Server in running status. If not, start the server. Click **Auto-detect** under **General Settings in Configuration**, and select the HTTP proxy server available in your environment.

2. Click the URL provided in the Status window to launch the SAS Markup Matcher Server in a separate window on the browser.

3. In this example, you can use the web page http://www.allanalytics.com/archives.asp?blogs=yes, from which you can easily extract content from specific sections of the page into separate fields in a document. www.allanalytics.com is an online community sponsored by SAS for topics such as analytics, information management, and business intelligence. Users of this website post articles, blogs, and other resources related to any of these topics in general.

4. For this demonstration, you can simply use the blogs section of the website. See Display 2.16, which shows the blog archives from the website ordered by recent to oldest. As you can see, there are many blogs posted by various authors on different topics. Each topic has a title, name of the author, date on which it was published, and a brief description of what that article is all about. If you intend to perform web crawling on this web page similar to how you did in scenario 1, you will capture all of the content in a single document. However, all of these blog posts are different from one another. Hence, it makes sense to create a single document for each post, and then store the information in separate fields. You can create an XML file with tag elements representing the title, author, date, and description of the blog posts.

**Display 2.16: Partial Screen Capture of www.allanalytics.com Web Page Listing Latest Blogs**

5. In SAS Markup Matcher, select **Testing ▶ Download Test Document** to open a window. In that window, enter the URL http://www.allanalytics.com/archives.asp?blogs=yes as the address, and click **OK**. You will see that the HTML code of this web page is opened in the Test Document pane.

6. Click **Edit ▶ Add Document Creation Rule** to create a new document creation rule for the content that you are looking to extract from this web page. This creates a new document in the Rules folder in the Matcher pane. Enter the name "AllAnalyticsBlogs" in the Document Creation Rule pane. For this scenario, the XPath expression of the document rule is very important because it decides how many documents can be extracted.

7. Click **Edit ▶ Add Field Creation Rule** four times to create four fields for this document. Once you select each field in the Matcher pane, you will see the corresponding Field Creation Rule pane. Enter the names "Title," "Author," "Date," and "Description" for these four fields, respectively.

8. Define the XPath expressions for these four fields using the HTML code of the web page that you imported into SAS Markup Matcher as a test document. To do this, you need to locate the first occurrence of the first blog title on the web page. From Display 2.16, you can see that **24 Extra-Large Bunny Suits** is the first title listed on the page. Click **Edit ▶ Find**, and enter the following search term in the pop-up window—**24 Extra-Large Bunny Suits**. Click **Next**. It will take you to the section of the HTML page where this title appears for the first time.

9. Follow these steps for each of the four fields to create the XPath expressions:
   a. Move your pointer and highlight the specific HTML tag that contains the content. In this case, you need to click on the **<span** portion of the tag that encapsulates the content. (See Display 2.17.)
   b. The XPath expression in the Field Creation Rule pane for that field should be populated automatically. (See Display 2.17.) If it is not automatically populated, then you must have either clicked on a wrong tag or a wrong field in the Test Document pane.
   c. The value of the field in the Test Results pane should be automatically populated. (See Display 2.17.)

**Display 2.17: Partial Screen Capture of www.allanalytics.com Web Page Listing Latest Blogs**

10. By now, you should have created the XPath expressions for all the fields and verified that they are populating the test results accurately. Click **File → Save Matcher** to save the matcher that you have created.

11. Click **File → Publish Matchers**, and then click **Upload** to upload this matcher. Enter a new matcher name such as "AllAnalytics," and click **OK** to finish uploading the matcher. The uploaded SAS Markup Matcher should be available within the markup_matcher document processor in the pipeline server.

12. Go back to the SAS Information Retrieval Studio interface, and click **Projects**. Create a new project named "AllAnalyticsCrawl." Click the SAS Web Crawler, and change the following under **General Settings**, keeping the default values for everything else:

    a.  HTTP proxy: Use **Auto-detect** to find a proxy server that is local to your environment.
    b.  Quota (files): 100
    c.  Quota (megabytes): 100
    d.  Limit traversal depth: Yes
    e.  Maximum traversal depth: 2
    f.  Number of downloader threads: 1
    g.  Sleep interval (seconds): 1
    h.  Timeout (seconds): 60
    i.  Maximum number of retries: 3
    j.  Retry delay (seconds): 60
    k.  Find links in Javascript and Flash: Yes

13. Enter the URL http://www.allanalytics.com/archives.asp?blogs=yes as the entry point, and add the same URL in the Scope with a **Match Type** of **Prefix** and an **Action** of **Allow**. Change **Quota (files)** to 100 in the Entry Points section. Click **Apply Changes** to make sure the settings that you changed are saved for the web crawler.

14. Go to the pipeline server, and click **Add** to open the list of available document processors. From this list, select **markup_matcher**, and click **Next**. Provide the name of the matcher that you saved

previously (AllAnalytics) in this window, and click **Finish**. You will see the markup_matcher document processor added to the pipeline server.

15. Now that you have the markup_matcher processor in place to extract only the relevant fields from the web pages, add another document processor, export_csv, to the pipeline server. In the Document Processor: export_csv window, make the following changes as shown in Display 2.18:

    a. Filename: Allanalytics.csv
    b. Drop document from pipeline
    c. Edit the field names under **CSV Columns** to ensure that only the following fields are listed: **Title**, **Author**, **Date**, and **Description**. Click **Finish**.

16. Add another document processor, export_to_files, to the pipeline server. Under **Configuration**, make the following changes:

    a. Document XML root tag: blogs
    b. Drop document from pipeline: Yes
    c. Add the following field names in the **Included Fields** pane to ensure that only these fields are listed: **Title**, **Author**, **Date**, and **Description**. Click **Finish**.

17. You should now have markup_matcher, export_csv, and export_to_files listed as document processors in the pipeline server (in that order). Click **Apply Changes**.

**Display 2.18: export_csv Document Processor Configuration Window**

18. The markup_matcher document processor parses the content on the web and extracts only the four specific fields that you defined in the matcher AllAnalytics. Both export_csv and export_to_files document processors should create the CSV and XML files with those four fields.

19. Go to the SAS Web Crawler, and click **Start** to start the web crawling process on the web pages. You should see the SAS Web Crawler in running status. Go to the proxy server and monitor the status of documents flowing through the proxy server into the pipeline server. Once the web crawler finishes, go to the SAS Information Retrieval Studio installation directory to find the AllAnalytics.csv file created

with the extracted content. Open this file to see the four fields that you defined using the SAS Markup Matcher listed as different columns. Also, the values of the relevant fields are populated in the corresponding cells. (See Display 2.19.)

**Display 2.19: CSV file Created by the export_csv Document Processor**

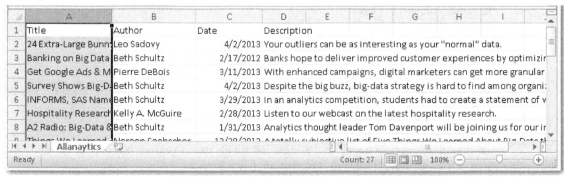

20. In the `\work\export-to-files` folder in the SAS Information Retrieval Studio installation path, you will find individual XML files created for each blog post separately. Select any one file, and open it in the XML editor to view the fields or values populated in the tag or element pairs. (See Display 2.20.)

**Display 2.20: Snapshot of an XML File Created by export_to_files Document Processor**

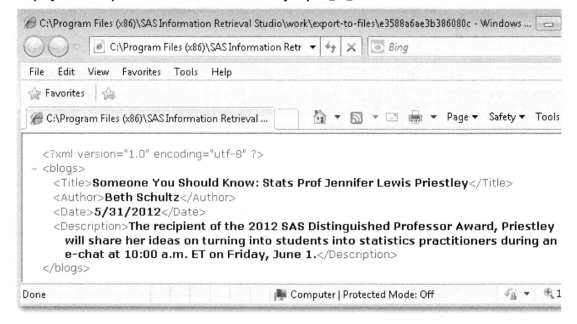

21. In Chapter 4, you will see how to easily use SAS XML Mapper or the SAS Text Miner Text Import node for mapping and exporting the fields or values in CSV or the tags or elements in XML to columns or rows in SAS data sets. Once the data is exported into a SAS data set, it can be used in SAS Text Miner for your analysis.
    **Note:** You can view the logs for all of the server components in SAS Information Retrieval Studio with an option to highlight specific text of interest in the log data. Unless otherwise mentioned, it is important to remember that you should click **Apply Changes** in any server component to ensure that any changes made to the configuration settings are put into effect.

## Summary

In this chapter, we showed you how to configure and use SAS Web Crawler and SAS Search and Indexing, configure some of the prebuilt document processors in the pipeline server, and build XPath expressions using SAS Markup Matcher for custom crawling. Using the integrated capabilities of these features, you can connect to any data source on the web or on your file share systems and grab content in a systematic and consistent way. In Case Study 7, we demonstrate how you can integrate content categorization and concept extraction in SAS Information Retrieval Studio to perform crawling, automatically classify documents, and extract facts or entities simultaneously. With this feature enabled, you can perform faceted searches on the indexed files and intuitively navigate through the search results. The SAS Information Retrieval Studio framework stands out as a key component to extract, parse, normalize, and manipulate data, which is a precursor to any text analysis exercise. In Chapter 3, we discuss how you can use either the file import functionality in SAS Text Miner or the %TMFILTER SAS macro to retrieve unstructured information from the web or from a local file system. These tools were primarily designed to gather data for the text mining process.

## References

Blum, T., Keislar, D., Wheaton, J., and Wold, E. 1998. "Writing a Web Crawler in the Java Programming Language." Muscle Fish LLC.

Gantz, J. and Reinsel, D. 2011. "Extracting Value from Chaos". IDC iView, June 2011. Available at: http://www.emc.com/collateral/analyst-reports/idc-extracting-value-from-chaos-ar.pdf.

Lieberman, H., Fry, C., and Weitzman, L. 2001. "Exploring the Web with Reconnaissance Agents". *Communications of the ACM*, 44(8): 69-75.

Manning, C. D., Raghavan, P., and Schutze, H. 2008. *Introduction to Information Retrieval.* New York, NY: Cambridge University Press.

McCallum, A. K., Nigam, K., Rennie, J., and Seymore. K. 2000. "Automating the Construction of Internet Portals with Machine Learning". *Information Retrieval.* 3(2): 127-163.

Mearian, L. 2011. "World's Data Will Grow by 50X in Next Decade, IDC Study Predicts." Computerworld, June 2011. Available at: http://www.computerworld.com/s/article/9217988/World_s_data_will_grow_by_50X_in_next_decade_IDC_study_predicts.

Mitchell, B. 2008. "UNC - Universal Naming Convention - Windows UNC." About.com Guide: Wireless Networking. Available at: http://compnetworking.about.com/od/windowsnetworking/g/unc-name.htm.

Pant,G. and Menczer,F. 2003. "Topical Crawling for Business Intelligence". *Proceeding of 7th European Conference on Research and Advanced Technology for Digital Libraries (ECDL 2003).* Trondheim, Norway, 2003.

Pant, G., Srinivasan, P., and Menczer, F. 2004. "Crawling the Web". *Web Dynamics: Adapting to Change in Content, Size, Topology and Use.* New York: Springer-Verlag, 153–78.

SAS® Information Retrieval Studio, Release 12.1. Administrator's Guide. Cary, NC: SAS Institute Inc.

SAS® Information Retrieval Studio, Release 12.1. User's Guide. Cary, NC: SAS Institute Inc.

SAS® Information Retrieval Studio, Release 12.1. Quick Start Guide. Cary, NC: SAS Institute Inc.

Srinivasan, P., Mitchell, J., Bodenreider, O., Pant, G., and Menczer, F. 2002. "Web Crawling Agents for Retrieving Biomedical Information". *NETTAB 2002: Agents in Bioinformatics*, Bologna, Italy, 2002.

---

[i] By default, the virtual indexing server and the virtual query server are not enabled. If they are turned on, they work hand in hand with back-end indexing servers and query servers to distribute the document load. Enabling these components can help improve the overall performance and optimal disk space or memory utilization of the system in cases where your crawling process can potentially retrieve documents in the range of a few petabytes or more by volume.

# Chapter 3   Importing Textual Data into SAS Text Miner

This chapter starts with a brief introduction to SAS Enterprise Miner and SAS Text Miner. We later discuss how SAS Text Miner can be used to import raw data available in various file formats into SAS and how to import data from the web using a web crawler. We briefly discuss and demonstrate the powerful use of Perl regular expressions for cleaning textual data before it is used for text mining. This chapter includes a brief introduction to the rich collection of character functions available in SAS for manipulating textual data.

## An Introduction to SAS Enterprise Miner and SAS Text Miner

SAS Enterprise Miner lets you build predictive and descriptive data mining models based on large volumes of data. The easy-to-use interface simplifies many common tasks associated with applied data mining, including text mining. Text mining results can be easily integrated with the data mining model process flow to enhance model performance. SAS Enterprise Miner offers secure analysis management and provides a wide variety of tools with a consistent graphical interface. You can customize it by incorporating your choice of analysis methods and tools that you want to analyze. The interface window is divided into several functional components.

- The *menu bar* and corresponding *shortcut buttons* perform the usual Windows tasks, in addition to starting, stopping, and reviewing analyses.
- The *Project panel* manages and displays data sources, diagrams, results, and project users.
- The *Properties panel* enables you to view and edit the settings of data sources, diagrams, nodes, results, and users.
- The *Help panel* displays a short description of the property that you select in the Properties panel. Extended help can be found in the Help Topics option in the Help main menu.
- In the *Diagram workspace*, process flow diagrams are built, edited, and run. The workspace is where you graphically sequence the tools that you use to analyze your data and to generate reports. The Diagram workspace contains one or more *process flows*.

A process flow starts with a data source and sequentially applies SAS Enterprise Miner tools to complete your analytic objective. A process flow contains several nodes. *Nodes* are SAS Enterprise Miner tools connected by arrows to show the direction of information flow in an analysis. The SAS Enterprise Miner tools available to your analysis are contained in the *tools palette*. The tools palette is arranged based on the SAS process for data mining, SEMMA (Sample, Explore, Modify, Model, and Assess). Additional tools are available in the Utility group. Other specialized group tools (for example, Credit Scoring, Applications, Text Mining, and Time Series) are available if licensed. Each group of tools contains specialized tools for doing appropriate analysis. For example, the Text Mining group contains the following tools: Text Cluster, Text Filter, Text Import, Text Parsing, Text Topic, and Text Rule Builder.

**Display 3.1: SAS Enterprise Miner Interface**

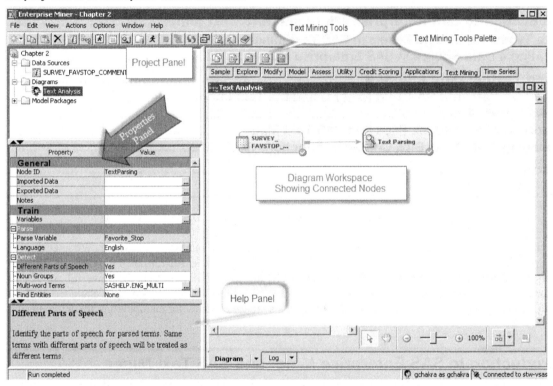

## Data Types, Roles, and Levels in SAS Text Miner

SAS Text Miner provides a set of tools that enable you to collect textual data and analyze the data to extract information. In addition, because you can embed the SAS Text Miner node in a SAS Enterprise Miner process flow diagram, you can combine quantitative variables with unstructured text in the mining process. As a result, you are incorporating text mining with other traditional data mining techniques. Therefore, data types in SAS Text Miner accommodate both structured (numeric) and unstructured (textual) data.

Data roles are typically different based on data types. For unstructured (textual) data that you want to analyze, the role is usually set to **Text** and the measurement level to **Nominal**. For numeric data, there are many data role options depending on your analysis objectives. For example, you can set the role of a numeric variable to **ID** when you want to use that numeric variable to reflect the unique identification number for each value in the text data. You can set the role of a numeric variable to **Time ID** when you want to use that numeric variable to reflect the time stamp associated with each value in the text data. You can set the role of a numeric variable to **Input** when you want to use that variable as an input variable in any subsequent nodes. For example, if you created a flag variable to indicate the presence or absence of a graphics file for each value in your text, and you want to use this flag variable in subsequent analysis, then you can assign the flag variable a role of **Input**. You can set the role of a numeric variable to **Target** when you want to use that variable as a target variable in any subsequent modeling nodes such as Regression. For example, if you have a numeric variable that indicates positive, negative, or neutral sentiments expressed in text data, and you want to predict this sentiment variable using the text data, you can assign the sentiment variable a role of **Target**. For other data roles available for numeric data, see the online Help for SAS Enterprise Miner.

For numeric data, there are five possible measurement levels, depending on the measurement scale used in creating or collecting the data. The level *Unary* is used when a variable has the same value for every observation in the data set. The level *Binary* is used when a categorical variable has two distinct values. The level *Nominal* is used when a categorical variable has three or more values and the measurement property reflects that the observations with different values simply represent different categories. The level *Ordinal* is used when a categorical variable has three or more values and the measurement property of the attribute reflects that different values present a natural ordering, but on which no arithmetic-like operations can be performed. The level *Interval* is used when the variable has many numeric values and the measurement property of the attribute reflects meaningful and equal intervals in the scale.

## Creating a Data Source in SAS Enterprise Miner

Before any data can be processed via SAS Text Miner, the data has to be brought into a SAS Enterprise Miner project. The following discussion takes you through the creation of a SAS Enterprise Miner project, the creation of a SAS library, the creation of a diagram, and the creation of a data source.

### (a) Create a SAS Enterprise Miner Project

The exact procedure to access and start SAS Enterprise Miner depend on the type of installation. A commonly used procedure to access the welcome window is to select **Start ▶ Programs ▶ SAS ▶ Analytics ▶ SAS Enterprise Miner** (or, double-click the SAS Enterprise Miner icon if it is on the desktop).

**Display 3.2: SAS Enterprise Miner Welcome Window**

To define a project, you specify a project name and the location of the project on the SAS Foundation Server.

- Select **File ▶ New ▶ Project** from the main menu or click **New Project** in the welcome window.

The Create New Project wizard opens at Step 1. Enter a project name and specify a path to a folder on your machine where you want the project to be saved.

**Display 3.3: SAS Enterprise Miner Create New Project Wizard**

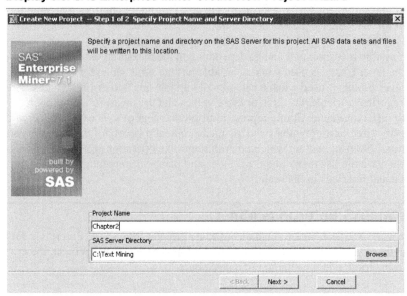

- Click **Next**. To finish defining the project, click **Finish** in the next step.

The SAS Enterprise Miner client application opens the project that you created.

**Display 3.4: SAS Enterprise Miner Window after Project Creation**

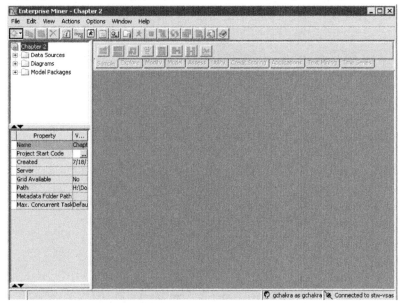

## (b) Create a SAS Library

A SAS library connects SAS Enterprise Miner with the raw data sources, which are the basis of your analysis. To define a SAS library, you need to know the name and location of the data that you want to link with SAS Enterprise Miner.

- Select **File ▶ New ▶ Library** from the main menu.

**Display 3.5: SAS Enterprise Miner Library Wizard—Step 1**

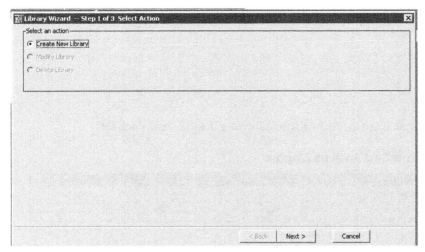

- Click **Next**. In Step 2, enter a name for the library (book) and specify a path to the folder where your data is located.

**Display 3.6: SAS Enterprise Miner Library Wizard—Step 2**

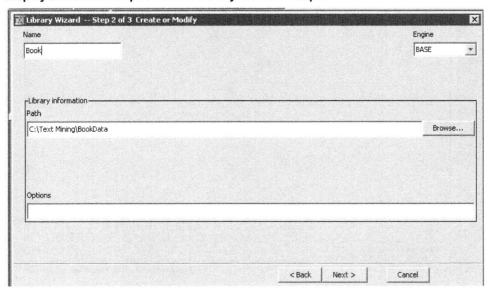

- Click **Next**. Click **Finish**.

Alternatively, with the project name selected, open the Project Start Code dialog box in the Properties panel, enter the following code, and run it to create a library:

```
libname Book 'C:TextMining\BookData';
```

All data available in the SAS library can now be used by SAS Enterprise Miner.

## (c) Create a SAS Enterprise Miner Diagram

All analyses in SAS Enterprise Miner are usually performed by adding nodes to a diagram.

- To create a diagram, select **File ▶ New ▶ Diagram** from the main menu.

**Display 3.7: SAS Enterprise Miner Create New Diagram**

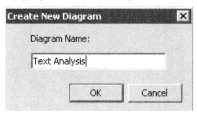

- Enter an appropriate name (such as Text Analysis) for your diagram, and click **OK**.

**Display 3.8: SAS Enterprise Miner Text Analysis Diagram**

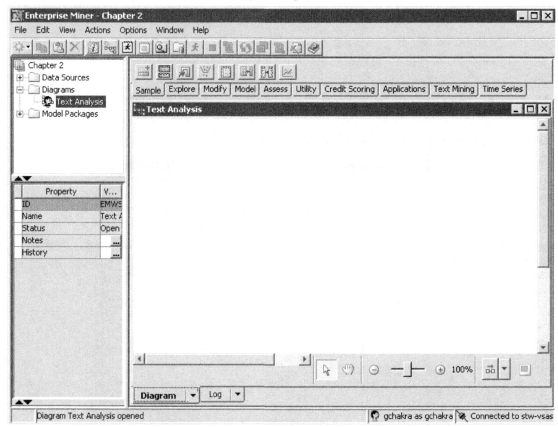

## (d) Create a SAS Enterprise Miner Data Source

A data source links SAS Enterprise Miner to an existing data table. In this example, you create a data source that has comments from customers of a national fuel chain. These customers completed a survey that has closed-ended questions (with numeric categories as responses) and open-ended questions (where customers entered their comments). For this example, you use only the data that has comments from one question in the survey, "Why do you consider this store a favorite stop?" The SAS data set name is Survey_FavStop_Comments.

- Select **File** ⇨ **New** ⇨ **Data Source** from the main menu to create a data source. Alternatively, you can right-click on the Data Sources folder in the Project panel, and select Create Data Source to start the Data Source Wizard.
- In Step 1 of the Data Source Wizard, click **Next**.
- In Step 2 of the Data Source Wizard, click **Browse**.
  - ○ Double-click to open the library where the data is located (Book).

     ◦   Select the table or data that you want to analyze (Survey_FavStop_Comments).

     ◦   Click **OK**.

     ◦   Click **Next**.

- Step 3 of the Data Source Wizard provides some basic information about the data set such as the number of variables, number of observations, etc. Click **Next**.
- Step 4 of the Data Source Wizard starts the metadata definition process. SAS Enterprise Miner assigns initial values to the metadata based on characteristics of the selected SAS table. The **Basic** setting assigns initial values to the metadata based on variable attributes such as the variable name, data type, and assigned SAS format. The **Advanced** setting assigns initial values to the metadata in the same way as the **Basic** setting, but it also assesses the distribution of each variable to better determine the appropriate measurement level. This is where it becomes important for you to know the modeling role and appropriate measurement level of each variable in the source data set. Click **Next** to use the **Basic** setting.
- Step 5 of the Data Source Wizard displays its best guess for the metadata assignments.

**Display 3.9: SAS Enterprise Miner Text Analysis Diagram**

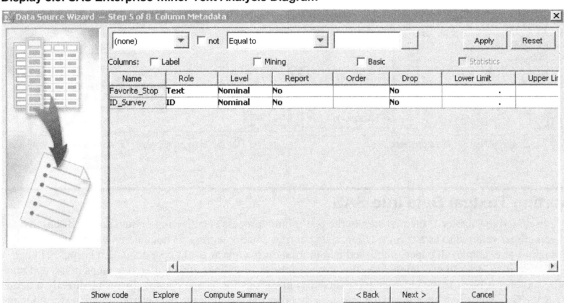

The variable **ID_Survey** is an identification number and is assigned a role of **ID** and measurement level of **Nominal**. The variable **Favorite_Stop** is text data and is assigned a role of **Text** and measurement level of **Nominal**. You can click next to each variable's role or level and change it to another value if the metadata guess is incorrect. In this case, it appears that the metadata guesses are correct.

- Click **Next** to accept these role and level assignments.
- Step 6 of the Data Source Wizard gives you the option to create a sample data set. Click **Next** to accept the default (no sample data set) option.
- Step 7 of the Data Source Wizard enables you to set a role for the data source and to add descriptive comments about the data source definition. For most analyses, a table role of Raw is acceptable. Click **Next** to accept the default options.
- The final step in the Data Source Wizard provides summary details about the data source that you created. Click **Finish**.

The Survey_FavStop_Comments data source is added to the Data Sources folder in the Project panel and is now ready to be analyzed.

**Display 3.10: SAS Enterprise Miner Text Analysis Diagram**

## Importing Textual Data into SAS

As discussed in Chapter 1, the first step in the text mining process is collecting textual data and setting up what is sometimes referred to as a corpus. Often, a text corpus usually represents documents from a particular domain. For example, all papers published in a journal during a year could be a document corpus. Though collecting data looks like a simple task, this is one of the most tedious and challenging steps in the text mining process. This is because the unstructured data exists in various forms that cannot always be directly processed by SAS Text Miner. A data conversion step usually takes place before the data is used in the text mining task unless the data is readily available as a SAS data set.

The data collection step heavily depends on the business problem. This involves answering simple questions such as the following:

- *What data is needed for this project?*
    - ✓ Customer reviews, application documents, news articles, e-mails, etc.
- *Where is the data available?*
    - ✓ Client directory, database, data warehouse, web, etc.
- *How is it available?*
    - ✓ PDF files, XML pages, Word documents, web pages, etc.
- *When should the data be collected (when collecting non-static data)?*
    - ✓ Before or after a specific event, data from wikis or blogs that are frequently updated.

Once the data that is needed to solve the current problem is identified, the next challenge is to collect the data and convert it for SAS Text Miner to process. It is quite possible that you will face any of the following situations in your project:

1. Data is readily available as a SAS data set.
2. Data is available as textual files (PDF, XML, HTML, Word, etc.) in a directory or in a database.
3. Data needs to be collected from the Internet.

The text mining process in SAS requires that the data be available as a SAS data set. This does not mean that the source data has to exist as a SAS data set (see situations 2 and 3 for other formats). In the first situation, you can directly create a data source and perform text mining. In the latter two situations, the source data has to be converted into a SAS data set. There are various ways to create a SAS data set from the data available in commercial databases using SAS data access features. Files in common formats such as comma-separated values (CSV) or Microsoft Excel can be easily imported into SAS using the SAS Enterprise Guide Data Import Wizard or the File Import node in SAS Enterprise Miner. The challenging part is to create SAS data sets from textual files. SAS Text Miner has the capability to create SAS data sets dynamically from textual files available in a directory or on the web. This is accomplished using the Text Import node in SAS Text Miner.

In Chapter 2, you learned how to collect data from the web and local files using SAS Information Retrieval Studio. Data collected using SAS Information Retrieval Studio is fed into SAS Text Miner for further text mining accomplishments. The following sections discuss in detail the different ways to collect textual data using SAS Text Miner.

## Importing Data into SAS Text Miner Using the Text Import Node

The Text Import node in the Text Mining collection of nodes can be used to import source data into SAS Text Miner as a SAS data set. The source data can exist in a directory in any proprietary file format type. The Text Import node requires that the SAS Document Server be installed and running for performing document conversion. More than 100 file formats are supported by the Text Import node. Some of them include the following:

- Microsoft Word (.doc, .docx)
- Microsoft Excel (.xls, .xlsx)
- Microsoft PowerPoint (.ppt, .pptx)
- Rich Text Format (.rtf)
- Adobe Acrobat (.pdf)
- ASCII (.txt)

The node requires a source directory (or Import File directory) that contains all of the source files and a destination directory to save all of the converted files. SAS Text Miner traverses the source directory, extracts the text from the files in this directory, and creates plain text files in the destination directory. Any source file in a form not readable by SAS Text Miner is omitted from the document conversion. Similarly, if the number of characters in each file exceeds 32,000, then the file is truncated. If source data is collected from the web, the node crawls the websites, retrieves files, and puts them in the source directory before the extraction process. The node can easily identify the language of the document and then transcode the document into the session encoding. Analysts who prefer programming can perform the same text import tasks. They can automate the process using the %TMFILTER macro. We discuss the macro in the next section.

### (a) Extract Text from PDF Files and Create a SAS Data Set

Suppose you want to analyze thousands of documents that are available in PDF format and stored on your desktop or on a network location. You want to represent each PDF file as a unit of analysis (i.e., the contents of one PDF file should reside in exactly one cell in your table or SAS data set as shown in Display 3.11).

**Display 3.11: Mapping Each PDF Document to Each Cell in a SAS Data Set**

In this example, you can use the Text Import node to convert 1,000 SAS Global Forum conference papers available in PDF format to a SAS data set. After this step, you will have a SAS data set of 1,000 observations in which each observation corresponds to one SAS Global Forum paper. A SAS data set cell can hold a maximum of 32,000 characters. Hence, it is not possible to include all of the contents of a single paper in one cell. In this situation, a cell can take only the first 32,000 characters in the paper. The Text Import node provides a facility to capture the links to the actual papers instead of the content of the actual papers. The output data set contains a variable that tells the successor nodes to use the document path information to access the full content of the papers. If you think you need only a certain number of bytes of text from each paper, you can set the node properties to fetch only that number of bytes of text. Only text that is present in the documents is imported. To extract tables, check boxes, or any other objects from the documents requires additional preprocessing. If the PDF is produced by a scanner, you first need to convert the images to textual documents by using any optical character recognition (OCR) software.

Let's see in detail how you can perform these operations using the Text Import node.

We start by creating the two necessary folders for the Text Import node to work right. We have a folder named Source created on our desktop that contains 1,000 SAS Global Forum papers, all available as PDF files. We have another folder named Destination in the same parent folder to save the filtered files. Drag and drop a Text Import node onto the Diagram workspace in SAS Enterprise Miner. This node does not require any predecessor nodes. Set the Train properties of the node as shown in Display 3.12.

Import File Directory: C:\TextMining\Source
Destination Directory: C:\TextMining\Destination

**Display 3.12: Text Import Node Train Property Settings**

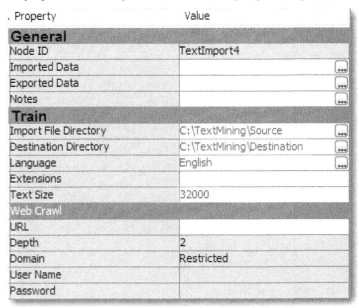

| Property | Value |
|---|---|
| **General** | |
| Node ID | TextImport4 |
| Imported Data | |
| Exported Data | |
| Notes | |
| **Train** | |
| Import File Directory | C:\TextMining\Source |
| Destination Directory | C:\TextMining\Destination |
| Language | English |
| Extensions | |
| Text Size | 32000 |
| **Web Crawl** | |
| URL | |
| Depth | 2 |
| Domain | Restricted |
| User Name | |
| Password | |

The Extensions property is used to restrict the types of files to extract. For example, if you provide HTML as the value for this property, the node will filter only files with an HTML extension. If you leave this value blank, the node will filter all of the file types that are supported by the SAS Document Conversion server. The Text Size property is used to control the size of the text that you want to extract from each document into the TEXT variable in the output data set. This will serve as a snippet when the size is small. We have set it to 32,000 to extract the maximum possible text.

The Language property sets the possible choices that the language identifier uses when assigning a language to each document. Select one or more languages in the Languages dialog box (this dialog box can be opened by clicking the ellipsis button next to the property). Only languages that are licensed can be used. We have only a license for English as shown in Display 3.13.

**Display 3.13: Dialog Box to Select Language Choices**

The results from running the Text Import node with these settings produces the results shown in Display 3.14.

**Display 3.14: Text Import Node Results Window**

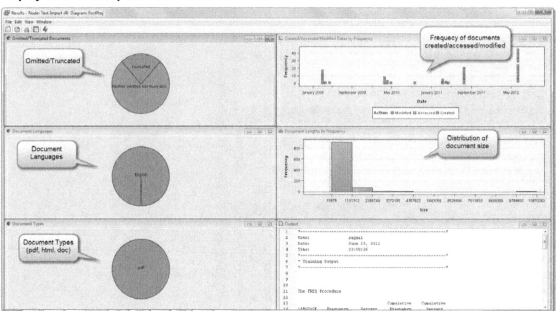

The Results window of the Text Import node shows various distribution plots that report the number of documents that were omitted or truncated, document languages, types, sizes, and a time series plot of when the document was created, modified, and accessed. These plots give a basic idea of how successful the text import task was. The Output window (lower right corner in Display 3.14) reports the PROC FREQ results on document languages and a cross tabulation of omitted versus truncated documents. In Display 3.15 , there are three documents for which the Text Import node was not able to correctly identify the language. The cross tabulation shows that no documents were omitted, but 236 of them were truncated. You can explore the Destination folder for the TXT files that were extracted from PDF files.

**Display 3.15: PROC FREQ Results of the Document Import Process**

```
The FREQ Procedure

                                  Cumulative   Cumulative
LANGUAGE      Frequency   Percent  Frequency    Percent
-------------------------------------------------------
English          997      100.00      997      100.00

Frequency Missing = 3


Table of TRUNCATED by OMITTED

TRUNCATED      OMITTED

Frequency|
Percent  |
Row Pct  |
Col Pct  |       0|   Total
---------+--------+
       0 |    764 |     764
         |  76.40 |   76.40
         | 100.00 |
         |  76.40 |
---------+--------+
       1 |    236 |     236
         |  23.60 |   23.60
         | 100.00 |
         |  23.60 |
---------+--------+
Total        1000      1000
            100.00    100.00
```

The variables in the data set that was exported by the Text Import node are shown in Display 3.16.

**Display 3.16: Variables in the Output Data Set from the Text Import Node**

Properties - EMWS1.TextImport4_TRAIN

| Table | Variables |

Columns: ☐ Label ☐ Mining ☐ Basic

| Name | Role | Level |
| --- | --- | --- |
| ACCESSED | Time ID | Interval |
| CREATED | Time ID | Interval |
| EXTENSION | Input | Nominal |
| FILTERED | Text Location | Nominal |
| FILTEREDSIZE | Input | Interval |
| LANGUAGE | Input | Nominal |
| MODIFIED | Time ID | Interval |
| NAME | Input | Nominal |
| OMITTED | Input | Interval |
| SIZE | Input | Interval |
| TEXT | Text | Nominal |
| TRUNCATED | Input | Interval |
| uri | Web Address | Nominal |

It is usually the Text Parsing node that is connected to the Text Import node in the process flow. There are two columns that can be used by the Text Parsing node to identify the text: FILTERED and TEXT. In the Text Parsing node, when you set the Use property of the TEXT variable to Yes, and the Use property of the FILTERED variable to No, only the text that is contained in the TEXT variable will be used for parsing. Otherwise, the FILTERED variable is always given the preference when both of the variables have the Use property set to Yes.

Exploring the exported data of the Text Import node will give you a sense of the data that will be used for parsing. The TEXT variable captures a snippet of the text that is extracted. From the Explore window of the exported training data set, sort the data by the Language column so that the table will show the records with a missing value for the Language column on the top. As shown in Display 3.17, the first three rows of the TEXT variable contain just letters and symbols without any words or sentences. This is because the three papers in these three rows were encoded in a form the document conversion tool cannot identify.

**Display 3.17: Explore Window of the Exported Data Set from the Text Import Node**

## (b) Retrieve Data from the Web and Create a SAS Data Set

Analyzing the data available on the web has become a competitive advantage for almost any organization. A treasure trove of information about brands, customer preferences, product reviews, stock prices, movie reviews, etc., can be found on the World Wide Web. This information from the Internet, in combination with internal data, can make your analysis more powerful and current. However, the challenging part of the task is to find a way to get this data into SAS. This can be easily achieved using a web crawler.

A web crawler is a sophisticated computer program that crawls the web in a methodical and automated manner. Web crawling is used for various purposes by organizations. Internet search engines use web crawlers to index web pages and stay current. Others use crawlers for automating maintenance tasks on websites. The web crawling process starts with a list of starting URLs or seed pages. The pages identified with these URLs are downloaded and any hyperlinks on these pages are extracted for further downloading.

A web crawler undertakes the following four tasks:

1. Starts with an URL from candidate URLs.
2. Downloads the associated web pages.
3. Extracts any hyperlinks on these web pages.
4. Updates the candidate list with these hyperlinks for further downloading.

There are different methods available in SAS to use the web crawler functionality. You can write a simple DATA step program or use the Text Import node in SAS Enterprise Miner. SAS Crawler is a sophisticated tool with more advanced web crawler capabilities as discussed in Chapter 2. In this chapter, you will see how you can use the Text Import node for downloading web pages and creating SAS data sets for text mining. SAS Crawler enables you to download different types of files such as HTML, PDF, PPT, DOC, etc., from the web from which text can be extracted.

For example, we show you how to download the web pages from the website www.allanalytics.com. This website is a great resource for analytics professionals. It brings together professionals, researchers, and students working in the field of analytics to share their experiences and know-how on business intelligence, advanced analytics, and data management. This website is a good start if you want to understand the latest happenings in the field of analytics and what has happened in the recent past. However, numerous blogs and message boards on this website make it tough to read each post. An easy solution is to download the web pages and perform text mining to see what different topics were discussed and what was the trend for these topics.

The Text Import node properties that need to be set for web crawling are shown in Display 3.18. Similar to importing PDF files (as seen in an earlier section), you need two folders, a Source folder for storing the downloaded HTML pages, and a Destination folder for storing the textual extracts. The paths to these two folders are set as property values for the Import File directory and the Destination File directory settings:

Import File Directory: C:\TextImport\Source

Destination Directory: C:\TextImport\Destination

**Display 3.18: Text Import Node Web Crawling Property Settings**

| Property | Value |
| --- | --- |
| **General** | |
| Node ID | TextImport |
| Imported Data | |
| Exported Data | |
| Notes | |
| **Train** | |
| Import File Directory | C:\TextImport\Source |
| Destination Directory | C:\TextImport\Destination |
| Language | English |
| Extensions | |
| Text Size | 1000 |
| ⊟ Web Crawl | |
| URL | www.allanalytics.com |
| Depth | 2 |
| Domain | Restricted |
| User Name | |
| Password | |

The use of the property settings **Language**, **Extensions**, and **Text Size** is the same as discussed in the previous section. Setting the **Text Size** property to **1000** will show a maximum text value of 1,000 characters for each extracted web page. The **URL** property is where you specify the initial web page to crawl. In this example, we have assigned the name of the website www.allanalytics.com. The **Depth** property is used to specify the extent of the number of levels of links to crawl from the initial URL page. A web page can contain links to many other pages, where each page, again, is linked to hundreds of other pages. These links can belong to the same website domain or to a different website. For example, you can see that the page www.allanalytics.com has links to other pages in the domain and links to a resource page on the www.sas.com website as shown in Display 3.19.

**Display 3.19: Home Page of the www.allanalytics.com Website**

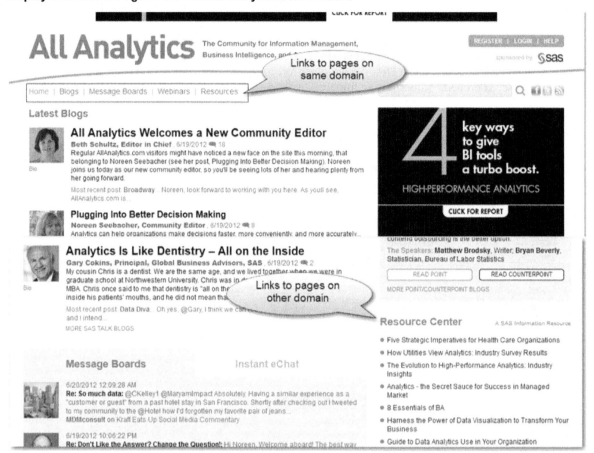

A value of 1 for the Depth property would extract all of the files linked to from the initial page. Hence, the web pages for the links on the home page of the allanalytics.com site, such as **Blogs**, **Message Boards**, **Webinars**, and **Resource Center**, are extracted. If you leave this property value set to its default (i.e., 2), the web pages of the links available in **Blogs**, **Message Boards**, **Webinars**, **Resources**, and **Resource Center** are also extracted. In this way, you can control the level up to which you want to extract web pages. You need to be cautious when using this property. As you increase the depth, the number of pages extracted grows exponentially.

Most of the time, you want to restrict your analysis only to web pages from one particular domain or website. For example, when analyzing the allanalytics.com web page, you do not want to extract web pages from other domains such as www.sas.com using the links that are contained in the allanalytics.com web pages. The property setting **Domain** is used to control this behavior. A value of **Restricted** prevents the crawler from processing documents outside the domain of the initial web page. If this is not a concern in your study, then leave this property value set to **Unrestricted**. **User Name** and **Password** properties are used when the URL input refers to a secured website and requires a user name and password to access the content on that domain.

After the Text Import node run is complete, the Results window generates the same types of plots as seen in the previous section. You can look in the Destination and Source folders for the files that are downloaded from the web and for the text files that were filtered from these downloaded files. The downloaded files are named File1, File2, File3, and so on. With the current settings, it took 20 minutes to download around 1,800 web pages. This completely depends on the website that you are downloading from and the value set for the Depth property.

Display 3.20 shows a sample of the data set that is created by the Text Import node. You can browse the data using the Exported Data property from the Properties panel. The **Text** field shows a text snippet with the first

1000 characters from the downloaded web page. The **URL** column lists the source URL of the web page. This list shows that all of the pages are downloaded from the same domain, www.allanalytics.com.

**Display 3.20: Sample Output of the Data Set Created from Web Crawling**

| | Text Snippet | URL | | Path to filtered .txt file | |
|---|---|---|---|---|---|
| EMWS2.TextImport_TRAIN | TEXT | uri | NAME | FILTERED | LANGUAGE |
| 1 | MOBILE Home \| Blogs \| Message Boards \| Webinars \| Resources REGISTER \| LOGIN \| HELP </for m> Latest Blogs Bio All Analytics Welcomes a New Community Editor Beth Schultz, Editor in Chief , 6/19/2012 18 Regular AllAnalytics.com visitors might have noticed a new face on the site this m orning, that belonging to Noreen Seebacher (see her post, Plugging Into Better Decision Making) . Noreen joins us today as our new community editor, so you'll be seeing lots of her and hearing plenty from her going Forward. Most recent post: Broadway ... Noreen, look forward to working with you here. As youll see, AllAnalytics.com is... Bio Plugging Into Better Decision Making Noree n Seebacher, Community Editor , 6/19/2012 8 Analytics can help organizations make decisions fa ster, more conveniently, and more accurately... at least in theory. In practice, however, faster and easier decision making is often just an unrealized goal, and floods of data overwhelm the pe ople struggling to make sense of it. Most recen | http://www.allanalytics.com | file1.html | C:\TextImport\Destination\file1.html.txt | English |
| 1460 | MOBILE Home \| Blogs \| Message Boards \| Webinars \| Resources REGISTER \| LOGIN \| HELP </for m> Blogs Latest Content View Content by Month recent \| most commented \| bloggers Page 1 / 2 > > > All Analytics Welcomes a New Community Editor Beth Schultz 6/19/2012 14 comments Say hello to Noreen Seebacher, who joins our community today. Plugging Into Better Decision Makin g Noreen Seebacher 6/19/2012 8 comments Big-data is supposed to provide big answers. But fa ster and easier decision making remains an often unrealized goal, and floods of data overwhelm the people struggling to make sense of it. Analytics Is Like Dentistry All on the Inside Gary Cokin s 6/19/2012 2 comments BI and business analytics let organizations see inside and find ways to i mprove operations. A Perspective on Using Metrics Appropriately Tom Sattler 6/19/2012 7 comm ents Not all uses of metrics and measurements are beneficial, as this example shows. 'Big Brothe r in Arkansas' Is Watching (& Profiling) Us Beth Schultz 6/18/2012 | http://www.allanalytics.com/archives.asp?blogs=yes | file7.html | C:\TextImport\Destination\file7.html.txt | English |
| 773 | MOBILE Home \| Blogs \| Message Boards \| Webinars \| Resources REGISTER \| LOGIN \| HELP </for m> Blogs Latest Most Commented Content View Content by Month recent \| most commented \| bl oggers Page 1 / 2 > > Banking on Big Data Literally Beth Schultz 2/17/2012 103 comments Ban ks hope to deliver improved customer experiences by optimizing how they manage, mine, and an alyze the data they're amassing. Watson Is Going to Take My Job! Tricia Aanderud 5/25/2012 8 6 comments Some might think an intelligent, automated learning machine can fill the data scientis t skills gap, but not me. Tuning In to Google+ Pages & Web Analytics Pierre DeBois 11/29/2011 7 0 comments As companies rush to deploy Google+ Pages profiles, they shouldn't underestimate how availability affects Web analytics. Target Has You in Its Bullseye Beth Schultz 2/22/2012 64 | http://www.allanalytics.com/archives.asp?blogs=yes&p_fltr=mostcommented | file1721.html | C:\TextImport\Destination\file1721.html.txt | English |

The exported data is used by the successor nodes, primarily the Text Parsing node, for processing and use in the text mining process. Chapter 4 discusses in detail the functionality of the Text Parsing node.

## %TMFILTER Macro

The macro program that runs behind the Text Import node is %TMFILTER. Starting with SAS Text Miner 4.1, this macro is made available as a node. If you are using an earlier version of SAS Text Miner, the same operations of filtering and web crawling can be performed using this macro. It is supported in all operating systems for filtering and on Windows for web crawling. Here is the macro syntax:

```
%TMFILTER (DIR=path,
URL=path <,DATASET=<libref.>output-data-set <,
DEPTH=n,
DESTDIR=path,
EXT=extension1 <extension2  extension3...>,
FORCE=anything,
HOST=name | IP address,
LANGUAGE=ALL | language1 <language2 language3...>,
         NORESTRICT=anything,
     NUMBYTES=n,
     PORT=port-number,
     PASSWORD=password,
     USERNAME=username>)
```

All of the above options are available as properties of the Text Import node, except for FORCE=, HOST=, and PORT=. The FORCE option is used to let the macro run even if the Destination directory is not empty. If no value is set for the property, the node or the macro will terminate if the Destination directory is not empty. The HOST and PORT options are necessary for the SAS Document Conversion server to function properly. The HOST property specifies the name or IP address of the machine on which to run the %TMFILTER macro. If you do not specify a value, then the macro assumes that the SAS Document Conversion server will use its own defaults. The PORT option is used to specify the number of the port on which the SAS Document Conversion server resides. It is more convenient for troubleshooting if you are using the %TMFILTER macro for importing text. For more technical information and examples of how to work with the macro, see the SAS Text Miner User Guide.

## Importing XLS and XML Files into SAS Text Miner

In the previous sections, we demonstrated how you can extract data from documents with file extensions such as .pdf, .html, or .txt into SAS data  the Text Import node feature in SAS Text Miner to import the data, it grabs all of the text in one row of the output SAS data set. Data is truncated from the TEXT field if its length exceeds the value provided in the Text Size property for the Text Import node. However, this is not the output that you would expect to generate. It is important to import each of the review comments as separate rows in the output data set. Hence, you can use the File Import functionality in SAS Enterprise Miner, located on the **Sample** tab, to import data from an Excel file. Drag and drop the node onto the Diagram workspace, and change the **Import File** property to point to the location of the Excel file. (See Display 3.21.) Change the **Delimiter** property to **SP**, which represents that a space character separates the columns in the Excel spreadsheet. You can use "**,**" as the delimiter for CSV files.

**Display 3.21: File Import Node Properties Pane**

| Train | |
|---|---|
| Variables | [...] |
| Import File | C:\test\dir\TabletReviews.xls [...] |
| Maximum rows to import | 1000000 |
| Maximum columns to import | 10000 |
| Delimiter | SP |
| Name Row | Yes |
| Number of rows to skip | 0 |
| Guessing Rows | 500 |
| File Location | Local |
| File Type | xls |
| Advanced Advisor | No |
| Rerun | No |

Run the File Import node to extract data from the Excel file into the SAS environment. When the **Name Row** property is set to **Yes**, SAS identifies the first row in the file as column names, and it uses the column names to name SAS variables in the output data set. The **Guessing Rows** property is useful to assess the data in the first $n$ rows of the Excel file, and it estimates values for attributes such as Level, Type, Format, Informat, and Length of the SAS variables. In the Properties pane, click on the ellipsis button for **Exported Data**, select the **Train** data identified by the column **Role**, and click **OK**. The Properties window appears. Go to the **Variables** tab, and select the **Basic** check box to find all of the variables and their attributes. (See Display 3.22.)

**Display 3.22: SAS Output Data Set Variables and Its Data Attributes**

| Table | Variables | | | | | | |
|---|---|---|---|---|---|---|---|

Columns: ☐ Label    ☐ Mining    ☑ Basic    ☐ Statistics

| Name | Role | Level | Type | Format | Informat | Length |
|---|---|---|---|---|---|---|
| NumComments | Input | Interval | Numeric | BEST15.0 | | 8 |
| Review | Text | Nominal | Character | $16418.0 | $16418.0 | 16418 |
| StarRating | Input | Interval | Numeric | BEST10.0 | | 8 |
| Title | Text | Nominal | Character | $125.0 | $125.0 | 125 |
| Useful | Input | Interval | Numeric | BEST6.0 | | 8 |
| Useless | Input | Interval | Numeric | BEST7.0 | | 8 |

You can use PROC IMPORT to import data from an Excel file into a SAS data set. (See Program 3.1.) For this to work, you should have a license for SAS/ACCESS Interface to PC Files. This product enables SAS to treat an Excel spreadsheet as a database and a subset of its cells (also called a range) as a table. You can specifically point to and read a relevant worksheet in an Excel workbook. For example, the worksheet Sheet1 in TabletReviews.xls contains all of the review comments that we need. Data values in the first row of columns in the worksheet can be used to name the variables in the output data set.

**Program 3.1: SAS Code to Import Data from an Excel File**

```
PROC IMPORT
        /* Library and name of the data set in which the output is stored */

                OUT=work.TabletReviews
        /* Physical path of the input file to read data */
                DATAFILE="C:\test\dir\TabletReviews.xls"
        /* Data source identifier specifying the type of data to import */
                DBMS = XLS
        /* Overwrite SAS output data set, if it already exists */
                REPLACE;
        /* Name of the worksheet containing the data to import */
                SHEET = 'Sheet1';
        /* Data values in first row used to name variables in SAS data set */
                GETNAMES=YES;
    RUN;
```

If you run this code, you will find that the SAS data set TabletReviews has been created in your Work library with all of the review comments stored as individual rows. This is a very quick and efficient approach to import text data from an Excel file. You can import data from an Excel file using the SAS Import Wizard in SAS Enterprise Guide without writing any SAS code.

Data is often stored in XML files. XML stands for Extensible Markup Language. XML files are typically composed of tags and values. An XML file can either have a simple or complex structure, which mainly depends on how the tag hierarchies are defined. For example, a sample XML file, SAS_Books_Sample.xml, contains information about some of the SAS Press books. (See Display 3.23.) You can see information specific to each book enclosed in relevant tags. Tag **book id** appears as the parent tag under which all other tags (**author**, **title**, **publisher**, and **price**) are formatted in the same level. This is a classic example of a very simple XML file structure that many websites typically follow.

**Display 3.23: Sample XML File with Information about SAS Press Books**

```xml
<?xml version="1.0"?>
<catalog>
    <book id="1">
        <author>Norm O'Rourke and Larry Hatcher</author>
        <title>A Step-by-Step Approach to Using SAS for Factor Analysis and Structural Equation Modeling, Second
            Edition</title>
        <publisher>SAS Institute</publisher>
        <price>74.95</price>
    </book>
    <book id="2">
        <author>Larry Hatcher</author>
        <title>A Step-by-Step Approach to Using SAS for Factor Analysis and Structural Equation Modeling</title>
        <publisher>SAS Institute</publisher>
        <price>89.95</price>
    </book>
    <book id="3">
        <author>Norm O'Rourke, Larry Hatcher, and Edward Stepanski</author>
        <title>A Step-by-Step Approach to Using SAS for Univariate and Multivariate Statistics, Second Edition</title>
        <publisher>SAS Institute</publisher>
        <price>89.95</price>
    </book>
```

XML files can be read programmatically by SAS by reading the tags and feeding their values to columns in a SAS data set. However, this process is not very efficient, and you need to write a customized program if the structure of the XML file changes. SAS XML Mapper is an easy-to-use tool exclusively developed for the purpose of mapping XML tags and their values directly to columns and rows in a SAS data set. You can download it from http://support.sas.com for free. Using this tool, XML files of almost any complex structure can be parsed.

Start SAS XML Mapper, and select **File ▶ Open** to locate the XML file to be used for generating the XML map. Select the XML file named SAS_Books_Sample.xml, and click **Open** to load the file into SAS XML Mapper. Right-click on the root tag **book**, and click **AutoMap this branch** to move all of the tags into the map.

(See Display 3.24.) Select **File ▶ Save XMLMap As**, and save the map as SASBooks.map. You can click the **Table View** tab in the bottom half of the screen to see how various tags and values are transformed into rows and columns.

**Display 3.24: SAS XML Mapper Using SAS_Books_Sample.xml**

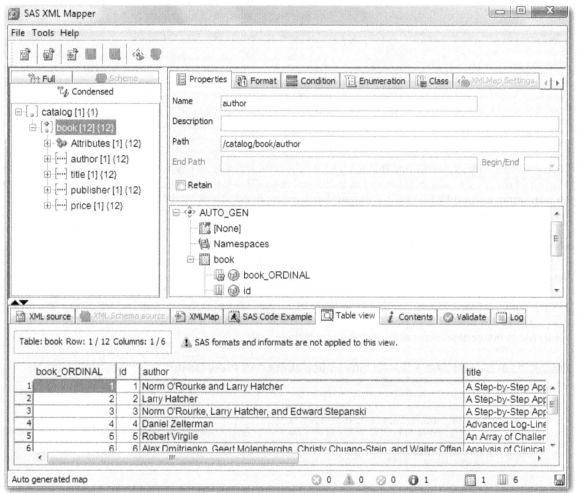

You can use this map to import data from the XML file through the XML LIBNAME engine. You can directly copy the program from the **SAS Code Example** tab to import data from the XML file. Uncheck all boxes except **Copy to WORK** to retain the code required for this purpose. (See Display 3.25.) Copy and run this code in Base SAS to generate the SAS output data set in the Work library. If you would like to create the output data set in a permanent library, replace the data set name **book** in the DATA step with **Lib.book**, where **Lib** is the name of permanent library.

**Display 3.25: SAS XML Mapper Showing SAS Code Generated Automatically**

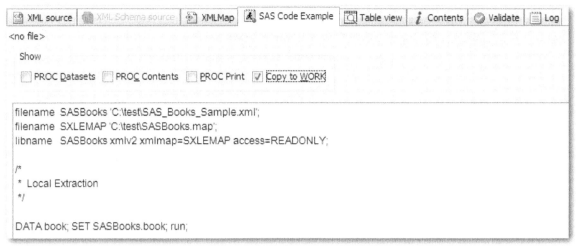

In the second maintenance release for SAS 9.3, the AUTOMAP option became available with the XMLV2 LIBNAME engine. This feature enables you to create default XMLMap files in SAS without the need for SAS XML Mapper. AUTOMAP= option accepts either REPLACE or REUSE as its value. If you use REPLACE, an XMLMap file with the same name is overwritten with the new file if it exists in the same location. The REUSE value uses the XMLMap file with the same name if already exists, otherwise, a new file is created. Using the sample XML file we used in the previous example SAS_Books_Sample.xml, you just need to run the following code to create the SAS data set automatically using the XMLV2 LIBNAME engine and the AUTOMAP option. The FILENAME option creates a temporary location for XMLMap file tempMap in this example.

**Program 3.2: SAS Code to Use XMLV2 LIBNAME Engine with the AUTOMAP Option**

```
libname out 'C:\test';
filename SASBooks 'C:\test\SAS_Books_Sample.xml';
filename tempMap temp;
libname SASBooks xmlv2 xmlfileref=out xmlmap=tempMap automap=replace;
data out.book; set SASBooks.book; run;
```

For a moment, let's move away from local files to websites where the information of interest is often found in news articles, blogs, user reviews, social media conversation, etc. Many websites storing massive volumes of unstructured data usually allow external applications to connect and access its data by means of an API (application programming interface). Social media giants Twitter and Facebook are the most popular websites offering APIs for data exchange. You can send your requests to these APIs and retrieve unstructured data that you can use in various text analytics applications, such as analyzing social media trends, topic discovery, and sentiment analysis.

Generally, you need custom-built SAS programs capable of querying APIs and parsing the returned responses in various formats. The most commonly used API response formats are Atom, XML, RSS, and JSON (JavaScript Object Notation). Data consumers face challenges to keep up with the ever-changing structures and formats of API responses. For example, Twitter, in its current version (API v1.1), offers support only in JSON format, and has deprecated old formats such as XML, Atom, and RSS. Fortunately, the SAS user community has been contributing and sharing conference papers, blogs, and articles and providing SAS programs to deal with most of the formats. In their paper, Badisa, Mandati, and Chakraborty (2013) discussed a SAS macro that they developed to extract movie reviews from a website and return data in JSON format. You need an API key and the name of a movie to execute this macro. It predominantly uses SAS Perl regular expression functions to meaningfully parse the returned responses and to store the reviews of that movie in a SAS data set.

In 9.4, SAS provides a new procedure, PROC JSON. The JSON procedure helps you export a SAS data set into an external file in JSON format. Here is sample code to generate an external file in JSON format based on the content in the SASBooks SAS data set that was created in the previous section using the XMLV2 LIBNAME engine.

**Program 3.3: SAS Code to Export a SAS Data Set into an External File in JSON Format**

```
proc json out="C:\test\BooksOut.json" pretty nosastags;
export out.book;
run;
```

The PRETTY option makes the output data easy to read, and the NOSASTAGS option suppresses metadata supplied by SAS. This helps keep it in a format similar to web service formats. Here is the sample output generated in the JSON format:

```
[
  {
    "book_ORDINAL": 1,
    "id": 1,
    "author": "Norm O'Rourke and Larry Hatcher",
    "title": "A Step-by-Step Approach to Using SAS for Factor
Analysis and Structural Equation Modeling, Second Edition",
    "publisher": "SAS Institute",
    "price": 74.95
  },
  {
    "book_ORDINAL": 2,
    "id": 2,
    "author": "Larry Hatcher",
    "title": "A Step-by-Step Approach to Using SAS for Factor
Analysis and Structural Equation Modeling",
    "publisher": "SAS Institute",
    "price": 89.95
  },
  {
    "book_ORDINAL": 3,
    "id": 3,
    "author": "Norm O'Rourke, Larry Hatcher, and Edward
Stepanski",
    "title": "A Step-by-Step Approach to Using SAS for
Univariate and Multivariate Statistics, Second Edition",
    "publisher": "SAS Institute",
    "price": 89.95
  },
]
```

## Managing Text Using SAS Character Functions

Many projects might require preprocessing textual data before any text mining is performed.

SAS programmers or analysts might have to validate text, replace text, and extract a substring from a string. These operations can be performed using powerful SAS character functions like SCAN, INDEX, SUBSTR, etc., and simple DATA step programs. Although SAS character functions are valuable for data cleaning, they might not be very efficient when the text is too long.

Many projects involving textual data require disguising a piece of text or a set of words because of confidentiality reasons or simply to make the data look clean. A typical example is eliminating or disguising all swear words in the text that are commonly found when dealing with data from social media websites. TRANWRD, TRANSLATE, and TRANSTRN functions can be appropriately applied to perform this task. Functions like COMPBL, COMPRESS, and SUBSTR can be extensively used to eliminate the noise from the text. Consider a data set that has the following Tweets about a general store:

1. #Valuemart clearing off shelf space for $599 #HP #TouchPad? http://pulsen.wk/1QQL7 :)
2. Value-mart has a Blu-ray Value Bundle #Deal Choice of 2 #Movies for $15 Pick 2 from a list of 48. RT
3. VALUE MART Coupons- Get Your Name In For The $1000 Gift Card We Will Be Giving Away At The End Of The Month **DEAL**

The following DATA step with SAS character functions replaces all occurrences of the word "ValueMart" or its variants, eliminates double spaces, and eliminates special characters like * . : , and ) from the text.

**Program 3.4: SAS Code with Examples Using TRANWRD, COMPBL, and COMPRESS Functions**

```
data valuemart1;
      set valuemart;

/*Compresses successive blanks to a single blank*/
text = compbl(text);

/*Replace 'Valuemart' with 'Store' */
text = tranwrd(text, 'Valuemart','Store');

/*Replace 'Value-mart' with 'Store' */
text = tranwrd(text, 'Value-mart','Store');

/*Replace 'VALUE MART' with 'Store' */
text = tranwrd(text, 'VALUE MART','Store');

/*Remove specified characters in the second argument from text */
text = compress(text, '#*:()&@!-?+');

run;
```

The DATA step transforms the three Tweets in the following example:

*Store clearing off shelf space for $599 HP TouchPad http//pulsen.wk/1QQL7*

*Store has a Bluray Value Bundle Deal Choice of 2 Movies for $15 Pick 2 from a list of 48. RT*

*Store Coupons Get Your Name In For The $1000 Gift Card We Will Be Giving Away At The End Of The Month DEAL*

Character functions can significantly improve the quality of the textual data. There are various other functions available in SAS that can be wisely applied for manipulating textual data. In the previous example, we had to write three different TRANWRD statements to replace the three variants of the string "Valuemart." In the next section, you learn how to accomplish this task with just one statement by defining patterns using Perl regular expressions.

## Managing Text Using Perl Regular Expressions

SAS character functions lack flexibility and tend to be not very effective with dynamic text. In situations like these, regular expressions and routines give the programmer the power to define, locate, and extract patterns from highly unstructured text easily. They can make a complicated string manipulation task look very simple by combining most of the steps, if not all, into one expression.

SAS provides two types of regular expressions for pattern-matching: SAS regular expressions (the RX functions) and Perl regular expressions (the PRX functions). Regular expressions are popularly used to easily process large amounts of textual data. These expressions help in defining patterns that can be searched and extracted in a character string. Regular expressions are nothing but a string of standard characters and special characters, called metacharacters, which tell the program what types of patterns to find in the text. It is strongly suggested to comment any regular expression that you write because it is highly possible that it will be very difficult to interpret the ones that you wrote yourself.

Perl is a powerful text processing language widely popular amongst programmers. In this chapter, we briefly discuss how Perl regular expressions can be used for effectively preprocessing the data. A modified version of Perl regular expressions is implemented in SAS via a set of functions called PRX functions. Only Perl regular expressions are available in SAS, not the entire Perl language. You do not need to be an advanced SAS programmer to write Perl regular expressions. Basic knowledge of DATA step programming is sufficient. The examples discussed in the following sections can be easily modified to meet your requirements.

## Using Perl Regular Expressions to Match Exact Text in a Document

The simplest use of Perl regular expressions is to locate a piece of text or a pattern in a textual document. You can write a regular expression to check whether a document contains text in the form (XXX) XXX – XXXX, indicating a phone number. Here, Xs are digits. You can use a Perl regular expression to search for specific characters and other classes of characters (digits, letters, non-digits, etc.).

Perl helps in defining and extracting custom patterns in text, and SAS Text Miner contains a built-in functionality to extract common types of text like phone number, address, and SSN, and entities like names of people, places, companies, etc., during text parsing. A more advanced type of entity extraction can be done by defining concepts in text using SAS Content Categorization Studio. We discuss these topics in Chapter 7.

Consider the corpus of web pages downloaded from the allanalytics.com website in the previous section. From the entire corpus, if you want to consider only those articles that talk about big data, you can use a Perl regular expression to filter the articles that you need.

Here is a simple DATA step that can meet this requirement:

**Program 3.5: SAS Code Showing Usage of PRXPARSE and PRXMATCH Functions in a SAS DATA Step**

```
data bigdata;
      set work.allanalytics;
if _N_=1 then pattern= PRXPARSE ("/big data /");
retain pattern;
Locate =PRXMATCH(pattern,text);
if locate;
drop pattern locate;
run;


Log:
NOTE: There were 1788 observations read from the data set WORK.ALLANALYTICS.
NOTE: The data set WORK.BIGDATA has 140 observations and 15 variables.
NOTE: DATA statement used (Total process time):
      real time          0.25 seconds
      cpu time           0.17 seconds
```

In the program, we create the data set **allanalytics** from the web pages available in the **WORK** library. The **PRXPARSE** function is used to create a regular expression. Because you want to use the pattern for each observation in the data set, the pattern has to be available for each iteration of the DATA step. Instead of compiling it for each observation, you can execute the statement with PRXPARSE only once, and then use the **RETAIN** statement with _N_ to make it available for each iteration. The **PRXPARSE** function returns a code that is a sequential number generated by SAS whenever a regular expression is compiled. The **PRXMATCH** function is used to return the first position of the string "big data" in the given text. A value of 0 is returned when no match is found. The **PRXMATCH** function takes two arguments: the return code from **PRXPARSE** and the string to be searched.

The string "big data" might be represented in the articles in multiple forms: big data, Big Data, big-data, BIG DATA, etc. You can include all of these different forms of the string "big data" in the regular expression using a | (OR) condition as shown:

```
pattern= PRXPARSE("/big data | big-data | Big Data | BIG DATA /");
```

The search can be made more compact and complex to include other possible forms of the string "big data" by writing it the following way:

```
pattern= PRXPARSE("/[Bb]ig( |-)?[Dd]ata /i");

Log:
NOTE: There were 1788 observations read from the data set WORK.ALLANALYTICS.
NOTE: The data set WORK.BIGDATA has 989 observations and 15 variables.
NOTE: DATA statement used (Total process time):
            real time            0.10 seconds
            cpu time             0.04 seconds
```

This pattern will match the strings "big data," "Big Data," "BIG DATA," "big-data," "Big-Data," "bigdata," "BigData," "BIGDATA," and "BIG-DATA." In the log, there are more documents filtered because the new regular expression contains more forms of the search word. In this regular expression, the letters "[Bb]ig" match either Big or big. Similarly, the expression in parenthesis ( |- ) is used to search for a space or a hyphen after the letters "[Bb]ig." To include the words "BIGDATA" or "bigdata" (no space or hyphen between the two words) in the search, we use the ? metacharacter, which matches the previous sub-expression zero or one time. The letter "i" after the last slash is used to match all types of cases (small, upper, and proper case) of the string.

## Using Perl Regular Expressions to Replace Exact Text

In many situations, the business application demands to anonymize certain words in your text before you use it in text mining. Perl regular expressions are of great help in performing text substitutions. The Perl substitution function, which is in a PRXPARSE function, looks like this:

```
PRXPARSE("s/ pattern to find / text to substitute /");
```

You need to specify the substitution operator **s** before the first slash in the regular expression. In the previous example, if you want to replace different forms of the string "big data" with the string "Big Data," you can do it using PRXPARSE and PRXCHANGE functions. You can anonymize all of these words with a different keyword if required. Here is the DATA step for completing this task:

**Program 3.6: SAS Code Showing Usage of PRXPARSE and PRXCHANGE Functions in a SAS DATA Step**

```
data bigdata;
      set bigdata;
if _N_=1 then pattern= PRXPARSE(" s/[Bb]ig( |-)?[Dd]ata / Big Data /i");
retain pattern;
call PRXCHANGE (pattern, -1, text);
drop pattern;
run;
```

The PRXCHANGE function is used to substitute one string for another. The function takes the return code from PRXPARSE as the first argument. The second argument is used to indicate the number of times to search for and replace a string. A value of -1 replaces all matches. The third argument is the string to be replaced. We have explained only the required arguments of the function in this example. For more information about this function, see the online SAS Help.

The substitution approach can be heavily used for cleaning the data. If you can define the pattern of noise in the data to the best extent possible, you can create regular expressions and replace the matches with a blank space. This helps in generating cleaner data for further text mining.

## Using Perl Regular Expressions to Extract Text between Two Keywords

The ability to define patterns and locate those patterns in text using regular expressions makes the Perl language very powerful in handling textual data. Organizations have ample textual data at their disposal. We have the computer power to use all textual data for text mining. Remember that textual data is first transformed into numbers before any traditional data mining tasks are performed. The more text used for text mining, the more complex the processing. Instead, in many situations, it is sufficient to use a piece or part of a document for performing text mining.

For example, consider you want to analyze conference papers to understand the trend of topics represented in various conferences. Considering the entire paper in your analysis will likely add a lot of noise because an entire paper can include tables, images, references, SAS programs, etc., which are problematic and might not add much value in text mining for topic extraction. The abstract of a paper is the most appropriate text because it captures the detailed objective of a paper and does not contain extraneous items. A typical conference paper has the following layout:

**Display 3.26: Typical Layout of a Conference Paper**

We define two regular expressions that give the positions of the keywords ABSTRACT and INTRODUCTION in the conference paper. The text between the two positions (that represents the abstract portion of the paper) is extracted for text mining. However, not all authors will follow the same layout. For this example, we assume we are dealing with papers that strictly follow the above layout. This problem is dealt with in detail in Case Study 1. The DATA step program that can solve this requirement is shown in Program 3.7:

**Program 3.7: SAS Code Showing Usage of PRXPARSE and PRXMATCH Functions to Extract a Portion of Text**

```
data abstracts;
length text $ 3000;
      set work.sasgf;
if _N_=1 then do;
            pattern1 = PRXPARSE ("/Abstract/i");
            pattern2 = PRXPARSE ("/Introduction/i");
end;
retain pattern1 pattern2;
            position1=PRXMATCH (pattern1, text);
            position2=PRXMATCH (pattern2, text);
```

```
if position1 gt 0 AND (position2-position1) gt 0
then   text= SUBSTR(text,position1+9,position2-position1-9);
else   text= SUBSTR(text,150,2850);
drop pattern1 pattern2 position1 position2;
run;
```

After locating the positions of the keywords using the **PRXMATCH** function, a **SUBSTR** function is used to fetch the text between the two positions. In the case that you do not find a match for the two keywords, we extract text with a length of 2,850 characters after the first 150 characters, assuming that the first 150 characters represent the title and author details, which are unnecessary for the study. Display 3.27 shows sample results of abstracts extracted using the above code.

**Display 3.27: Sample Output of the Abstracts Extracted from SAS Global Forum Papers**

| | text |
|---|---|
| 3 | SAS® Enterprise Guide® empowers organizations exploiting the power of SAS by offering programmers, business analysts, statisticians and end-users with powerful built-in wizards to perform a multitude of reporting and analytical tasks, access multi-platform enterprise data sources, deliver data and results to a variety of mediums and outlets, perform important data manipulations without the need to learn complex coding constructs, and support data management and documentation requirements quickly and easily. Attendees learn how to use the graphical user interface (GUI) to access tab-delimited and Excel input files; subset, group, and summarize data; join two or more tables together; flexibly export results to HTML, PDF and Excel; and visually manage projects using flowcharts and diagrams. |
| 4 | Last's year's presentation "Sending email from the DATA step" drew a lot of attention. But receiving email in the DATA step, i.e. using the DATA step as a POP3 mail client is also possible. This is a simple low-cost solution for companies that require branches to send in information for handling at the head office. This paper explains how to access a POP3 mail server and save the messages in a SAS® data set and also how to make it run repeatedly. |
| 5 | This paper describes how SAS® jobs are automated under UNIX at the BC Ministry of Health Services. A job is a set of SAS program(s) and/or Perl modules to be run in sequence. Job steps can be listed in a text file for the Perl script to run. Having a well-tested perl script run the show has some advantages. Notification of job success or failure is reliable. Jobs can be restarted at the point of error. Logs are automatically date- and time-stamped. It is easy to migrate SAS code from development to production. |
| 6 | Some of the common measures used to monitor rater reliability of constructed response items include Percent Exact Agreement, Percent Adjacent Agreement, Simple Kappa Coefficient, Weighted (Cicchetti-Allison, Fleiss-Cohen) Kappa Coefficient, Pearson Correlation Coefficient, Mean Difference, Standardized Mean Difference, and Matched-Pair T-Test. This paper discusses the development of a Windows application using Visual C#.NET and SAS® to calculate various measures of rater reliability and produce custom reports. Users can select which measures to include in their reports and set threshold values to flag items with low rater reliability values. |
| 7 | An old maxim holds that every nontrivial computer program has bugs. As a developer, you want to minimize the number and severity of the bugs that appear in your released application. When a bug is reported, you need to be able to identify and resolve it as efficiently as possible. Effective debugging requires a scientific approach and an understanding of the tools available to help you identify, understand, and solve the problems that cause bugs. |
| 8 | This paper illustrates how to use ODS markup to create PivotTable and PivotChart reports. You can use these reports to help your organization with business requirements such as analyzing expense trends over time or planning for inventory. Users of all experience levels will learn how to specify options in ODS statements that enable them to perform the following tasks • customize PivotTable layouts • customize cell format • specify statistics for the analysis • modify the display of the analysis • generate a PivotTable report and a PivotChart report for each worksheet • format tables that you create • customize worksheets |
| 9 | As a fast growing large public metropolitan research institution, the Office of University Analysis and Planning Support at the University of Central Florida (UCF) is challenged with transforming very large data sets into information that will support strategic planning and decision making. This paper describes a process that utilizes SAS® macro language and integration technologies with Microsoft Excel®, to generate a series of highly customized Excel reports in an automated approach. This methodology allows SAS to process a data set of over 400,000 records, consolidate the data, open pre-formatted Excel templates, and write the data to designated data ranges. The integration technology used, called Dynamic Data Exchange (DDE), reduces report generating time by up to 90%, compared to prior methods that used Excel PivotTables and manual copying and pasting of results, while improving report accuracy by reducing human error. Additionally, SAS macro language enables programmers to customize the report to suit the needs of a variety of end users. |

The examples described in this section are very basic and are intended to illustrate the simple functionality of Perl regular expressions available in SAS. There are other functions like PRXPOSN, PRXNEXT, PRXPAREN, etc., which make processing unstructured data much easier and powerful. Using these functions cleverly can help you locate, extract, and replace any type of text. Whether it is extracting sales figures from annual reports, extracting ZIP codes or phone numbers from application documents, or cleaning customer comments from Internet public forums, anything can be done using Perl regular expressions. For more information about using Perl regular expressions in SAS, see http://support.sas.com or http://sascommunity.org.

## Summary

In this chapter, you learned various techniques for importing textual data that exists in a wide variety of file formats: .html, .txt, .pdf, .xlsx, .xml, .doc, etc. In many situations, you might have to combine text data in these different forms for text mining purpose. SAS Enterprise Miner and SAS Text Miner have easy-to-use capabilities to import data from these files and to prepare SAS data sets. The strong web crawling functionality in SAS that is available through the Text Import node makes it easier to retrieve data available from the web. Every data importing task in SAS can be performed either via a user interface or via SAS programming. The powerful data management tools and techniques in SAS help in combining, transforming, and cleaning data. This extracted textual data can be processed before text mining analysis using SAS character functions. Although most of the text validation, extraction, and replacement tasks can be performed using these functions, more text processing capabilities are available using Perl regular expressions. All of these features empower an analyst with innumerable capabilities from data extraction, consolidation, and transformation that are generally performed before starting text mining analysis.

# References

Badisa, G., Mandati, S., and Chakraborty, G. 2013. "%GetReviews: A SAS® Macro to Retrieve User Reviews in JSON Format from Review Websites and Create SAS® Datasets". *Proceedings of the SAS Global Forum 2013 Conference*. SAS Institute Inc., Cary, NC. Available at:http://support.sas.com/resources/papers/proceedings13/342-2013.pdf

Cassell, D. L. 2005. "PRX Functions and Call Routines." *Proceedings of the Thirtieth Annual SAS Users Group International Conference*. SAS Institute Inc., Cary, NC. Available at: www2.sas.com/proceedings/sugi30/138-30.pdf.

Cody, R. 2004. "An Introduction to Perl Regular Expressions in SAS 9." *Proceedings of the Twenty-Ninth Annual SAS Users Group International Conference*. SAS Institute Inc., Cary, NC. Available at: www2.sas.com/proceedings/sugi29/265-29.pdf

Cody, R. 2004. *SAS Functions by Example, 1ˢᵗ ed.* Cary, NC: SAS Institute Inc.

Francis, L. 2010. "Text Mining Handbook." *CAS E-Forum: Publication of the Casualty Actuarial Society*. Available at: www.casact.org/pubs/forum/10spforum/Francis_Flynn.PDF.

McGowan, K. and Lagle, B. 2011. "A Better Way to Search Text: Perl Regular Expressions in SAS." *Proceedings of the SAS Global Forum 2011 Conference*. SAS Institute Inc., Cary, NC. Available at: http://support.sas.com/resources/papers/proceedings11/006-2011.pdf.

McNeill, B. 2013. "The Ins and Outs of Web-Based Data with SAS®." *Proceedings of the SAS® Global Forum 2013 Conference*. SAS Institute Inc., Cary, NC. Available at:

http://support.sas.com/rnd/papers/sasgf13/024-2013.pdf.

Miller, T. W. 2005. *Data and Text Mining: A Business Applications Approach*. Upper Saddle River, New Jersey: Pearson Prentice Hall.

Pant, G., Srinivasan, P., and Menczer, F. 2004. "Crawling the Web". *Web Dynamics: Adapting to Change in Content, Size, Topology and Use*. New York: Springer-Verlag, 2004.

Text Analytics Using SAS® Text Miner. Course Notes. SAS Institute Inc., Cary, NC.

Course information: https://support.sas.com/edu/schedules.html?ctry=us&id=1224.

Weiss S, Indurkhya N, Zhang T, Damerau F. 2005. *Text Mining: Predictive Methods for Analyzing Unstructured Information*. Springer-Verlag.

Zhang, S. 2007. "Use Perl Regular Expressions in SAS®." *Proceedings of the 20th Annual NorthEast SAS Users Group Conference*. Available at: www.nesug.org/proceedings/nesug07/bb/bb18.pdf.

# Chapter 4  Parsing and Extracting Features

## Introduction

In this chapter, we discuss the next step and perhaps the most important step in the text mining process flow—text parsing. In Chapters 2 and 3, we have seen how various methods collect and process textual documents. The next task is to convert the collected text documents (in unstructured form) to a vector representation (a structured form). Fundamentally, parsing is the first step in converting unstructured text to the familiar spreadsheet (structured) format for ease of analysis. Typically, this process involves tokenization, normalization of tokens (via lemmatization or stemming), part-of-speech (POS) tagging, and so on.

There are two broad-based approaches to analyzing text. The first is the bag-of-words method, where the basic assumption is counting words in the text, plus understanding how these words are syntactically (structurally) related to each other in regard to laws of grammar, etc. This method is sufficient to summarize and classify text documents. The second approach, which is linguistic, posits that to truly understand and classify text, you have to move beyond syntax (structure) to semantics (meaning of words). For a detailed discussion of the similarities and differences between these approaches, as well as the philosophical debate over Chomsky's empirical approach versus the behaviorist's approach in the NLP field, see Manning and Schutze (2000) and Dale and Somers (2000). In order to apply either approach, the text document is first parsed to find the words contained in it.

Term parsing was originally derived from the Latin word "pars orationis," meaning the identification of parts of speech. Parsers are now being extensively used in various disciplines. In the field of computer science, a computer program is parsed to identify the syntax of the code before it is compiled. In computational linguistics, parsing is used to identify the grammatical structure in regard to a structured language. In both cases, the text is first broken into pieces called tokens (for example, words), and then the structural relationships among the tokens (words) within the text are delineated. In this chapter, we demonstrate how to use the Text Parsing node for parsing a textual corpus.

## Tokens and Words

Parsing begins by taking a stream of characters (such as a sequence of sentences in a text document) and breaking it down to tokens (to units, where a unit is either a word, a number, or a punctuation mark). This process is called tokenization. A term identified through tokenization might not be just a single word, but can be a group of words (for example, noun groups). Typically, the tokenization process involves applying commonly accepted delimiters such as a period, space, tab, new line, and so on, to unstructured text, and then figuring out instances (occurrences) of each token. Consider sentence, "I love eating chocolate but I worry about how many calories are in each bite." In that sentence, there are 15 tokens, but the token "I" appears twice. Strictly speaking, linguists refer to each token as an instance of a type. In that sense, we have two instances of the type "I" in the sentence. In this book, we use frequency of type and frequency of token interchangeably.

In addition to identifying tokens, the purpose of various punctuation characters is deciphered during parsing. Although this process might sound simple for humans trained in a language, it is a difficult task for computers because of the complexities of the language and the variations based on contextual use. For example, a comma might often be a delimiter, unless it appears in the middle of a string of numbers, in which case the comma is often placed as a separator for thousands, millions, and so on, to make it easier for humans to quickly grasp the number. A period might be at the end of a sentence or a part of an abbreviation (such as Mr.). A dash between two numbers might be a symbol of a subtraction operation or a separator as in a telephone number, 744-5000. Parsing faces problems with how to handle single apostrophes. For example, is "I'll" a single token, or should it be considered two tokens, "I" and "will"? There are no easy answers to any of these problems.

## Lemmatization

After tokens are found in a text document, the next step is often the normalization of tokens to reduce complexity. In any text, you will find different forms of the same word such as "play," "plays," and "playing" used for grammatical reasons. For computational purpose, it is useful to treat all variants of a word similarly. This process involves identifying the root terms in different variants of a token. Lemmatization and stemming are typically used for this purpose. These two methods have the same goal of reducing inflected or derived words to their stem or root forms, but they are slightly different in their approach. Stemming follows a crude heuristic process that chops off the ends of the words. Lemmatization uses a standard vocabulary and morphological analysis of words for reducing an inflectional form of the word to its base or dictionary forms, which is known as lemma.

The basic idea is that by reducing the frequency of the number of distinct types in a corpus, you can reduce the complexity of the document summarization task without much loss of information. This is possible because word categories are systematically related by morphological processes such as the singular and plural forms of nouns or different tenses of the same verb. One example is using the root term (lexeme) such as "run" to represent terms such as "ran," "runs," "running," and so on. Another example of stemming is to use the singular form of a noun (such as car) as the root term for plural forms of the same noun (such as cars).

When normalization is restricted to the grammatical variants of the tense forms or singular and plural, it is called inflectional stemming in morphological analysis. Derivation in morphological analysis occurs when a word is changed more radically from one syntactic category to another. An example is adding a suffix such as "er" to the verb form of the word "teach" to convert it to the noun form "teacher." However, stemming should not be restricted to grammatical variants. It can be extended to include semantically equivalent words (or synonyms). For example, a synonym-based lemmatization might use the word "car" as the root form for semantically equivalent words such as automobile. Compounding in morphological analysis refers to the formation of a new word by merging two words. The most common examples are noun-noun compounds such as DVD disk, flash drive, college degree, and so on. In parsing, compounds can often be handled as entities or through phrase recognition.

## POS Tags

Once text is split into tokens and the tokens are normalized, the next logical step is to determine the POS for each token to perform additional linguistic analyses and to extract more sophisticated features. POS tagging (often called just tagging) is the task of labeling each token (word) in a sentence with its appropriate part of speech. That is, for each word in a sentence, the algorithm decides whether it is a noun, verb, adjective, adverb, preposition, conjunction, and so on. A commonly known tag set in English is the Brown tag set based on the American Brown corpus of about one million English words (Brown et al., 1991). The Brown tag set is quite large with 179 tags. It is very common that the same word can be identified as a different part of speech based on the context. In this situation, each combination of word and POS tag can be represented as a unique term. For example, the term "address" tagged as noun is different from the term "address" tagged as verb. Hence, a large set of tags causes the term space to grow vastly. In a computational world, often a smaller tag set (such as Penn Treebank) is used. The Penn Treebank is a smaller subset of the Brown tag set, and it is constructed from the Wall Street Journal corpus, available from the Linguistic Data Consortium (LDC) at www.ldc.upenn.edu. Some of the Penn Treebank categories are shown in Table 4.1.

**Table 4.1: Partial List of POS Tags Used in the Penn Treebank Tag Set**

| Number | Tag | Description |
|--------|-----|-------------|
| 1. | CC | Coordinating conjunction |
| 2. | CD | Cardinal number |
| 3. | DT | Determiner |
| 4. | EX | Existential *there* |
| 5. | FW | Foreign word |
| 6. | IN | Preposition or subordinating conjunction |
| 7. | JJ | Adjective |
| 8. | JJR | Adjective, comparative |
| 9. | JJS | Adjective, superlative |
| 10. | NN | Noun, singular or mass |
| 11. | NNS | Noun, plural |
| 12. | POS | Possessive ending |
| 13. | RB | Adverb |
| 14. | RBR | Adverb, comparative |
| 15. | RBS | Adverb, superlative |
| 16. | UH | Interjection |
| 17. | VB | Verb, base form |
| 18. | VBD | Verb, past tense |
| 19. | VBG | Verb, gerund or present participle |
| 20. | VBN | Verb, past participle |

Here is an example of a tagged sentence ("The student put books on the table.") where the Brown tags for each word is shown in boldface:

The-**AT** student-**NN** put-**VBD** books-**VBZ** on-**IN** the-**AT** table-**NN**

## Parsing Tree

Tagging a sentence is often limited in scope compared to a full parsing of a sentence. In the full parsing of a sentence, each word is represented usually in a tree form that depicts the structure of the sentence in which the word appears. For example, consider a simple sentence, "Bob is educating students." A complete parse of this sentence in a tree form using a commonly available set of POS tags might look like the following:

**Display 4.1: Sample Tree Form of a Parsed Sentence with Tagged Parts of Speech**

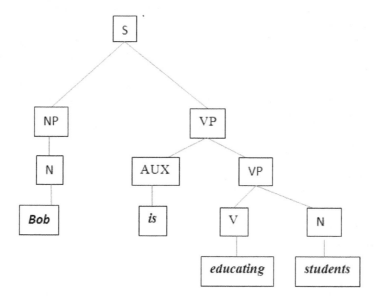

In the tree depiction, the root (top) node has a label of **S**, denoting a sentence. The leaf (terminal) nodes of the tree are the words of that sentence. The internal (intermediate) nodes represent the grouping of the words in phrases (such as **VP** for verb phrase) and POS tags. Even for a simple sentence such as this one, there are different representations possible. Another slightly different representation of the same sentence might look like the following:

**Display 4.2: A Different Tree Form of a Parsed Sentence with Tagged Parts of Speech**

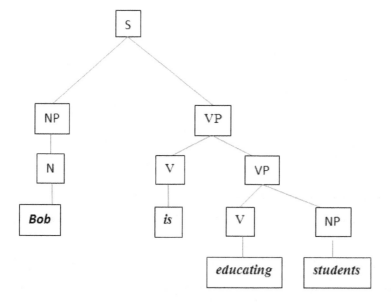

It is easy to see that as a sentence becomes more complex and longer, there can be many different parses possible for the same sentence, which leads to ambiguous interpretation. Fortunately, many algorithms exist that have been developed and trained using a large corpora of documents that can be used to select the most relevant parsing. SAS Text Miner incorporates many of these algorithms and features.

## Text Parsing Node in SAS Text Miner

In the process flow, the Text Parsing node must be preceded by a node that exports a training data source. Typically, it is either a Data Source node or a Text Import node. A SAS Code node can act as a data source when using the %TMFILTER macro for data collection. The input data should contain at least one variable with Role Text, Text Location, or Web Address. In the previous chapter, there were two data collection methods. In one method, the whole document is captured in a single cell. In the other method, a snippet of the document text is captured. To process all of the text in the latter case, you should use the FILTERED variable that captures the path to the full document. The role of the variable, FILTERED, must be set to Text Location in the Text Parsing node.

Let's go back to the data set that was created from the SAS Global Forum papers using the Text Import node in Chapter 3. Display 4.3 shows the variables accessible from the Text Parsing node when it is connected to the Text Import node. When this node is run, the text of all SAS Global Forum papers is accessed using the path to the file available in the FILTERED variable.

**Display 4.3: Input Variables for Text Parsing**

| Name | Use | Report | Role | Level |
|------|-----|--------|------|-------|
| FILTERED | Default | No | Text Location | Nominal |
| TEXT | Default | No | Text | Nominal |
| uri | Default | No | Web Address | Nominal |

The Text Parsing node starts with breaking down the text stream into tokens (or words). Various actions are performed on the identified word list based on the property values set in the Properties panels. These settings significantly impact the number of terms that will be used for analysis in the nodes. Remember that the larger the number of terms that are used for analysis, the more complicated the text mining task turns into. The Text Parsing node features are invaluable in controlling the number of terms that are extracted for text mining analysis. Most of the tasks performed through this node act as a preliminary method of reducing the dimensions of a document-by-term matrix. An in-depth discussion on the mathematical techniques for dimension reduction is provided in the next chapter.

## Stemming and Synonyms

Text mining mainly deals with the co occurrence of the terms in a document and across documents. Stemming and synonyms play key roles in identifying the relationships between the documents. For example, by treating the terms "airlines," "aviation," and "aero" as similar, we can say that the documents that contain any of these words are related. The stemming functionality enables SAS Text Miner to automatically assign synonyms to terms that can be represented by their root word forms. However, in some cases, the unchecked application of stemming rules might group words that are semantically different. For example, the words "university" and "universal" might be grouped together by a stemmer. This will lead to a potential loss of information. A closer look at the results from the stemming process is required for any application of text mining. The algorithm that does stemming is known as a stemmer. A stemmer functions based on rules or a dictionary. Rules identify and strip suffixes from the words and deduce stems. For example, a rule might be the following, "If the word ends in "ed," then remove "ed."). SAS Text Miner uses a dictionary-based stemmer, which can significantly reduce the number of terms. Stemming can be turned on (**Yes**) or off (**No**) using the **Stem Terms** property as shown in Display 4.4. The default value is Yes.

**Display 4.4: Stemming Property Setting in Text Parsing Node**

| | |
|---|---|
| Find Entities | None |
| Custom Entities | |
| **Ignore** | |
| Ignore Parts of Speech | 'Aux' 'Conj' 'Det' 'Interj' 'Par[...] |
| Ignore Types of Entities | [...] |
| Ignore Types of Attributes | 'Num' 'Punct' [...] |
| **Synonyms** | |
| Stem Terms | Yes |
| Synonyms | SASHELP.ENGSYNMS [...] |
| **Filter** | |
| Start List | [...] |
| Stop List | SASHELP.ENGSTOP [...] |

Display 4.5 shows the Terms window from the results of the Text Parsing node being run with default property settings.

**Display 4.5: Results of Text Parsing Node with Default Property Settings**

| Term | Role | Attribute | Freq | # Docs | Keep | Parent/Child Status | Parent ID | Rank for Variable numdocs |
|---|---|---|---|---|---|---|---|---|
| + forum | Noun | Alpha | 11193 | 995 Y | | + | 519 | 1 |
| global | Adj | Alpha | 11546 | 995 Y | | | 518 | 1 |
| + global forum | Noun Group | Alpha | 10871 | 993 Y | | + | 94 | 3 |
| + sas institute | Company | Entity | 42266 | 990 Y | | + | 221207 | 4 |
| + paper | Noun | Alpha | 4528 | 988 Y | | + | 284 | 5 |
| + be | Verb | Alpha | 76710 | 987 N | | + | 357 | 6 |
| other | Adj | Alpha | 5740 | 980 N | | | 550 | 7 |
| inc | Prop | Alpha | 4216 | 974 Y | | | 296 | 8 |
| information | Noun | Alpha | 7268 | 974 Y | | | 554 | 8 |
| + name | Noun | Alpha | 7916 | 974 N | | + | 706 | 8 |
| + institute | Noun | Alpha | 4185 | 973 Y | | + | 1539 | 11 |
| + other | Noun | Alpha | 2683 | 971 N | | + | 556 | 12 |
| + product | Noun | Alpha | 3900 | 967 Y | | + | 1568 | 13 |
| + indicate | Verb | Alpha | 2035 | 966 Y | | + | 1457 | 14 |

In the Terms window, the **Term** column displays all of the terms extracted from parsing. A + sign before the word identifies the word as a parent term. The same information can be derived from the **Parent/Child Status** column, where a + sign is populated for all of the parent terms. A parent term is the root word that is created out of the stemming process or is defined as a parent word in the synonyms dictionary.

As shown in Display 4.4, the Properties panel has a facility to use a synonyms list for parsing. Click the ellipsis button next to the **Synonyms** property to view, add, or delete synonyms. The default synonyms list, ENGSYNMS, available in the SASHELP dictionary, has only one entry—the first entry—as shown in Display 4.6. To add custom entries, click [icon] at the top left corner of the window. For example, you want to treat the first three terms in Display 4.6 as synonyms with "global forum" as the parent term. You want to treat the terms "same" and "similar" as synonyms with "similar" as the parent term. Display 4.6 shows the new synonyms added to the list.

**Display 4.6: Creating Synonyms**

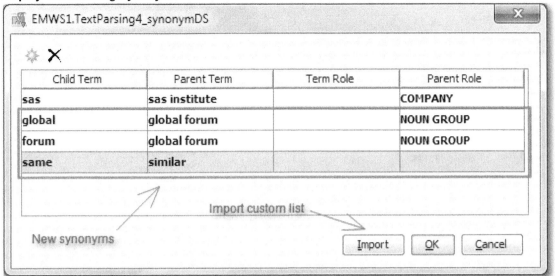

Each term is given a role that represents either the part of speech of the term, entity classification of the term, or the value **NOUN GROUP**. The role controls how the synonyms are applied when the part-of-speech and entities property settings are enabled or disabled. We discuss more about roles in the following sections when dealing with parts of speech and entities. A custom data set of synonyms can easily be imported and used for parsing. (See Display 4.6.) The data set should contain the following columns:

- TERM
- TERMROLE
- PARENT
- PARENTROLE

The imported data set will replace the default ENGSYNMS data set for the analysis. Any custom-created synonyms in the ENGSYNMS data set (as shown in Display 4.6) will not be used. If you do not want to use any synonyms in the analysis, check the **No data set to be specified** check box in the import window as shown in Display 4.7.

**Display 4.7: Import Window Settings**

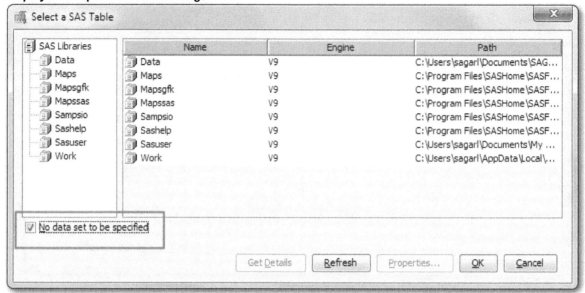

The terms that were combined from stemming and synonym groupings can be viewed using the interactive results viewer feature of the Text Filter node. The Text Filter node is connected to the Text Parsing node and must be run to access the interactive filter viewer window. Display 4.8 shows the interactive filter viewer window. You can see that the words "use," "used," "using," and "uses" are represented by one root term, "use." Similarly, you no longer see individual occurrences of the terms "global" and "forum."

**Display 4.8: Interactive Filter Viewer Window from the Text Filter Node**

Terms

| TERM | FREQ ▼ | # DOCS | KEEP | WEIGHT | ROLE | ATTRIBUTE |
|---|---|---|---|---|---|---|
| ⊞ be | 76711 | 987 | ☐ | 0.0 | Verb | Alpha |
| ⊞ sas institute | 42266 | 990 | ☑ | 0.07 | Company | Entity |
| data | 42215 | 936 | ☑ | 0.067 | Noun | Alpha |
| ⊟ use | 20055 | 908 | ☐ | 0.0 | Verb | Alpha |
| uses | 1186 | 464 | | | Verb | Alpha |
| used | 5137 | 821 | | | Verb | Alpha |
| use | 5332 | 775 | | | Verb | Alpha |
| using | 8400 | 874 | | | Verb | Alpha |
| ⊞ have | 14949 | 909 | ☐ | 0.0 | Verb | Alpha |
| ⊞ variable | 12599 | 684 | ☑ | 0.139 | Noun | Alpha |
| ⊞ not | 11494 | 879 | ☐ | 0.0 | Adv | Alpha |
| ⊞ value | 11047 | 768 | ☑ | 0.12 | Noun | Alpha |
| ⊟ global forum | 10871 | 993 | ☑ | 0.031 | Noun Group | Alpha |
| global forum | 10865 | 993 | | | Noun Group | Alpha |
| global forums | 6 | 5 | | | Noun Group | Alpha |
| ⊞ do | 10012 | 857 | ☐ | 0.0 | Verb | Alpha |
| ⊞ create | 9961 | 838 | ☑ | 0.091 | Verb | Alpha |

Display 4.9 shows the results of the synonym grouping for the terms "similar" and "same." Because the **Term Role** and **Parent Role** are left blank in the synonym definition (see Display 4.6), the grouping is done for

different parts of speech. Essentially, each different part of speech for the same word is treated as a different word. We explore part-of-speech settings in detail in the next section.

**Display 4.9: Results of Custom Synonym Definition**

| TERM ▲ | FREQ | # DOCS | KEEP | WEIGHT | ROLE | ATTRIBUTE |
|---|---|---|---|---|---|---|
| simila | 1 | 1 | ☐ | 0.0 | Noun | Alpha |
| ⊟ similar | 4863 | 796 | ☑ | 0.08 | Adj | Alpha |
| same | 3704 | 749 | | | Adj | Alpha |
| similar | 1159 | 493 | | | Adj | Alpha |
| ⊟ similar | 15 | 13 | ☑ | 0.635 | Prop | Alpha |
| same | 3 | 2 | | | Prop | Alpha |
| similar | 12 | 11 | | | Prop | Alpha |
| ⊟ similar | 9 | 7 | ☑ | 0.727 | Adv | Alpha |
| same | 9 | 7 | | | Adv | Alpha |
| similar | 0 | 0 | | | Adv | Alpha |
| similar access | 1 | 1 | ☐ | 0.0 | Noun Group | Alpha |

In a typical text mining project, using a custom synonym list is just a starting point. The synonym list grows by adding more terms that are identified from the extracted terms. Therefore, you end up creating a synonym list for each specific type of data analysis. For example, you have a synonym list for analyzing customer complaints, a synonym list for analyzing claims filings, etc. The creation of synonym lists on the fly can be done using the interactive filter viewer window. In this window (as shown in Display 4.10), identify and highlight the terms that you want to be treated as synonyms. Then, right-click and select the option **Treat as Synonyms.** In the following window, you select one of the terms as the parent term. As shown in Display 4.10, the terms "procedure" and "proc" are treated as similar, and the term "procedure" is selected as the parent term, as shown in Display 4.11. This is reasonable for this corpus because the term "proc" in SAS Global Forum papers is referring to a procedure.

**Display 4.10: Creating Synonyms Using the Interactive Filter Viewer Window**

| | TERM | FREQ | # DOCS | KEEP ▼ | WEIGHT | ROLE | ATTRIBUTE |
|---|---|---|---|---|---|---|---|
| ⊞ | procedure | | | | | | Alpha |
| ⊞ | global forum | 6 | | Add Term to Search Expression | | Noun Group | Alpha |
| ⊞ | data | 25 | | Treat as Synonyms | | | Alpha |
| ⊞ | information | 3 | | Remove Synonyms | | | Alpha |
| ⊞ | product | 1 | | Toggle KEEP | | | Alpha |
| ⊞ | register | | | View Concept Links | | | Alpha |
| | usa | 1 | | Find | | | Alpha |
| ⊞ | indicate | 1 | | Repeat Find | | | Alpha |
| ⊞ | value | 9 | | | | | Alpha |
| ⊞ | trademark | 1 | | Clear Selection | | | Alpha |
| ⊞ | abstract | | | Print... | | | Alpha |
| ⊞ | set | 9793 | 447 | ✓ | 0.092 | | Alpha |
| ⊞ | contact | 873 | 446 | ✓ | 0.023 | | Alpha |
| ⊞ | service | 1236 | 445 | ✓ | 0.153 | | Alpha |
| ⊞ | country | 885 | 445 | ✓ | 0.145 | | Alpha |
| ⊞ | registration | 476 | 444 | ✓ | 0.016 | | Alpha |
| ⊞ | run | 7181 | 439 | ✓ | 0.063 | | Alpha |
| ⊞ | service name | 442 | 438 | ✓ | 0.0090 | Noun Group | Alpha |
| ⊞ | create | 6532 | 435 | ✓ | 0.079 | | Alpha |
| ⊞ | proc | 10518 | 425 | ✓ | 0.079 | | Alpha |
| | respective | 487 | 424 | ✓ | 0.028 | | Alpha |

**Display 4.11: Selecting the Parent Term When Creating Custom Synonyms**

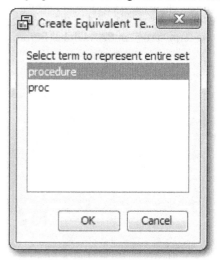

When you are finished with your changes, you can save your results or create a synonym data set by selecting **Export Synonyms** from the **File** menu. This is one of the time-consuming tasks in parsing in which you have to manually explore all of the terms and treat similar terms in your study as synonyms.

## Identifying Parts of Speech

After the extraction of tokens from the text, SAS Text Miner identifies parts of speech for each token or term based on both the definition of the term and the context. This is complicated because a word can represent multiple parts of speech depending on the context. This type of ambiguity can be because of the lexical definition of the term or the syntactic usage. For example, the word "institute" is lexically ambiguous because it can either be a noun or a verb. A syntactic ambiguity occurs in a sentence such as, "I made her duck." The sentence can mean the following:

- I cooked waterfowl for her.
- I caused her to quickly lower her head or body.
- I waved my magic wand and turned her into a waterfowl.

A deep understanding of the adjacent terms in the sentence or in the paragraph is required to resolve the ambiguities and correctly identify the part of speech for a term. The written word loses the discriminating features of the spoken language, such as volume and inflection. The POS tagging algorithms overcome these challenges to a certain extent by using predefined rules. The tagger algorithms determine whether the word is a common noun, verb, adjective, proper noun, adverb, and so on. The goal of POS tagging is to create a group of syntactically related terms.

Using the part-of-speech feature results in a much larger bag of words for analysis. As seen in Display 4.9, the word "similar" is identified as an adjective, proposition, and adverb, which makes it three different words. SAS Text Miner provides an option to turn the part-of-speech feature off completely or to ignore terms assigned to specific part-of-speech tags. This depends on the analyst's choice and the type of the study. One way to overcome the lexical ambiguity for a term is to forcibly assign a part of speech to the term. For example, in our corpus, the term "author" is identified as a noun and as a verb. You can force the tool to treat the term "author" as a noun only by adding the entry (as shown in Table 4.2) to the synonym list. This enforces the condition that the term "author," whenever parsed as a verb, is treated as a noun.

**Table 4.2: Creating a Custom Synonym for the Term "author"**

| Term | Term Role | Parent | Parent Role |
|------|-----------|--------|-------------|
| Author | Verb | Author | Noun |

The other way to fix the ambiguity is to use the **Treat as Synonyms** option from the interactive filter viewer window as discussed in the previous section. Given that there are thousands of terms extracted, correcting the part-of-speech tagging for the terms using these types of approaches requires substantial human intervention. By now, you can understand why it might be sometimes preferable to disable part-of-speech tagging to reduce the complexity, even though that might result in loss of information. You can safely ignore identifying parts of speech in situations where you have grammatically incorrect textual data (for example, in online chat comments, text messages or places where incomplete sentences are often used).

In addition to POS tagging, SAS Text Miner can identify noun groups based on linguistic relationships that exist within sentences. For example, the words "data" and "model" are parsed as individual terms and are also identified as a noun group, "data model." A noun group acts as a single term along with the other parsed single terms. If stemming is enabled, noun groups are also stemmed. For example, the term "data models" is parsed as "data model." You can either enable or disable the Noun Groups feature for your analysis.

**Display 4.12: Property Settings for POS Tagging and Noun Groups**

| Detect | |
|---|---|
| Different Parts of Speech | Yes |
| Noun Groups | Yes |
| Multi-word Terms | SASHELP.ENG_MULTI |
| Find Entities | None |
| Custom Entities | |
| Ignore | |
| Ignore Parts of Speech | 'Aux' 'Conj' 'Det' 'Interj' 'Par |
| Ignore Types of Entities | |
| Ignore Types of Attributes | 'Num' 'Punct' |

By default, the **Different Parts of Speech** and **Noun Groups** properties are enabled (**Yes**) as shown in Display 4.12. To disable these features, set these property values to **No**. Results in Display 4.5 show the second column (**Role**) with either a part-of-speech type or value **Noun Group** if the term is a noun group.

The **Ignore Parts of Speech** property provides a facility to select different types of parts of speech that you want to ignore in parsing. Click the ellipsis button next to the property setting to access the selection window. Display 4.13 shows the part-of-speech types that are ignored by default. You can select others or deselect the defaults. These two properties of parts of speech come in very handy in reducing the number of terms. The **Different Parts of Speech** property can be set to **No**, which does not treat the same word as multiple types. At the same time, you can ignore a type of part of speech so that the terms for that particular part of speech are not extracted.

For example, for the term "institute," Table 4.3 shows the term frequencies for three different types of settings. In any text mining project, the outcome of analysis largely depends on both quantity and quality of the terms that are extracted in text parsing. Hence, it is always suggested to play with the different options and explore the extracted terms.

**Table 4.3: Term Frequencies for the Term "institute" under Different Part-of-Speech Settings**

| Default | Different Part of Speech: No | Different Part of Speech: No Ignore: Prop, Verb |
|---|---|---|
| Noun: 4185 Prop: 21 Verb: 6 | 4212 | 4185 |

**Display 4.13: Parts of Speech Ignored in Parsing By Default**

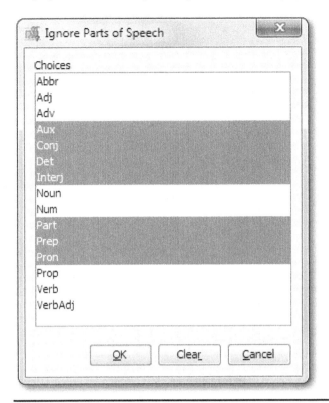

## Using Start and Stop Lists

The complexity of the text mining job increases with the number of terms parsed from the textual documents. There are always terms that are parsed that barely add any value to analysis. Examples of these terms are frequently occurring words in the English language such as, "a," "the," "be," "of," "in," "at," "to," etc. In most of the document corpora, these terms can account for almost 50% of the terms extracted. When dealing with documents from a specific field or topic, terms related to the topic occur most frequently. For example, in the SAS Global Forum paper corpus, the most occurring words are "global forum," "sas," "institute," etc. Display 4.14 shows the frequency distribution of the terms parsed from the SAS Global Forum papers without ignoring any type of part of speech. Clearly, the most frequently occurring words dominate the list. Each term is given a rank based on the occurrence of the term in the number of documents. Frequently occurring terms in many documents can simply add noise to the task of discriminating or classifying documents.

**Display 4.14: Results of Text Parsing with Default Part-of-Speech Option Values**

| Term | Role | Attribute | Freq | # Docs | Keep | Parent/Child Status | Parent ID | Rank for Variable numdocs |
|------|------|-----------|------|--------|------|---------------------|-----------|---------------------------|
| + global forum | Noun Group | Alpha | 10871 | 993 Y | + | | 94 | 1 |
| and | Conj | Alpha | 82945 | 992 N | | | 309 | 2 |
| the | Det | Alpha | 220862 | 991 N | | | 332 | 3 |
| + sas institute | Company | Entity | 42266 | 990 Y | + | | 221862 | 4 |
| + paper | Noun | Alpha | 4537 | 988 Y | + | | 284 | 5 |
| + be | Verb | Alpha | 76711 | 987 N | + | | 375 | 6 |
| + be | Aux | Alpha | 31516 | 986 N | + | | 196861 | 7 |
| in | Prep | Alpha | 57419 | 986 N | | | 289 | 7 |
| of | Prep | Alpha | 79958 | 986 N | | | 324 | 7 |
| or | Conj | Alpha | 17327 | 986 N | | | 479 | 7 |
| for | Prep | Alpha | 38874 | 984 N | | | 342 | 11 |
| other | Adj | Alpha | 5740 | 980 N | | | 599 | 12 |
| all | Det | Alpha | 4490 | 976 N | | | 381 | 13 |
| to | Prep | Alpha | 79583 | 975 N | | | 328 | 14 |
| inc | Prop | Alpha | 4216 | 974 Y | | | 298 | 15 |
| information | Noun | Alpha | 7268 | 974 Y | | | 603 | 15 |
| + name | Noun | Alpha | 7916 | 974 N | + | | 763 | 15 |
| + institute | Noun | Alpha | 4185 | 973 Y | + | | 1645 | 18 |
| a | Det | Alpha | 69346 | 971 N | | | 290 | 19 |

If ignoring specific parts of speech is one method of filtering terms from parsing, another easy way to filter words without ignoring part-of-speech type is by using a start or stop list. A start or stop list helps control the terms that are used in text mining analysis. A stop list consists of stop words that have little or no value in identifying a document or in comparing documents. Standard stop lists contain stop words that are articles (the, a, this), conjunctions (and, but, or), and prepositions (of, from, by). A custom stop list can be created to identify low information words, like the word "computer" in a collection of articles about computers. In contrast, a start list contains the words that you want to include in the analysis. When using a start list, only the terms in the start list are included in the analysis.

Start lists are mainly used when documents are dominated by technical jargon or in situations where adequate domain expertise is available. On the other hand, stop lists are useful when documents are loosely related like news articles, business reports, and Internet searches or in situations where domain expertise is not available.

Display 4.15 shows the Properties panel of the Text Parsing node with start and stop list properties.

**Display 4:15: Text Parsing Node Properties Panel**

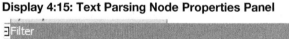

| Filter | |
|--------|--|
| Start List | |
| Stop List | SASHELP.ENGSTOP |
| Report | |

A default stop list (ENGSTOP) is automatically included with SAS Text Miner. New terms can be added or existing terms can be deleted from this list. A custom start or stop list can be imported the same way a synonym list is imported. This list should contain the column **Term**, which represents the term to include or exclude, and an additional **Role** column that contains the role of the term. A role is a part of speech, an entity classification, or the value Noun Group. Enabling the part-of-speech setting controls which form of a word is excluded or included. For example, if you have the term "duck" with the role Verb in your stop list, and you set the **Different Parts of Speech** property to **Yes**, then the word "duck" is excluded from parsing only when it is used as a verb. If no role is defined for the word "duck," then any instance of the term "duck" is excluded from the parsing results. In this example, we have added custom terms to the stop list as shown in Display 4.16.

**Display 4.16: Adding Custom Entries to the Default Stop List**

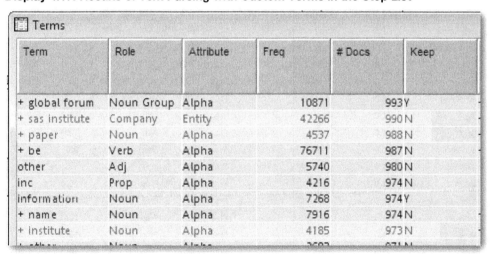

After text parsing with the modified stop list, the terms added to the stop list now have the value of **N** for the **Keep** column (as you can see in Display 4.17). A value of **Y** for this column indicates that the term is kept for use in subsequent nodes. Otherwise, it is dropped.

**Display 4.17: Results of Text Parsing with Custom Terms in the Stop List**

| Term | Role | Attribute | Freq | # Docs | Keep |
|------|------|-----------|------|--------|------|
| + global forum | Noun Group | Alpha | 10871 | 993 | Y |
| + sas institute | Company | Entity | 42266 | 990 | N |
| + paper | Noun | Alpha | 4537 | 988 | N |
| + be | Verb | Alpha | 76711 | 987 | N |
| other | Adj | Alpha | 5740 | 980 | N |
| inc | Prop | Alpha | 4216 | 974 | N |
| information | Noun | Alpha | 7268 | 974 | Y |
| + name | Noun | Alpha | 7916 | 974 | N |
| + institute | Noun | Alpha | 4185 | 973 | N |

It is very common that analyzing text requires processing and understanding words together. Multiple terms could include simple words like "as far as" or include non-decomposable idioms like "to kick the bucket." SAS Text Miner extracts and processes a multiple term as a single term. SAS uses a dictionary, a default multi-word list, for identifying the multi-word terms. User-defined multi-word terms created in the custom synonym list are added to the default multi-word dictionary. In the same way a start and stop list is defined and managed, the tool provides a facility to use custom multi-word lists. A default multi-word list is provided. Some of the multi-word terms extracted and processed in our example include "high quality," "stand out," "to the point," etc.

## Spell Checking

Misspelled words often account for a major part of the noise that comes with text. Even a professionally written document can contain misspelled words. This type of noise is even more frequent in informal text like customer comments, e-mails, online discussion forums, etc. After text parsing is performed and terms are extracted, a dictionary can be used to spell check and correct misspelled words. When no dictionary is used, SAS Text Miner compares words with each other in the corpus to identify misspelled words. Most frequently occurring words are considered to be correctly spelled, and they are used to correct the less frequently occurring words. When using a dictionary, any word that is in the dictionary is considered correctly spelled.

The algorithm calculates a customized distance score using the SPEDIS function from Base SAS. The distance score is the distance between the less frequently occurring word, called the child term, and all other frequently occurring words starting with the same first letter, called the parent term. The distance score is the measure of how close the misspelled word is to the parent term in terms of spelling. The smaller the distance score, the closer the terms. Spell checking can be performed using the Text Filter node or using the %TEXTSYN macro. The result of the spell checking task is a data set that can be used as a synonym list. In Case Study 3, we show the advantages of creating a custom synonym list using spell check results. Each misspelled word is a distinct word that increases the word count. Rerunning the Text Parsing node with the spell check results can phenomenally reduce the number of different words extracted. Display 4.18 shows the sample results of a spell check on the terms extracted from the SAS Global Forum paper corpus. The **term** column shows the misspelled word, and the **parent** column shows the correctly spelled word. The **minsped** column is the distance score calculated using the SPEDIS function.

**Display 4.18: Sample Spell Checking Results**

EMWS1.TextFilter3_spellDS

| | numdocs | term | childndocs | parent | termrole | parentrole | minsped | dict |
|---|---|---|---|---|---|---|---|---|
| 9 | 17.0 | workspaceserver | 3.0 | workspace server | Prop | NOUN_GROUP | 6.0 | |
| 10 | 17.0 | workspace-server | 1.0 | workspace server | Noun | NOUN_GROUP | 12.0 | |
| 11 | 9.0 | resourcerequirement | 1.0 | resource requirements | Prop | NOUN_GROUP | 10.0 | |
| 12 | 7.0 | configuration-directory | 1.0 | configuration directory | Noun | NOUN_GROUP | 8.0 | |
| 13 | 7.0 | user count | 1.0 | user account | NOUN_GROUP | NOUN_GROUP | 14.0 | |
| 14 | 775.0 | contract information | 1.0 | contact information | NOUN_GROUP | NOUN_GROUP | 4.0 | |
| 15 | 893.0 | other band | 1.0 | other brand | NOUN_GROUP | NOUN_GROUP | 10.0 | |
| 16 | 893.0 | productnames | 1.0 | product names | Noun | NOUN_GROUP | 8.0 | |
| 17 | 40.0 | achieving | 5.0 | achieving | Prop | Verb | 0.0 | |
| 18 | 259.0 | hig | 1.0 | high | Prop | Adj | 10.0 | |
| 19 | 259.0 | haigh | 1.0 | high | Prop | Adj | 12.0 | |
| 20 | 91.0 | availability | 1.0 | availability | Prop | Noun | 0.0 | |
| 21 | 91.0 | availabilit | 1.0 | availability | Prop | Noun | 2.0 | |
| 22 | 990.0 | sas™ | 1.0 | sas | Prop | Prop | 10.0 | |
| 23 | 990.0 | saas | 5.0 | sas | Prop | Prop | 8.0 | |
| 24 | 990.0 | sas8 | 1.0 | sas | Prop | Prop | 10.0 | |
| 25 | 990.0 | sasa | 1.0 | sas | Noun | Prop | 10.0 | |
| 26 | 990.0 | sask | 1.0 | sas | Prop | Prop | 10.0 | |
| 27 | 990.0 | sasl | 1.0 | sas | Noun | Prop | 10.0 | |
| 28 | 990.0 | sasr | 1.0 | sas | Prop | Prop | 10.0 | |
| 29 | 990.0 | sass | 1.0 | sas | Noun | Prop | 8.0 | |

Without a dictionary, it is possible that a correctly spelled word is falsely identified as misspelled. Custom dictionaries can be easily converted to SAS data sets and plugged in to SAS Text Miner for spell checking.

Consider the following method for creating a SAS data set as described in the online Help for SAS Text Miner:

OpenOffice.org has links to dictionaries for many languages that are available for free download from http://wiki.services.openoffice.org/wiki/Dictionaries.

To create an English dictionary data set that you can use with the Text Filter node or the %TEXTSYN macro, do the following:

1. Go to http://extensions.openoffice.org/en/project/en_US-dict to download the OpenOffice English US dictionary file.
2. Save the file to your local machine and unzip it when the download completes. The file extension can be different depending on your browser. However, you can easily unzip the file using most unzipping tools.
3. In SAS Enterprise Miner, select **View ▶ Program Editor**.
4. In the Program Editor that opens, enter and run the following SAS code. This code removes extraneous characters from the OpenOffice dictionary, assigns the Proper Noun part of speech to any term that is capitalized in the file, and creates an English dictionary data set with the required format.

Note: Be sure to change *<fileLocation>* to the path of the unzipped dictionary file. Also, the following code uses a library called **tmlib**. You will need to create this library with a LIBNAME statement before running this code. Or, you can change **tmlib** to another library that you have already created.

**Program 4.1: SAS Code to Create English Dictionary to use in SAS Text Miner**

```
data tmlib.engdict (keep=term pos);
   length inputterm term $32;
   infile '<fileLocation>en_US.dic'
      truncover;
   input linetxt $80.;
   i=1;
   do until (inputterm = ' ');
      inputterm = scan(linetxt, i, ' ');
      if inputterm ne ' ' then do;
            location=index(inputterm,'/');
            if location gt 0 then
            term = substr(inputterm,1,location-1);
            if location eq 0 then
            term = inputterm;
         if lowcase(term) ne term then pos = 'Prop';
         term = lowcase(term);
         output;
         end;
      i=i+1;
      end;
   run;
```

The type and quality of the data determines the need for spell checking. A text corpus of news articles and journal papers would rarely require spell checking. However, an analysis of customer comments, discussion forums, and social media data would definitely require spell checking. Display 4.19 shows sample spell check results on a data set of Tweets with the hashtag "usa" (#usa) using a dictionary (prepared as described above). You can see in Display 4.19 that correctly spelled words like "produce" and "guilt" are identified as misspelled. Even with a dictionary, it is possible that a correctly spelled word could be identified as misspelled. Hence, it is always recommended to carefully review the results from spell checking and discard the ones that are wrongly classified.

**Display 4.19: Sample Spell Check Results on Tweets with #usa**

| numdocs | term | childndocs | parent | termrole | parentrole | minsped | dict |
|---|---|---|---|---|---|---|---|
| 33.0 | canad | 1.0 | canada | | | 12.0 | Y |
| 13.0 | australian | 1.0 | australia | | | 6.0 | Y |
| 7.0 | produce | 1.0 | producer | | | 8.0 | Y |
| 4.0 | incident | 3.0 | incidente | | | 6.0 | Y |
| 5.0 | russian | 1.0 | russia | | | 10.0 | Y |
| 6.0 | german | 1.0 | germany | | | 10.0 | Y |
| 12.0 | manny | 5.0 | many | | | 12.0 | Y |
| 4.0 | guilt | 1.0 | guilty | | | 12.0 | Y |
| 5.0 | spring | 2.0 | springs | | | 10.0 | Y |
| 4.0 | cuban | 2.0 | cubano | | | 12.0 | Y |
| 8.0 | closet | 1.0 | close | | | 12.0 | Y |
| 6.0 | amricans | 1.0 | americans | | | 6.0 | N |
| 5.0 | trials | 1.0 | trial | | | 6.0 | N |
| 7.0 | clase | 1.0 | case | | | 12.0 | N |
| 4.0 | occup | 1.0 | occupy | | | 6.0 | N |
| 22.0 | amurica | 1.0 | america | | | 14.0 | N |

## Entities

Entity detection is one of the many key features offered in SAS Text Miner. An entity is a piece of useful information contained in the textual data that is usually different from the general text. SAS Text Miner can identify various types of entities that are prebuilt into the software. Factual information such as a street name or postal address, name of a person, phone number, name of an organization, e-mail address, name of a city, etc., are examples of some of the many standard entities that SAS Text Miner is capable of detecting automatically. If SAS Text Miner detects an entity that has one or more words, then each one of the words is analyzed by the software to find entities. The entity extraction feature is available within the Text Parsing node in SAS Text Miner. Currently, SAS Text Miner is capable of detecting the standard entities as outlined in Table 4.4.

**Table 4.4: Default Standard Entity Types in SAS Text Miner (ordered alphabetically)**

| Standard Entity | Description | Example |
|---|---|---|
| ADDRESS | Postal address or a street name | 2526 Carywood Drive |
| COMPANY | Name of a company | SAS Institute Inc. |
| CURRENCY | Currency or Currency Expression | $450,000 or 6000 USD |
| DATE | Full Date, Year, Month, or Day | 1st Oct 1949 or January or 1983 |
| INTERNET | URL of a website or e-mail address | http://www.sas.com or support@sas.com |
| LOCATION | Name of a city, state, country, or any other geographical place or region | Istanbul or Malaysia or Utah |
| MEASURE | Measurement or measurement expression | 100 miles or 25 sec |
| ORGANIZATION | Name of a government or service agency | FBI – Federal Bureau of Investigation or SEC – Securities and Exchange Commission |
| PERCENT | Percentage or percent expression | 45% or 20 PERCENT |

| Standard Entity | Description | Example |
|---|---|---|
| PERSON | Name of a person | Abraham Lincoln or John F. Kennedy |
| PHONE | Phone number in a standard format | 1-XXX-XXX-XXXX or (XXX)XXX-XXXX |
| PROP_MISC | Proper noun with an ambiguous classification | US President George Bush |
| SSN | Social Security number in a standard format | ZZZ-ZZ-ZZZZ |
| TIME | Time or time expression | 2:45 pm 14:45 |
| TIME_PERIOD | An expression of measured time | 10 days or 5 years |
| TITLE | Title of a person or a position | Mr. or Ms. or Dr. |
| VEHICLE | Make, name, model, year, and color of a motor vehicle | Honda Accord EX 2005 Black |

Raw entities found during the analysis are sometimes transformed into their normalized forms. SAS Text Miner performs normalization on the entities in certain instances. Entities in short form are normalized to return them to their full form. For example, when the Text Parsing node finds "NASA" in the text, it is looked up against the dictionary, and its complete name "National Aeronautics and Space Administration" is returned. A prebuilt dictionary with a fixed set of organization and company names is used to identify entities and associate them each with a parent. As a generally accepted norm, the entity expanded in its most precise expanded form is used as the parent form. Entities found in the ISO (International Standardization for Organization) standard are normalized to replace the original term instead of keeping the parent–child (original term) relation. Entities returned by SAS Text Miner can be changed by tweaking the synonym list. The original entity is placed in the **Term** column, the entity in the appropriate form in the **Parent** column, and the entity category in the **Category** column before rerunning the node to achieve the results. Custom entities can be derived using either SAS Concept Creation for SAS Text Miner or SAS Contextual Extraction Studio. The compiled custom entity file (with an .li extension) should be imported to use in a Text Parsing node. In this chapter, we limit our discussion about how to use the custom entities built using SAS Contextual Extraction Studio in SAS Text Miner with a simple example. All of the features and capabilities of SAS Content Categorization Studio and SAS Contextual Extraction Studio are discussed in greater detail in Chapter 7.

***Important Note:*** Ensure that you specify April 12, 2011 as the compatibility date when compiling the custom entity files to use in SAS Text Miner.

Let's see the different types of entities that are extracted from our example corpus, the SAS Global Forum papers. Use the **Standard** option in the **Find Entities** property setting in the Text Parsing node. Run the node to identify standard entities within the data set. As shown in Display 4.20, the parsed terms are tagged as **Entity** for the attribute type with roles such as **Currency** and **Date**. Other entity roles that are identified from this corpus include Person, Organization, Location, Internet, etc. The role Internet is assigned to e-mail IDs and website names.

**Display 4.20: Partial Output from the Text Parsing Node Showing Identified Standard Entities**

| Term | Role ▲ | Attribute | Freq | # Docs | Keep | Parent/Child Status |
|------|--------|-----------|------|--------|------|---------------------|
| $660 billion | Currency | Entity | 1 | | 1Y | |
| $69 | Currency | Entity | 1 | | 1Y | |
| $7,500 | Currency | Entity | 2 | | 1Y | |
| $750,000 | Currency | Entity | 2 | | 1Y | |
| $85 billion | Currency | Entity | 1 | | 1Y | |
| august | Date | Entity | 4 | | 4Y | |
| february | Date | Entity | 5 | | 4Y | |
| july | Date | Entity | 4 | | 4Y | |
| june | Date | Entity | 4 | | 4Y | |
| april | Date | Entity | 3 | | 3Y | |
| january 1, 1960 | Date | Entity | 8 | | 3Y | |
| may | Date | Entity | 3 | | 3Y | |
| tuesday | Date | Entity | 3 | | 3Y | |
| wednesday | Date | Entity | 3 | | 3Y | |

## Concept

A simple concept can be defined as a piece of information such as an entity (like name of a place, person, organization, location, etc.). An example of a simple concept is a place like New York or an organization such as SAS Institute Inc. A relational concept can be two or more entities or terms in relation with each other. For example, CEO of SAS Dr. Jim Goodnight is a relational concept. One of the important features of SAS Content Categorization Studio is concept extraction. There are primarily two types of concept definitions that you can create in SAS Content Categorization Studio.

Classifier – A classifier concept is based on an independent word or string identified in the text. It can use regular expressions for pattern matching. Boolean rules can be written to perform concept extraction with disambiguation to distinguish between two contexts in which the same word is used. For example, Bing can be referred to as a search engine as well as a heap or pile in British dialect.

Grammar – A grammar concept predominantly works based on the part-of-speech tags of the terms and special symbols to identify precise matches for patterns.

## Building Custom Entities Using SAS Contextual Extraction Studio

One of the several important features that SAS Contextual Extraction Studio offers is context-sensitive matching using complex advanced linguistic rules called LITI (or Language Interpretation/Text Interpretation). In this feature, concepts are matched with a specific context. For example, names of people are identified by SAS Text Miner automatically when the **Find Entities** property is set to **Standard**. However, suppose you are interested in only finding names of authors provided after the sub-header **CONTACT INFORMATION** in the text. (See Display 4.21.) In this case, SAS Text Miner is not able to exclusively find those names. You have to use SAS Contextual Extraction Studio and write a LITI rule to find matching entities with context.

**Display 4.21: Partial Screen Capture of the Data Set Showing the Highlighted Context and Author Names**

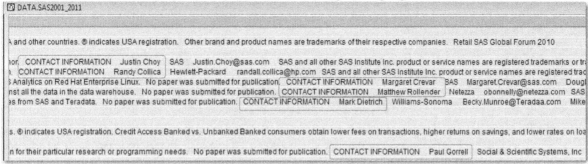

In SAS Contextual Extraction Studio, define a classifier for the keywords **CONTACT INFORMATION** in the **PREFIX** concept as shown below. Also, define a **AUTHOR** concept using the concept in context (**C_CONCEPT**) and using the **PREFIX** concept.

| Concept Name | Entry |
|---|---|
| PREFIX | CLASSIFIER: CONTACT INFORMATION |
| AUTHOR | C_CONCEPT: PREFIX _c {_cap _cap} |

**_cap** represents the symbol to match a word starting with a capital letter. **_c** is a context marker symbol to identify only the concepts in context. Here, it means to identify the names of people following the keywords **CONTACT INFORMATION**. In this way, only those authors whose names are specified after the CONTACT INFORMATION sub-header in the text will be identified.

**Display 4.22: Concepts Tested in SAS Contextual Extraction Studio Using Sample Text**

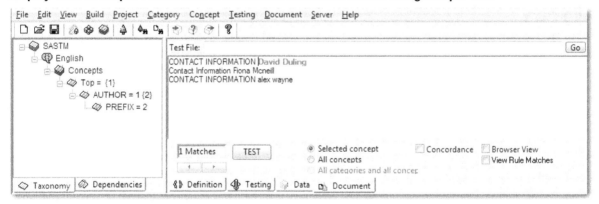

You can test the LITI rules using sample text populated with both false positive and true positive cases to ensure that they are working correctly. (See Display 4.22.) After compiling the concepts, an .li file is generated in the SAS Contextual Extraction Studio project directory. You now have the custom entity file (with an .li extension) generated from SAS Contextual Extraction Studio. This file can be used with SAS Text Miner to identify custom entities. In the Text Parsing node Properties panel, change the **Find Entities** property to **Custom** and update the **Custom Entities** field with the complete path to the .li file, as shown in Display 4.23. Rerun the Text Parsing node.

**Display 4.23: Text Parsing Node Properties to Extract Custom Entities**

| Detect | |
|---|---|
| Different Parts of Speech | Yes |
| Noun Groups | Yes |
| Multi-word Terms | SASHELP.ENG_MULTI |
| Find Entities | Custom |
| Custom Entities | C:\SASTM\CCS\TK240 Projects\SASTM\English.li |

The SAS Text Miner Text Parsing node now generates the output with the custom entities extracted from the text based on the LITI rules matching with the context that we developed earlier. (See Display 4.24.) In this way, you can define LITI rules using SAS Contextual Extraction Studio and use them to extract a wide range of custom entities.

**Display 4.24: Partial Output of the Text Parsing Results Showing Custom Entities Extracted**

| Terms | | | | | | |
|---|---|---|---|---|---|---|
| Term | Role | Attribute ▼ | Freq | # Docs | Keep | Parent/Child Status |
| contact information | PREFIX | Entity | 403 | 403 | Y | |
| david duling | AUTHOR | Entity | 1 | 1 | Y | |
| fiona mcneill | AUTHOR | Entity | 1 | 1 | Y | |
| john sall | AUTHOR | Entity | 1 | 1 | Y | |
| randy betancourt | AUTHOR | Entity | 1 | 1 | Y | |

## Summary

In this chapter, we covered in detail the text parsing task, a key and fundamental step in the text mining process. You have seen the advantages and disadvantages of various linguistics operations performed in this process, such as stemming, part-of-speech tagging, entity extraction, etc. These operations significantly impact the number and types of terms that are extracted from the corpus. A higher number of terms poses analytical and computational challenges. Hence, these operations should be chosen very carefully. In many situations, the business problem will influence the property selections.

Entity and concept extraction in SAS Text Miner were also discussed in this chapter. A detailed discussion on these topics using SAS Content Categorization Studio is provided in Chapter 7. The next chapters cover tasks that follow text parsing: text filtering, text clustering, and topic mining.

## References

Abney, S. 1997. "Part-of-Speech Tagging and Partial Parsing." *Corpus Based Methods in Language and Speech Processing*, Kluwer, Dordrecht.

Brown, Peter F., Della Pietra, Stephen A., Della Pietra, Vincent J., and Mercer, Robert L. 1991. "Word-Sense Disambiguation Using Statistical Methods". *Proceedings of the 29th Annual Meeting of the Association for Computational Linguistics*, 264-270.

Ciravegna, F. and Lavelli, A. 1999. "Full Text Parsing Using Cascades of Rules: An Information ExtractionPerspective". *Proceedings of the Ninth Conference of the European Chapter of the Association of Computational Linguistics: EACL '99*.

Dale, R. Moisl, H. and Somers, H. 2000. *Handbook of Natural Language Processing*. New York, NY: Marcel Dekker.

Gupta, V., Lehal, G.S. 2009. "A Survey of Text Mining Techniques and Applications." *Journal of EmergingTechnologies in Web Intelligence.* 1(1): 60-76.

Koppel, M., Schler, J., and Zigdon, K. 2005. "Determining an Author's Native Language by Mining aText forErrors". *KDD 05: Proceedings of the Eleventh ACM SIGKDD International Conference on Knowledge Discovery in Data Mining,* 624-628.

Kraft, R. and Zien, J. 2004. "Mining Anchor Text for Query Refinement". *Proceedings of the ThirteenthInternational Conference on World Wide Web (WWW-2004),* 666-674.

Manning, C. D. and Schutze, H. 1999. *Foundations of Statistical Natural Language Processing.* Cambridge,Massachusetts: The MIT Press.

Nivre, J. 2005. "Two Notions of Parsing". *Inquiries into Words, Constraints and Contexts: Festschrift forKimmo Koskenniemi on his 60th Birthday.* CSLI Studies in Computational Linguistics ONLINE. Stanford, CA: CSLI Publications. Available at: http://www.stanford.edu/group/cslipublications/cslipublications/koskenniemi-festschrift/.

Nivre, J. 2006. "Two Strategies for Text Parsing". *A Man of Measure, Festschrift in Honour of FredKarlsson,* 440-448.

"Part-of-Speech Tagging." Wikipedia. Retrieved Jun 20, 2012 from http://en.wikipedia.org/wiki/Part-of speech_tagging.

SAS® Content Categorization Studio 5.2: User's Guide. Cary, NC: SAS Institute Inc. 2011.

SAS® Contextual Extraction Studio 5.2: User's Guide. Cary, NC: SAS Institute Inc.2011.

SAS® Concept Creation for SAS® Text Miner: User's Guide. Cary, NC: SAS Institute Inc.. 2011.

SAS® Content Categorization and Contextual Extraction: Course Notes. Cary, NC: SAS Institute Inc.Cary, NC.2010.

Text Analytics using SAS® Text Miner. Course Notes. Cary, NC: SAS Institute Inc.Course information: https://support.sas.com/edu/schedules.html?ctry=us&id=1224

# Chapter 5 Data Transformation

## Introduction

You saw in Chapter 4 that the first task of text mining analysis is to break down the text into a bag of words or tokens. Then, you apply various linguistic rules to identify the parts of speech, synonyms, noun groups, attributes, etc. Even before doing this exercise, a good understanding of what is being talked about in a corpus can be obtained by looking at the counts of the words extracted from the corpus. A current popular technique to visually represent prominent terms in text is using a word cloud or text cloud. This is an easy and visually appealing way of presenting the frequency of words in text as a weighted list. The font size of the words represents the frequency of the word. High frequency words appear in a bigger font size. There are various online resources available for free that can be used to generate a word cloud. Display 5.1 shows a word cloud for text from five random SAS Global Forum papers from our SAS Global Forum corpus. It is clearly evident that the terms "data," "SAS," and "bivariate" appear frequently in this corpus.

**Display 5.1: Word Cloud**

Source: http://worditout.com

## Zipf's Law

The most widely celebrated theory on the distribution of words in a large corpus is given by George Zipf, popularly known as Zipf's law. This law was first stated by Estoup (1916), and later popularized by Zipf (1949). Manning and Schutze (1996) evaluate Zipf's law using Mark Twain's *Tom Sawyer*. According to Zipf's law, if you rank order the frequency of occurrence of the words in a corpus, you can observe an approximate mathematical relationship between the position of the word (or the rank) and the frequency of occurrence of the word, as stated below:

$$f \propto \frac{1}{r} \quad or \quad f \cdot r = k \qquad \text{(Equation 5.1)}$$

In this equation, $f$ is the frequency of occurrence of the word, $r$ is the rank of the word, and $k$ is a constant.

Zipf's law was widely applied to study different types of human behavior. In many cases, it was observed that the law holds true only for small corpora within processing capabilities during that time. When this law was applied to the SAS Global Forum corpus, we saw a big deviation in the constant term (frequency*rank) as shown in Table 5.1.

**Table 5.1: Applying Zipf's Law to SAS Global Forum Papers**

| Term | Frequency | Rank | Frequency*Rank |
|------|-----------|------|----------------|
| + be | 77920 | 1 | 77920 |
| + sas institute | 42292 | 2 | 84584 |
| + data | 42224 | 3 | 126672 |
| + use | 25155 | 4 | 100620 |
| + variable | 15218 | 5 | 76090 |
| -------------- | | | |
| + method | 4009 | 87 | 348783 |
| + add | 3991 | 88 | 351208 |

| Term | Frequency | Rank | Frequency*Rank |
|---|---|---|---|
| + order | 3931 | 89 | 349859 |
| + test | 3905 | 90 | 351450 |
| + specify | 3833 | 91 | 348803 |
| -------------- | | | |
| + complex filter | 3 | 19926 | 59778 |
| + opportunity cost | 3 | 19928 | 59784 |
| data lineage | 3 | 19929 | 59787 |
| auditability | 3 | 19930 | 59790 |
| monolithic | 3 | 19931 | 59793 |

However, Zipf's law continues to generate interest even today. It is universally accepted that Zipf's law gives a rough distribution of the frequency of words in a corpus. You can always identify a small set of words that account for almost 50% of the words in the corpus and identify a large set of words that occur very infrequently. You will find that these rare words account for a considerable portion of the text. The law tends to explain two forces that are controlled by the speaker and the hearer. The speaker is trying to minimize his effort by using fewer different types of words most often, and the hearer is trying to minimize his effort by having a large number of rare words (Marie, 1992).

Instead of term frequency, SAS Text Miner uses the frequency of number of documents for a word to rank order the terms. Display 5.2 shows the ZIPF plot from the Text Filter node results for the SAS Global Forum corpus.

**Display 5.2: ZIPF Plot from Text Filter Node Results**

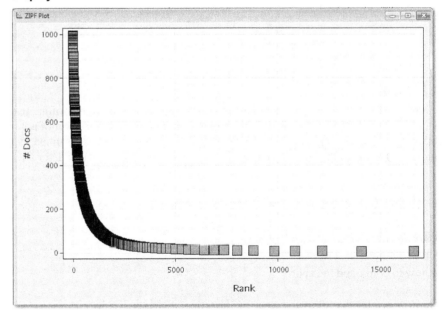

Mandelbrot (1954) has extensively validated Zipf's law. He identified that the law holds true only in a restricted range of the word distribution. He noted that the distribution according to the law is a bad fit, especially for high and low rank terms. Zipf's law belongs to the class of power laws ($y = kx^c$), with c=-1. According to a power law, the plot of frequency of the word and rank of the word on a logarithmic scale is approximately linear with slope -1. Display 5.3 shows the plot of frequency of the word and the rank on a logarithmic scale. You can clearly see a huge deviation from linear distribution for very high ranks. To achieve a better fit, Mandelbrot modified Zipf's law and suggested the relationship between rank and frequency as the following:

$$f = P\,(r + \rho)^{-B} \quad \text{or} \quad \log f = \log P - B \log(r + \rho) \qquad \text{(Equation 5.2)}$$

In this equation, $P$, $B$ and $\rho$ are parameters of the text. For B=1 and $\rho = 0$, Mandelbrot's formula simplifies to Zipf's law.

**Display 5.3: Term Frequency versus Term Rank (Log Scale)**

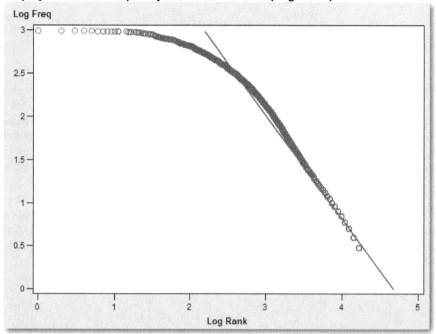

## Term-By-Document Matrix

The fundamental idea of applying classical data mining techniques to text mining relies on transforming text data (unstructured) to numbers (structured). This numerical representation of the text takes the form of a spreadsheet-like structure called a term-by-document matrix. In this matrix, dimensions are determined by the number of documents and number of terms in the corpus. Here is the same example from Chapter 1 for a term-by-document matrix.

Consider a collection of three reviews (documents) of a book as provided below.

Document 1: I am an avid fan of this sport book. I love this book.

Document 2: This book is a must for athletes and sportsmen.

Document 3: This book tells how to command the sport.

Parsing this document collection generates the following term-by-document matrix in Table 5.2.

**Table 5.2: Term-By-Document Matrix**

| Term/Document | Document 1 | Document 2 | Document 3 |
|---|---|---|---|
| I | 2 | 0 | 0 |
| am | 1 | 0 | 0 |
| an | 1 | 0 | 0 |
| avid | 1 | 0 | 0 |

| Term/Document | Document 1 | Document 2 | Document 3 |
|---|---|---|---|
| fan | 1 | 0 | 0 |
| of | 1 | 0 | 0 |
| this | 2 | 1 | 1 |
| book | 2 | 1 | 1 |
| love | 1 | 0 | 0 |
| is | 0 | 1 | 0 |
| a | 0 | 1 | 0 |
| must | 0 | 1 | 0 |
| for | 0 | 1 | 0 |
| athletes | 0 | 1 | 0 |
| and | 0 | 1 | 0 |
| sportsmen | 0 | 1 | 0 |
| tells | 0 | 0 | 1 |
| how | 0 | 0 | 1 |
| to | 0 | 0 | 1 |
| command | 0 | 0 | 1 |
| the | 0 | 0 | 1 |
| sport | 1 | 0 | 1 |

There are many theories that discuss what numbers should go into the cells of this matrix. In the simplest approach, each cell of the term-by-document matrix is allowed to contain the term frequency or the number of times a term appears in the document (as shown in Table 5.2). Another simple measure is the presence or absence of a term in a document. A detailed discussion about term weighting techniques appears in the next section. Once you have a table-like structure, traditional statistical methods can be comfortably applied to identify key inputs (terms) that differentiate or categorize the documents without having to use any grammar rules. Table 5.2 can be visualized as rows identified by documents and columns identified by terms. Even with a decent size corpus of a thousand documents, you can end up with thousands of unique terms, resulting in a huge dimensional matrix. The size of a term-by-document matrix can grow exponentially with more documents and more terms being extracted. As you can see from Table 5.2, there are many zeros in this table. This is because the rare words make up a significant portion of the text. Sparseness is a key characteristic of a term-by-document matrix compared to any typical table used for data mining. This is in-line with what is expected according to Zipf's law. The other two features of this table that make it unique are the absence of non-negative values and missing values. Algorithms that process text data have been developed to comply with the properties of a term-by-document matrix: sparseness, non-negative values, and high dimensions.

## Text Filter Node

In SAS Text Miner, the Text Filter node is primarily used for filtering documents and configuring the term weight and frequency weight settings. An important objective of text mining analysis is to differentiate the documents in the corpus. In this effort, the key task is to identify significant terms that can discriminate the documents. Research in the field of information retrieval has shown that using simple term frequencies does not help in document discrimination. Terms with low and high overall frequencies are found to be very bad discriminators. The best discriminators are those with neither too high nor too low frequencies–with document frequencies between $n/10$ and $n/100$ for a corpus of $n$ documents.

Various techniques are available that use term frequency, document frequency, number of documents that the term occurs in, and the size of the corpus to derive a weight for each term. Terms that occur very often in a document are assigned a higher weight and are considered important because they better describe the document. Terms that occur less frequently in the corpus are assigned a higher weight because they better discriminate the documents in the corpus. The value that goes into the term-by-document matrix is calculated using these weights. Two different types of weights are used. The first one is frequency weight, also called the local weight (within a document). Frequency weighting is the transformation of the frequency of the occurrence of the term in a document using a weighting function. The other one is term weight, also called global weight (corpus wide). A term weight is assigned to the term based on the overall frequency and document frequency. The cell value in a term-by-document matrix is a weighted frequency value derived by multiplying the local weight and the global weight. One of the most popular weighting schemes in the field of information retrieval is tf-idf (term frequency, inverse document frequency) (Salton and Buckley, 1988). The term frequency (tf) (or the local measure) is calculated with the following equation:

$$tf = \frac{tdf}{\max tdf}$$

In this equation, *tdf* is the document-specific frequency for a term, and *max tdf* is the frequency of the most frequent term in a document. Longer documents tend to have more terms (and, hence, a high tdf). These high frequencies are normalized using max tdf. The idf (Sparck Jones, 1972) portion of the weighting scheme is the global frequency measure, which is calculated with the following equation:

$$idf = \log\left(\frac{N}{df_t}\right)$$

In this equation, $N$ is the size of the collection, and $df_t$ is the number of documents in which the term $t$ appears. The tf-idf measure is simply the product of the local weight and global weight. The following section discusses different weighting schemes available in SAS Text Miner.

Let $f_{ij}$ be the raw frequency of $i$th term in the $j$th document. You first use a function $g(.)$ to transform the raw frequency, $g(f_{ij})$. Let $w_i$ be the weight of the $i$th term. The weighted frequency of the $i$th term for the $j$th document in a term-by-document matrix is given by $g(f_{ij}) * w_i$.

## Frequency Weightings

SAS uses three different local frequency weighting property settings: log, binary, and none.

**Log:** $g(f_{ij}) = \log(f_{ij} + 1)$

This is the default setting for this property and is used to control the effect of high frequency terms in a document.

**Binary:** $g(f_{ij}) = 1$

This is the equation if a term is present in a document. $g(f_{ij}) = 0$ is the equation if a term is absent in a document. This setting clearly doesn't differentiate between documents where a term is heavily present and between documents where a term rarely occurs.

**None:** $g(f_{ij}) = f_{ij}$

With this setting, the raw frequency value is used as is with no transformation done. This does not mean that no frequency weighting method is used.

## Term Weightings

With thousands of terms extracted from each document, the key question is, "Are all the terms in a document important?" Frequency weights help only in understanding the composition of a document. They do not help in identifying the terms that discriminate the documents. Each term is assigned a weight using any of the three different methods available in SAS. The fundamental assumption in assigning a weight to a term is that the term occurs in only a few documents, but it occurs many times in those few documents, and the term is significant in discriminating the documents. The three global frequency weighting methods available in SAS for assigning a weight to a term are entropy, mutual information, and inverse document frequency.

## Entropy

Entropy is a sophisticated metric derived from the concept of information gain in the field of information theory developed by Claude Shannon in the 1940s. Shannon used the concept of entropy from statistics to solve the problems of maximizing information transmission over imperfect channels of communication. In statistical parlance, entropy measures the average uncertainty in a single random variable, X, and is usually denoted as the following:

$$-\sum p(x) \log_2 p(x)$$

In this equation, *p(x)* is the probability that the random variable X takes the value of *x*. In text mining, this technique uses the random variable (which represents the distribution of words across the corpus) in the calculation of the weight for a term.

With the entropy method, the weight for term $w_i$ is calculated using the following equation:

$$w_i = 1 + \sum_{j=1}^{n} \frac{(f_{ij}/g_i).log_2\,(f_{ij}/g_i)}{log_2\,(n)} \qquad \text{(Equation 5.3)}$$

Here, $g_i$ is the number of times that term *i* appears in the document collection, and **n** is the number of documents in the collection. Because the logarithms of zero are undefined, the product in the numerator of the formula is 0, if $f_{ij}$=0.

Entropy weight for a term is zero if the term appears exactly once in all of the documents.

$$w_i = 1 + \sum_{j=1}^{n} \frac{(1/n).log_2\,(1/n)}{log_2\,(n)} = 1 + \frac{n*\frac{1}{n}*(-\log_2(n))}{(\log_2 n)} = 1\text{-}1 = 0$$

Entropy weight for a term is 1 if the term appears only once in onedocument.

$$w_i = 1 + \sum_{j=1}^{n} \frac{(1/1).log_2\,(1/1)}{log_2\,(n)} = 1 + \frac{1*0}{(\log_2 n)} = 1$$

For all other occurrences, the terms have a value between 0 and 1. Generally speaking, the terms that occur infrequently in a few documents get higher weights, meaning these terms are considered to be providing more information than terms with lower weights.

## Inverse Document Frequency (IDF)

It is essential to study the relevance of collection frequency and document frequency in identifying important terms. For example, it is obvious that the SAS Global Forum corpus is dominated by terms such as "sas institute," "paper," "data," and "information," which are present in all the documents. These terms definitely do not help discriminate the documents. As shown in Display 5.4, the terms "macro" and "information" occur with approximately the same frequency (collection frequency) in the SAS Global Forum corpus, whereas the document frequency is considerably different for the two terms. Hence, it makes more sense to use document frequency instead of collection frequency in calculating the scaling factor.

**Display 5.4: Term and Document Frequencies for the Terms "macro" and "information"**

| Term | Freq ▼ | # Docs |
|---|---|---|
| + macro | 7701 | 466 |
| + program | 7657 | 703 |
| + information | 7536 | 976 |
| + user | 7334 | 702 |

IDF is another measure used as a scaling factor in assigning an importance weight to a term. As the name implies, IDF calculates importance as the inverse of the frequency of occurrence of a term in documents. A term that appears infrequently is considered more important and is given a higher score, whereas a term with a high frequency of appearance is considered less important and is given a lower score.

$$w_i = log_2 \left(\frac{n}{df_i}\right) + 1 \qquad \text{(Equation 5.4)}$$

In this equation, $df_i$ is the document frequency or the number of documents that contain the $i$th term. $n$ is the number of documents in the collection.

If a term appears in every document, then the IDF weight for that term is the following:

$$w_i = log_2 \left(\frac{n}{n}\right) + 1 = 0+1 = 1$$

The maximum IDF weight for a term occurs when the term appears in exactly one document. However, no upper limit exists for the maximum IDF weight because it depends on the number of documents and is equal to $1+log_2 n$.

An IDF weight value of zero indicates that the word appears in all documents and has an insignificant effect on discriminating the documents.

## Mutual Information

Mutual information can be derived as the difference between the entropy of variable X and the entropy of variable X with the condition that X knows Y. That is, the mutual information is the reduction in uncertainty of X due to the knowledge of variable Y. In statistical parlance, mutual information between two variables works out to a symmetric non-negative measure of commonality between two variables. In text mining, this can be thought of as a measure of how much the presence of one term in a document tells us about whether the document belongs to a particular category. In SAS Text Miner, using this metric requires that your data contain a categorical target variable. In text mining problems where you have both structured and unstructured data, you will likely see a need for this metric because in most situations, you will have a target variable. This metric is defined as the following:

$$w_i = max_{C_k} \left[log \left(\frac{P(t_i, C_k)}{P(t_i) P(C_k)}\right)\right] \qquad \text{(Equation 5.5)}$$

In this equation, $P(t_i)$ is the proportion of documents that contain the term $t_i$. $P(C_k)$ is the proportion of documents that belong to category $C_k$. $P(t_i, C_k)$ is the proportion of documents that contain the term $t_i$ and belong to category $C_k$. $log(.)$ is 0 if $P(t_i, C_k) = 0$ or $P(C_k) = 0$.

The weight is proportional to the similarity of the distribution of documents that contain the term to the distribution of documents that are contained in the respective category. That is, in general, a higher mutual information weight for a term implies that the term tends to occur more in documents with the same target category. We demonstrate the benefits of this technique in Case Study 2.

## None

With this setting, no term weight is applied ($w_i = 1$).

Term weights are used to scale the frequency weighting for a term in a document. Although the basic purpose of the previous methods is the same, it is difficult to make a definitive statement about which weight is the best metric. In general, entropy and IDF are most widely used in text mining applications, with entropy being more effective for smaller documents and IDF for larger documents. In practice, you are advised to try all of the methods and compare the results. Display 5.5 shows the property settings for term weightings from the Text Filter node Properties panel.

**Display 5.5: Text Filter Node Properties Panel**

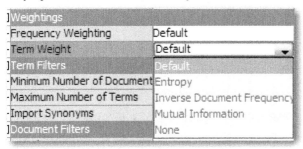

As the default setting, **Entropy** is used for the **Term Weight** property. In the presence of a categorical target, Mutual Information is used as the default setting. Let's explore the effect of these settings on the SAS Global Forum corpus. Display 5.6 shows the top 10 terms based on term weight. Display 5.7 shows the top 10 terms based on rank. Clearly, the terms that have the largest weights are those that occur in very few documents. The terms that have the lowest ranks have low weights due to their high frequencies. This is essentially what is expected because terms like "sas institute," "paper," and "global forum" occur in every document and, hence, are not good differentiators.

**Display 5.6: Top 10 Largest Weighted Terms**

Table: Number of Documents by Weight

| Term | Weight ▼ | Freq | # Docs | Rank |
|---|---|---|---|---|
| + calcium | 0.971 | 108 | 4 | 14095 |
| + sustainability | 0.961 | 119 | 4 | 14095 |
| + incomplete observation | 0.957 | 71 | 5 | 12178 |
| + blackberry | 0.954 | 85 | 4 | 14095 |
| + placemark | 0.951 | 96 | 4 | 14095 |
| + gadget | 0.949 | 114 | 5 | 12178 |
| + stripe | 0.948 | 56 | 5 | 12178 |
| residential | 0.947 | 90 | 4 | 14095 |
| automap | 0.944 | 66 | 4 | 14095 |
| + step boundary | 0.943 | 34 | 4 | 14095 |

**Display 5.7: Top 10 Highest Ranked Terms**

| Table: Number of Documents by Weight | | | | |
|---|---|---|---|---|
| Term | Weight | Freq | # Docs | Rank ▲ |
| + global forum | 0.031 | 10874 | 993 | 1 |
| + sas institute | 0.000 | 42292 | 991 | 2 |
| + paper | 0.000 | 5628 | 989 | 3 |
| + be | 0.000 | 77920 | 988 | 4 |
| + other | 0.000 | 8443 | 983 | 5 |
| + information | 0.084 | 7536 | 976 | 6 |
| + name | 0.000 | 10149 | 975 | 7 |
| + inc | 0.000 | 4266 | 974 | 8 |
| + institute | 0.000 | 4216 | 973 | 9 |
| + product | 0.083 | 4023 | 968 | 10 |

Display 5.8 shows the top 10 terms as ranked by IDF. This is a completely different list when compared with the top terms from the entropy weight setting. All terms that appear in the same number of documents get the same weight.

**Display 5.8: Top 10 Terms Based on IDF**

| Table: Number of Documents by Weight | | | | |
|---|---|---|---|---|
| Term | Weight ▼ | Freq | # Docs | Rank |
| complex analytics | 8.966 | 4 | 4 | 14095 |
| + grid deployment | 8.966 | 5 | 4 | 14095 |
| policy enforcement | 8.966 | 4 | 4 | 14095 |
| central component | 8.966 | 5 | 4 | 14095 |
| overall architecture | 8.966 | 5 | 4 | 14095 |
| + master list | 8.966 | 4 | 4 | 14095 |
| + dns | 8.966 | 47 | 4 | 14095 |
| + built-in capability | 8.966 | 4 | 4 | 14095 |
| location information | 8.966 | 6 | 4 | 14095 |
| uptime | 8.966 | 4 | 4 | 14095 |

## Filtering Documents

### Interactive Filter Viewer

The interactive filter viewer is a handy feature of the Text Filter node to refine the results from parsing and filtering. You have seen the utility of this tool in Chapter 4for creating a custom synonym list using the parsed terms. You have studied various strategies during text parsing to filter irrelevant terms to lessen the number of terms used in post-parsing text mining tasks. After applying all of these techniques, you can encounter useless terms being retained and important terms being dropped in the list of filtered terms. Using the interactive filter viewer, you can browse through all of the parsed terms and manually modify the list by dropping or keeping terms. This is the most time-consuming step in the text mining process flow. The results of the text mining analysis are very sensitive to the terms that are included in the analysis. Hence, it is essential to invest time and effort in refining the list of selected terms manually by using the interactive filter viewer.

Display 5.9 shows the Terms table of the interactive filter viewer. The **KEEP** column in this table indicates whether a term is used or dropped from analysis. A checked box indicates an included term.  To exclude the terms "data" and "global forum" from analysis, highlight the terms, right-click, and select **Toggle KEEP** (or you can directly clear the check box).

**Display 5.9: Terms Table from the Interactive Filter Viewer**

| TERM | FREQ ▼ | # DOCS | KEEP | WEIGHT | ROLE | ATTRIBUTE |
|------|--------|--------|------|--------|------|-----------|
| ⊞ be | 77920 | 988 | ☐ | 0.0 | | Alpha |
| ⊞ sas institute | 42292 | 991 | ☐ | 0.0 | Company | Entity |
| ⊞ data | 42224 | 936 | ☑ | 0.067 | | Alpha |
| ⊞ use | 25155 | 926 | ☐ | 0.0 | | Add Term to Search Expression |
| ⊞ variable | 15218 | 696 | ☑ | 0.139 | | Treat as Synonyms |
| ⊞ have | 15013 | 912 | ☐ | 0.0 | | Remove Synonyms |
| ⊞ set | 14293 | 836 | ☑ | 0.107 | | Toggle KEEP |
| ⊞ value | 13726 | 867 | ☑ | 0.109 | | View Concept Links |
| ⊞ not | 11643 | 879 | ☐ | 0.0 | | Find |
| ⊞ do | 11205 | 861 | ☐ | 0.0 | | Repeat Find |
| ⊞ run | 10959 | 793 | ☑ | 0.092 | | Clear Selection |
| ⊞ proc | 10936 | 635 | ☑ | 0.138 | | Print... |
| ⊞ global forum | 10874 | 993 | ☑ | 0.031 | Nou | |
| ⊞ table | 10760 | 690 | ☑ | 0.165 | | Alpha |
| ⊞ create | 10251 | 841 | ☑ | 0.09 | | Alpha |

When you close the interactive filter viewer, you can save your changes, and the modified data is used in further analysis. An easy way to omit low-frequency words is by using the **Minimum Number of Documents** property setting. To exclude all of the terms that occur in less than 10 documents, enter a value of 10 for this property. The other property that can be used to control the number of terms is **Maximum Number of Terms**. Using this property, you can set a maximum limit on the total number of terms that are kept in the analysis based on their frequency. These different types of filtering properties come in very handy in reducing the noise from the data.

## Document Search and Retrieval

In many situations when refining the terms and making a decision to include or exclude a term, you might want to quickly see the documents that contain that term. The interactive filter viewer contains a feature to search for documents in the corpus based on search expressions. A search expression can be a single term or a list of terms. Documents that match at least one of the terms are returned. The search results contain a relevance score for each document that indicates how well each document matches the search expression.

The matching documents are retrieved using the vector space model, which is the fundamental technique used in many information retrieval operations. In this model, all documents and search expressions are represented as vectors in the term space. The vector components are defined by the term weights given to the terms in the document. A similarity measure between two documents, *d1* and *d2*, is calculated using the dot product (or cosine similarity) of the two vectors, *V1* and *V2*.

$$similarity\,(d1, d2) = \frac{V_1 \cdot V_2}{|V_1|\,|V_2|} \qquad\qquad \text{(Equation 5.6)}$$

The numerator in the equation represents the dot product between the two vectors. The denominator accounts for the various sizes of the documents by considering the Euclidean length of the document defined as the following:

$$|V_i| = \sqrt{\sum v_i^2}$$

Consider the following unit vector:

$$v_i = \frac{V_i}{|V_i|}$$

Equation 5.6 can be rewritten as the following:

$$similarity\ (d1, d2) = \quad v_1 . v_2 \qquad \text{(Equation 5.7)}$$

The search expression, in itself, is considered a short document. Term weights are calculated (also known as tf-idf weights) treating the search expression as a whole corpus. The tf-idf weight is the multiplication of term frequency and inverse document frequency. It has the following characteristics:

1. The value of tf-idf is highest when the term occurs many times within a fewer number of documents (i.e., higher discriminatory ability).
2. The value of tf-idf is lowest when the term occurs in almost all documents (i.e., lower discriminatory ability).

Consider an example corpus with just two documents.

*Document 1:* A vector space model uses a dot product to calculate similarity. This model defines vectors in the term space.

*Document 2:* A giant model of a spacecraft is being built for a movie based on space.

Consider the search expression (or query) "vector space model" used to retrieve matching documents from the corpus. Let inverse document frequency be used as the technique to calculate term weights in the corpus. Sample weights for the terms in the query are calculated using frequency. Table 5.3 shows the common words and one non-common word, "movie," between the query and the documents.

**Table 5.3: Weights for Words in the Sample Documents and Search Expression**

| Term | Search Expression | | Document 1 | | | Document 2 | | |
|------|-----------|--------|-----------|----------|--------|-----------|----------|--------|
|      | **Term Freq** | **Weight** | **Term Freq** | **Doc Freq** | **TF*IDF** | **Term Freq** | **Doc Freq** | **TF*IDF** |
| vector | 1 | 0.4 | 2 | 1 | 4 | 0 | 1 | 0 |
| space | 1 | 0.4 | 2 | 2 | 2 | 2 | 2 | 2 |
| model | 1 | 0.4 | 2 | 2 | 2 | 1 | 2 | 1 |
| movie | 0 | 0 | 0 | 1 | 0 | 1 | 1 | 2 |

Similarity between the search expression (Q) and document 1 (Doc1) is calculated **using** equation 5.7.

Score (Q, Doc1) = (0.4 x 4) + (0.4 x 2) + (0.4 x 2) + (0 x 0) = 3.2

Similarly, the score between the search expression (Q) and document 2 (Doc2) is calculated.

Score (Q, Doc2) = (0.4 x 0) + (0.4 x 2) + (0.4 x 1) + (0 x 2) = 1.2

We can conclude that document 1 is a better match with the query.

In this example, the inverse document frequency method is used for calculating term weights. SAS Text Miner uses a different weighting scheme that is not configurable. A relevance measure, between 0 and 1, for every document searched with the search expression is calculated. A value of 1 indicates a best match in the collection.

In the interactive filter viewer, you can enhance the search using the following techniques:

- **+term** returns only documents that include the term.
- **-term** returns only documents that do not include the term.
- **+term1  term2** returns documents that include term1 and documents that include both term1 and term2. (Documents that contain term2 but not term1 are not returned.)
- **+term1 +term2** returns only documents that include both term1 and term2.
- **"text string"** returns only documents that include the quoted text.
- **string1*string2** returns only documents that include a term that begins with string1, ends with string2, and has text in between.
- **>#term** returns only documents that include the term or any of the synonyms that have been assigned to the term.

For example, to retrieve documents that contain the term "calcium" in the SAS Global Forum corpus, enter the term "calcium" or "+calcium" or ">#calcium" in the search expression. Or, you can locate the term "calcium" in the Terms table, right-click the term, and select **Add Term to Search Expression**. Display 5.10 shows the search results for the search expression ">#calcium."

**Display 5.10: Search Results for the Search Expression ">#calcium"**

Out of the four results in Display 5.10, the **RELEVANCE** column shows that one of the documents from the search results is a best match with a score of 1. The other documents have a very low score. This is very clear if you look at the title of the paper:

- An Analysis of a Calcium Nutrition Educational Intervention for Middle School Students in Las Vegas, Nevada Using SAS®
- Confessions of a Clinical Programmer: Dragging and Dropping Means Never Having to Say You're Sorry When Creating SDTM Domains
- Dealing with Lab Data – Stacking the Deck in Your Favor
- The Next Generation: SAS/STAT®

The first paper in the list, which got the best match score, refers and talks about calcium a lot. In the other three papers, the term "calcium" is referred to only occasionally and in examples. The relevance score is a useful measure in identifying irrelevant documents and excluding those documents from analysis.

To read the full contents of the cell, right-click on the results, and select **Toggle Show Full Text**. In our example, because the text variable does not contain the full text and is being referred to in the actual SAS Global Forum papers using the paths to the files, you can access the full paper by right-clicking and selecting **View Input File**. Selecting **View Original** allows you to access the source file whose path is recorded in the variable with the role **Text Location**. Display 5.11 shows the two windows, with a .txt file and a .pdf file, that pop up when you select **View Input File** and **View Original** for one of the documents from the documents table.

**Display 5.11: Viewing Source Documents from the Interactive Filter Viewer**

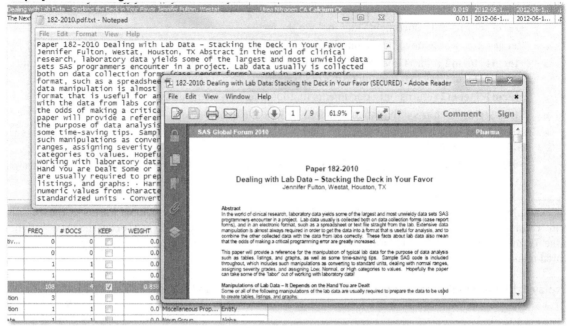

## Concept Links

Concept links help in understanding the relationships between words based on the co-occurrence of words in the documents. You can identify and define concepts in the text by selecting a keyword and exploring the words that are associated with this keyword using concept links. Concept links are intuitive and easy to understand because they are presented as interactive graphs. In the graph, the selected term is connected to other highly associated terms with a hub-and-spoke structure. The connected term can be expanded to display the terms associated with it. The structure can resemble a social network in which we can define multi-order associations. The width of the line between the center term and a concept link represents how closely the terms are associated. A thicker line indicates a closer association. The strength of association is calculated using binomial distribution. Here is the formula for calculating the strength measure as explained in the *SAS Text Miner: Reference Help*.

The strength of association between two terms, A and B in a corpus of r documents, is calculated as follows:

$$Strength = log_e(1/Prob_k)$$

$$Prob_k = \sum_{r=k}^{r=n} [n!/[r!(n-r)!]] \cdot p^r(1-p)^{(n-r)}$$

In this formula, *n* is the number of documents that contain term B.

*k* is the number of documents that contain both term A and term B.

*p=k/n* is the probability that term A and term B co-occur, assuming that they are independent of each other.

Concept links can be generated by right-clicking a term in the Terms tables in the interactive filter viewer, and selecting **View Concept Links**. Display 5.12 shows an example of the concept links for the term "calcium" from our example corpus.

**Display 5.12: Concept Links for the Term "calcium"**

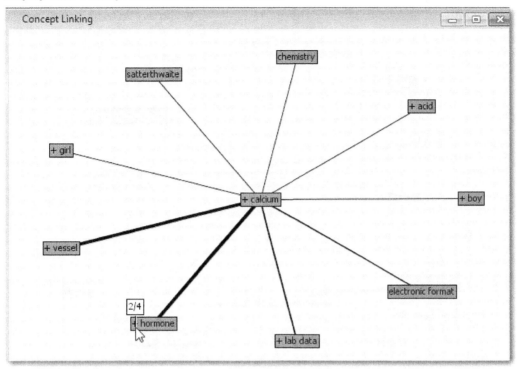

The term "calcium" is more strongly associated with the terms "vessel" and "hormone" than with the other words. If you hover over any term in the graph, you can see two numbers separated by a slash. The first number represents the number of documents in which the two terms co-occur, and the second number represents the total number of documents in which the specific term ("calcium") occurs. You can right-click on any child term, and select **Expand Links** to display the terms associated with the child term. Display 5.13 shows the expanded version of the graph in Display 5.12. In the context of text mining customer feedback, an organization can easily identify the good or bad words that are associated with the products being reviewed. Concept links is a great feature to investigate the relationships between words in the corpus.

**Display 5.13: Expanded Concept Links for the Term "calcium"**

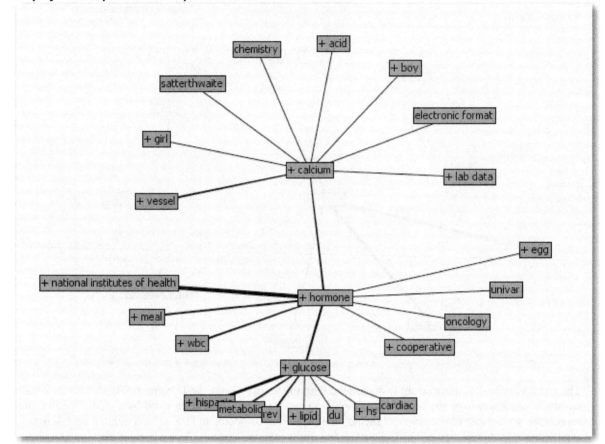

## Summary

In this chapter, we discussed the transformation methods available in the Text Filter node that can be applied to a term-by-document matrix. You should experiment with different frequency weighting (count, log, and binary) and term weighting (entropy, IDF, and mutual Information) methods. Term weighting methods help in achieving the primary objective of text mining analysis—to discriminate documents—by assigning weights to terms. The relationships between terms based on their co-occurrences in documents can be explored using the concept links feature in this node. The Text Filter node also has a capability to filter documents from analysis using search queries. Like any typical search engine functionality, you can retrieve documents for viewing or for dropping from analysis.

With a fully prepared term-by-document matrix, the next step in text mining analysis is applying traditional data mining techniques such as clustering or classification. Chapter 6 discusses in detail document clustering and topic extraction using SAS Text Miner.

## References

Aizawa, A. 2003. "An Information-Theoretic Perspective of tf-idf Measures." *Information Processing & Management.* 39 (1): 45-65.

Booth, A. D. 1967. "A 'Law' of Occurrences for Words of Low Frequency." *Information and Control.* 10 (4): 386-393.

Chen, Y. and Leimkuhler, F. F. 1987. "Analysis of Zipf's Law: An Index Approach*." Information Processing & Management*. 23(3): 171-182.

Chisholm, E and Kolda, T. G. 1999. "New Term Weighting Formulas for the Vector Space Method in Information Retrieval." Technical Report. Oak Ridge National Laboratory.

Collica, R. 2011. *Customer Segmentation and Clustering Using SAS® Enterprise Miner^TM. 2nd Ed.* Cary, NC: SAS Institute Inc.

Cummins, R., O'Riordan, C. 2006. "Evolving Local and Global Weighting Schemes in Information Retrieval." *Information Retrieval.* 9(3): 311–330.

Dumais, S. T. 1991. "Improving the Retrieval of Information from External Sources." *Behavioral Research Methods, Instruments, & Computers.* 23 (2): 229-236.

Le Quan Ha., Sicilia-Garcia, E.I., Ming, Ji, and Smith, F.J. 2002. "Extension of Zipf's Law to Words and Phrases." *Proceedings of COLING 2002: Proceedings of the 19th International Conference on Computational Linguistics*, Taipei, Taiwan.

Mandelbrot, B. 1953. "An Informational Theory of the Statistical Structure of Language." *Communication Theory*. New York: Academic Press, 486-502..

Manning, C. D., and Schutze, H. 1999. *Foundations of Statistical Natural Language Processing*. Cambridge, Massachusetts: The MIT Press.

Montemurro, M. A. 2001. "Beyond the Zipf-Mandelbrot Law in Quantitative Linguistics*." Physica A: Statistical Mechanics and Its Applications*. 300 (3-4): 567-578.

Salton, G., Buckley, C. 1988. "Term-Weighting Approaches in Automatic Text Retrieval." *Information Processing & Management.* 24(5): 513–523.

Salton, G., Yang. C. S., & Yu. C. T. 1975. "A Theory of Term Importance in Automatic Text Analysis." *Journal of the American Society for Information Science.* 26 (1): 33-44.

Sparck Jones, K. 1972. "A Statistical Interpretation of Term Specificity and Its Application in Retrieval." *Journal of Documentation.* 28 (1): 11-21.

Tesitelova, M. 1992. *Quantitative Linguistics*. Amsterdam: John Benjamins.

Text Analytics using SAS® Text Miner. Course Notes. SAS Institute Inc. Cary, NC. Course information: https://support.sas.com/edu/schedules.html?ctry=us&id=1224

Zipf, G.K. 1949. *Human Behavior and the Principle of Least Effort: An Introduction to Human Ecology*. Cambridge, MA: Addison-Wesley.

# Chapter 6 Clustering and Topic Extraction

## Introduction

In Chapters 1 through 5, you learned how to take a collection of documents and convert them into a vector space model that represents features of each document using numeric values. In this chapter, we discuss how to take that vector space model and assign each document to a small number of groups, called clusters. The basic idea is that documents within a cluster should be similar to each other, and documents in different clusters should be dissimilar to each other. The similarity between two documents is based on the similarity of features (such as terms or words) between documents in the vector space model. In this context, we discuss latent semantic indexing (LSI), which provides a method for determining the similarity of words and passages by the analysis of large text corpora. Then, we discuss the concept of topic extraction from a collection of documents. A topic is conceptualized as a collection of terms that capture the main themes or ideas in the document. Unlike cluster groups, where each document is assigned to only one cluster, the same document can be assigned to multiple topics, depending on how many ideas are represented in a document.

## What Is Clustering?

Clustering or cluster analysis is a generic name for a group of related techniques (such as unsupervised pattern recognition, unsupervised classification analysis, numerical taxonomy, typology constructions, Q-analysis, and so on) that automatically try to find natural groupings in the data. One crucial difference between clustering and a typical classification model is the absence of any target variable (where classes or groups are known *a priori*) in the data. In the context of textual data, this means that no labeled training examples are needed before documents can be clustered into groups. This is why clustering is often referred to as unsupervised classification.

As a conceptual activity, the assignment of objects into groups is something humans do routinely all through their lives to reduce the complexity of the environment that they have to work with. The natural grouping of objects and observations is extremely important to many disciplines (such as statistics, psychology, sociology, biology, engineering, economics, and business). Each of these disciplines, in turn, has used its own label to describe cluster analysis. Although the names might differ across disciplines, all disciplines share the fundamental concept of separating data suggested by the natural groupings in the data. In essence, cluster analysis attempts to group objects so that each object in a cluster is similar to the other objects in the same cluster. However, objects in different clusters are dissimilar to each other. In the context of textual data, objects are the documents that must be assigned to clusters so that within a cluster, documents are similar, but between clusters, documents are different.

The term-by-document matrix has been presented in previous chapters with terms in rows and documents in columns. For clustering, it often helps to visualize the transpose of that matrix, where documents are in rows (representing observations or objects) and terms are in columns (representing variables). The idea in clustering is to put documents (rows or observations) into groups so that within the groups, documents (observations) are similar and between the groups, documents (observations) are dissimilar.

## Similarity Metrics

To identify any natural groupings in the vector space model of textual data, you must first define similarity between the documents. In statistics, similarity is often measured via three broadly defined methods: distance, correlation, and association. Of these methods, distance- and correlation- based methods typically require metric data, whereas association measures can work on nonmetric data.

In the distance-based methods, dissimilarity is conceptualized as the distance between objects. That is, if two things are similar, the distance between them must be small. If two things are dissimilar, the distance between them must be large. There are many distance metrics in statistics literature. Two commonly used ones are Euclidean distance and Mahalanobis distance. The Euclidean and Mahalanobis distance metrics focus on the magnitude of the values and portray objects as similar that are close together in the variable space. The Mahalanobis distance (also called the generalized distance) is defined as the following:

$$\sqrt{(X_i - X_j)^T \sum{}^{-1} (X_i - X_j)}$$

In this equation, $X_i$ and $X_j$ are the vectors of values of variables for cases $i$ and $j$, and $\Sigma$ is the pooled within group variance-covariance matrix. Unlike the Euclidean distance, the Mahalanobis distance takes into account covariances among variables when calculating the metric. In fact, if all of the variables are uncorrelated with each other, then the Mahalanobis distance reduces to the Euclidean distance. In the SAS Text Cluster node, the Mahalanobis distance is used in the EM algorithm to measure the distance between a document and a cluster. The Euclidean distance is used to measure the distance between clusters in hierarchical clustering.

In correlation-based methods (such as cosine similarity introduced in chapter 5), high correlations indicate similarity (correspondence of patterns across variables) between objects (documents). Low correlation indicates dissimilarity between objects (documents). Correlational measures represent patterns rather than magnitudes. These measures are often used in document search and retrieval in SAS Text Miner as discussed in Chapter 5.

The association-based measures of similarity are typically used for non-metric variables. An association metric generally assesses the degree of agreement or matching between two pairs of observations. The simple matching coefficient (that measures the percentage of times that the two objects match in variables) is an example of an association measure. Although these metrics are often intuitively appealing, for most practical document clustering problems, the term-by-document matrix is too large and too sparse to effectively apply these metrics.

## Clustering Algorithms

Broadly speaking, clustering algorithms can be divided into four groups: hierarchical, non-hierarchical (or partitional), probabilistic (or spectral density), and neural network (SOM/Kohonen). For more information about advantages and disadvantages of each algorithm, refer to Cluster Analysis.

In hierarchical clustering, the algorithm iteratively groups objects into cascading sets of clusters so that clusters within a step are nested within a cluster from a prior (or subsequent) step. This can be achieved in a top-down (divisive) or bottom-up (agglomerative) manner. Within each of the hierarchical algorithms, often there are many variants (such as single linkage, complete linkage, average, centroid, and Ward's method) that differ based on how distances between clusters are calculated. In SAS Text Miner, the hierarchical clustering method uses Ward's minimum variance method to calculate the distance between two clusters. However, all variants of hierarchical clustering algorithms are available via SAS codes in SAS Enterprise Miner environment.

Unlike hierarchical algorithms, partitioning algorithms (such as k-means) are non-incremental and simultaneously assign all observations to clusters based on the distance of each observation to a cluster center. The term "$k$-means" implies that $k$ clusters are used. Typically, the algorithm selects $k$ cluster centers in the data space. Then, it assigns all observations that are closest to each center. The centers are updated to reflect the assignments of observations (that is, a new center becomes the average of all observations assigned to each center). The algorithm iterates through this process of assigning observations that are closest to the centers, updating centers based on assignments until a convergence criterion is reached. Although $k$-means is a very popular and widely used algorithm for clustering numeric data and is available as a node in SAS Enterprise Miner, it is not currently used in SAS Text Miner.

Probabilistic (or spectral density) clustering can be viewed as identifying dense regions of the data space. An efficient representation of the probability density function is the mixture model, which asserts that the data can be thought of as a combination of $k$ component densities corresponding to k clusters. The expectation-maximization (EM) algorithm is one of the most well-known and effective techniques for estimating mixture models. The EM algorithm assumes a probability density function (typically joint normal distribution) of the variables in each cluster. Then, the algorithm applies an iterative optimization to estimate the probabilities for each observation to belong to each cluster. The EM algorithm consists of two steps: the expectation (E) step and the maximization (M) step. In the expectation (E) step, input partitions are selected similar to the k-means technique. In this step, each observation is given a weight or expectation for each partition. In the second step, maximization (M), the initial partition values are changed to the weighted average of the assigned observations, where weights are identified from the E step. This cycle is repeated until the partition values do not change significantly as identified by the log likelihood of the iteration.

Neural networks are most widely known for their use in supervised classification (where a target variable with known classes exists) or prediction-type problems. Another variant of this type of network, the self-organizing map (SOM), originally developed by Teuvo Kohonen, is often used in unsupervised classification or clustering. The Kohonen network typically contains two layers. The input layer consists of $p$-dimensional observations. An output layer (represented by a grid) consists of $k$ nodes for $k$ clusters, each of which is associated with a $p$-dimensional weight. Each output node is initially assigned a random weight. These random weights are modified as the network learns the pattern of input data. Input observations are presented to the network, and each observation is provisionally assigned to one of the nodes based on the shortest Euclidean distance between each observation and each node. The node weights are updated based on which input observations are assigned to each node. Then, the process repeats itself, and starts presenting input observations to the nodes containing revised weights. The iterative process eventually stabilizes with weights corresponding to cluster centers in such a way that clusters that are similar to one another are located in close proximity on the map (grid). The technique is appealing for large dimensional data because it does two things simultaneously. It finds similarity among observations and represents them in a lower dimensional space on the map. The SOM/Kohonen method is available in SAS Enterprise Miner, but not currently used in SAS Text Miner.

## Singular Value Decomposition and Latent Semantic Indexing

For most document collections, the term-by-document matrix is often very large and very sparse. This makes it difficult to use this matrix directly in clustering or in any other algorithms. What's needed is a way to reduce the dimensionality of the data, yet retain most of the meaningful information. Addressing this curse of dimensionality is an age-old problem in statistics and data mining. Plenty of techniques have been developed to handle this issue. Some of the most commonly used and well known of these approaches are principal component analysis (PCA), factor analysis (FA), partial least squares (PLS), and latent semantic indexing (LSI). PCA, FA, and PLS are available as nodes in SAS Enterprise Miner. In SAS Text Miner, a form of LSI is used.

LSI (sometimes referred to as latent semantic analysis (LSA)) is a dimensionality reduction technique that typically operates on the term-by-document matrix (or weighted term-by-document matrix using weights discussed in Chapter 5). It uses a well-known mathematical matrix decomposition technique called singular value decomposition (SVD) to break down the original data into linearly independent components. These components are, in a sense, an abstraction away from the noisy correlations in the original data to sets of values that best approximate the underlying structure of the data. The majority of the components usually have small values and can be ignored, which results in dimensionality reduction. Therefore, the LSI approach is very similar to factor analysis. Just as factor analysis is used to extract underlying dimensions (factors) from multi-item questions measuring multidimensional constructs via surveys, so is LSI applied to extract underlying dimensions from large text corpora. Essentially, LSI combines surface information (the pattern of occurrences of words across the text corpora) into a deeper abstraction (the latent semantic dimensions) that captures the mutual implications of words and documents. Thus, LSI provides dimensions with semantic meaning so that features in the same dimension are often topically related. This is why LSI is very attractive for analyzing text.

## Mathematics of SVD

Cluster algorithms rely on SVD to transform the original weighted term-by-document matrix into a dense, but reduced, dimensional representation. Mathematically, a full SVD does the following. Consider that A (mxn) is the term-by-document matrix with m>n (more terms than documents) and where the entries in the matrix are real numbers (such as presence or absence of a term, entropy weight, etc.). SVD computes matrices **U**, **S**, and **V** so that the original matrix can be re-created using the formula A = USVT. In this formula, the following is true:

- **U** is the matrix of orthogonal eigenvectors of the square symmetric matrix $\mathbf{AA}^T$.
- **S** is the diagonal matrix of the square roots of eigenvalues of the square symmetric matrix $\mathbf{AA}^T$.
- **V** is the matrix of orthogonal eigenvectors of the square symmetric matrix $\mathbf{A}^T\mathbf{A}$.

The mathematical proof of the formula is available in any matrix algebra book.

## A Numerical Example of SVD

Assume that you have the following set of four text documents:

D1: I love iPad.
D2: iPad is great for kids.
D3: Kids love to play soccer.
D4: I play soccer at OSU.

In creating the term-by-document matrix from these documents, we have ignored the underlined stop words.

**Table 6.1: Term-by-Document Matrix for the Sample Four Text Documents**

| Term/Document | D1 | D2 | D3 | D4 |
|---|---|---|---|---|
| I | 1 | 0 | 0 | 1 |
| Love | 1 | 0 | 1 | 0 |
| iPad | 1 | 1 | 0 | 0 |
| Is | 0 | 1 | 0 | 1 |
| Great | 0 | 1 | 0 | 0 |
| Kids | 0 | 1 | 1 | 0 |
| Play | 0 | 0 | 1 | 1 |
| Soccer | 0 | 0 | 1 | 1 |
| OSU | 0 | 0 | 0 | 1 |

The term-by-document, which is **A** (9x4) matrix, looks like the following:

1.000 0.000 0.000 1.000
1.000 0.000 1.000 0.000
1.000 1.000 0.000 0.000
0.000 1.000 0.000 1.000
0.000 1.000 0.000 0.000

0.000 1.000 1.000 0.000
0.000 0.000 1.000 1.000
0.000 0.000 1.000 1.000
0.000 0.000 0.000 1.000

The transpose of the **A** matrix, which is **A**<sup>T</sup> (4x9) matrix, looks like the following:

1.000 1.000 1.000 0.000 0.000 0.000 0.000 0.000 0.000
0.000 0.000 1.000 1.000 1.000 1.000 0.000 0.000 0.000
0.000 1.000 0.000 0.000 0.000 1.000 1.000 1.000 0.000
1.000 0.000 0.000 1.000 0.000 0.000 1.000 1.000 1.000

**The A<sup>T</sup>A** matrix is the document-by-document matrix. It is created by the multiplication of the matrices **A**<sup>T</sup> and **A** as shown below. Note that it is a square symmetric (4x4) matrix.

3.000 1.000 1.000 1.000
1.000 4.000 1.000 1.000
1.000 1.000 4.000 2.000
1.000 1.000 2.000 5.000

The **AA**<sup>T</sup> matrix is the term-by-term matrix. It is created by the multiplication of the matrices **A** and **A**<sup>T</sup> as shown below. Note that it is also a square symmetric (9x9) matrix.

2.000 1.000 1.000 1.000 0.000 0.000 1.000 1.000 1.000
1.000 2.000 1.000 0.000 0.000 1.000 1.000 1.000 0.000
1.000 1.000 2.000 1.000 1.000 1.000 0.000 0.000 0.000
1.000 0.000 1.000 2.000 1.000 1.000 1.000 1.000 1.000
0.000 0.000 1.000 1.000 1.000 1.000 0.000 0.000 0.000
0.000 1.000 1.000 1.000 1.000 2.000 1.000 1.000 0.000
1.000 1.000 0.000 1.000 0.000 1.000 2.000 2.000 1.000
1.000 1.000 0.000 1.000 0.000 1.000 2.000 2.000 1.000
1.000 0.000 0.000 1.000 0.000 0.000 1.000 1.000 1.000

The nonzero eigenvalues of **A<sup>T</sup>A** are the following:

7.783, 3.511, 2.253, and 2.453

These values can be found by using the standard calculation of eigenvalues and eigenvectors for any square symmetric matrix.

The square roots of the eigenvalues are the following:

2.790, 1.874, 1.566, and 1.501

The SVD matrices based on the previous eigenvalues are shown below. Note that the eigenvectors of **A<sup>T</sup>A** make up the columns of **V**, the eigenvectors of **AA**<sup>T</sup> make up the columns of **U**, and the singular values in **S** are square roots of the eigenvalues from **AA**<sup>T</sup> or **A<sup>T</sup>A**. The singular values are the diagonal entries of the **S** matrix and are arranged in descending order. The singular values are always real numbers.

**U:**
0.355 0.120 -0.088 0.649

0.314 -0.056 0.677 0.205
0.265 -0.575 0.045 0.355
0.380 -0.164 -0.560 -0.069
0.145 -0.430 -0.214 -0.182
0.340 -0.340 0.205 -0.513
0.429 0.356 0.072 -0.220
0.429 0.356 0.072 -0.220
0.235 0.266 -0.346 0.112

**S:**
2.790 0.000 0.000 0.000
0.000 1.874 0.000 0.000
0.000 0.000 1.566 0.000
0.000 0.000 0.000 1.501

**$V^T$:**
0.335 0.405 0.542 0.655
-0.273 -0.806 0.168 0.498
0.405 -0.335 0.655 -0.542
0.806 -0.273 -0.498 0.168

If you multiply **U** with **S**, you get the following:

0.990 0.225 -0.138 0.974
0.876 -0.105 1.060 0.308
0.739 -1.078 0.070 0.533
1.060 -0.307 -0.877 -0.104
0.405 -0.806 -0.335 -0.273
0.949 -0.637 0.321 -0.770
1.197 0.667 0.113 -0.330
1.197 0.667 0.113 -0.330
0.656 0.498 -0.542 0.168

Next, if you multiply the result by **$V^T$**, then you get the complete A matrix (within round-off errors):

0.999 0.000 -0.001 0.999
1.000 0.000 0.998 -0.001
1.000 0.999 0.000 -0.001
0.000 0.999 0.000 0.999
0.000 1.000 0.001 0.000
0.001 1.000 1.001 0.001
-0.001 -0.001 0.999 1.000
-0.001 -0.001 0.999 1.000
0.000 0.000 0.001 1.000

In the full SVD method, there is no dimensionality reduction and, consequently, no loss of information. In practice, you typically use the first few of the eigenvalues (instead of all of the eigenvalues as in full SVD) so that the dimensionality is reduced. The eigenvalues are ordered from highest to lowest. These values can be plotted in a scree plot (against the eigenvalues) to identify an elbow in the plot for selecting a cutoff at the point where the eigenvalues fall off dramatically. Of course, not using the entire set of eigenvalues means some loss of information. The idea is we are willing to trade off some loss of information to get a more robust and simpler structure to represent the patterns in the data.

So, in effect, the application of SVD to a term-by-document matrix results in rearranging the original matrix so that the following is true:

- The values in each column after SVD are a linear-weighted combination of the values in all columns of the original matrix.
- Each column after SVD is uncorrelated with all other columns.
- The first SVD dimension provides the best approximation of the original matrix that is possible with one dimension. The addition of the second SVD dimension provides the best possible representation in two dimensions, and so on.
- The documents are represented in the SVD space by the column vector of the matrix $\mathbf{V}^T$.
- The terms are represented in the SVD space by the row vectors of the multiplication of matrix $\mathbf{U}$ and matrix $\mathbf{S}$.

In the context of our simple example, assume that you select to retain only the top two eigenvalues. This will result in a huge reduction of dimensionality from nine dimensions (terms) to only two dimensions (SVD space). The coordinates of the documents and the terms represented in the SVD space are shown in Table 6.2.

**Table 6.2: Final Coordinates for the Documents and Terms from the SVD Example**

| ID | Type | SVD1 | SVD2 |
|----|------|------|------|
| **D1** | Document | 0.335 | -0.273 |
| **D2** | Document | 0.405 | -0.806 |
| **D3** | Document | 0.542 | 0.168 |
| **D4** | Document | 0.655 | 0.498 |
| **I** | Term | 0.99 | 0.225 |
| **Love** | Term | 0.876 | -0.105 |
| **iPad** | Term | 0.739 | -1.078 |
| **is** | Term | 1.06 | -0.307 |
| **Great** | Term | 0.405 | -0.806 |
| **Kids** | Term | 0.949 | -0.637 |
| **Play** | Term | 1.197 | 0.667 |
| **Soccer** | Term | 1.197 | 0.667 |
| **OSU** | Term | 0.656 | 0.498 |

Plotting the documents in the SVD space (Display 6.1) shows that documents D1 and D2 are close together and documents D3 and D4 are close together. Although the task of separating the four documents in the two groups (D1, D2 versus D3, D4) seems easy to do for any English speaker, it is gratifying to see that it can be done reasonably well via SVD.

**Display 6.1: Two-Dimensional Plot of Document Coordinates from Table 6.2**

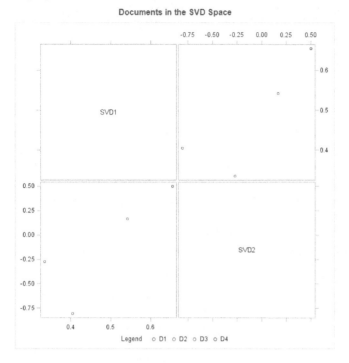

In the plot, SVD1 and SVD2 are shown as the X and Y axes. But, just like in factor analysis, these axes can be rotated to better align the documents with the axes. For example, if you rotate the axes (keeping the angle between them constant at 90 degrees or ensuring zero correlations between SVD1 and SVD2) by about 45 degrees in the plot, the documents align well with the rotated axes as shown in Display 6.2.

**Display 6.2: Rotated X and Y Axes for Dimensions in Display 6.1**

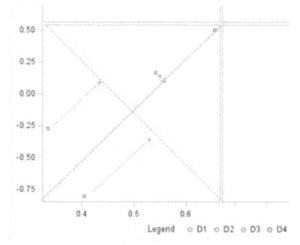

Recall that these SVDs are the latent semantic dimensions of the text documents. Plotting both documents and terms in the SVD space can provide more insight into how documents and terms are related in the SVD space and the meaning of the latent semantic dimensions. Labels for semantic dimensions are often arrived at subjectively by inspecting plots or by plotting other known and labeled variables in the same SVD space and noting the relationships among these variables and the SVDs. In this example, just by quickly reading the documents aligned closely with each dimension, you can see that one dimension seems to capture the semantic concept of the usefulness of the iPad, and the other dimension seems to capture the semantic concept of playing soccer.

## Text Cluster Node

In the process flow, the Text Cluster node must be preceded by the Text Parsing and Text Filter nodes as shown in the process flow diagram in Display 6.3.

**Display 6.3: SAS Global Forum Example Process Flow with Text Cluster and Text Topic Nodes**

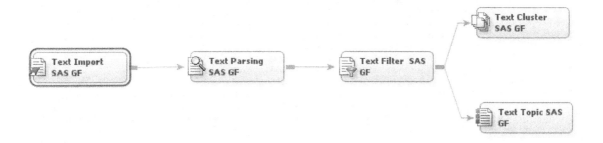

There are two algorithms available for clustering: the hierarchical algorithm and the expectation-maximization (EM) algorithm. The EM algorithm is the default option. In SAS Text Miner, EM clustering automatically selects between two versions of the EM algorithm—standard or scaled. The standard version of the EM algorithm analyzes all data at each iteration step and is used when the data size is small. The scaled version of the EM algorithm uses part of the input data in each iteration and is used when the data is large.

**Display 6.4: Text Cluster Node Properties Panel**

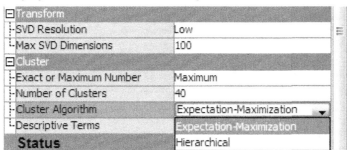

**SVD Resolution** is set to **Low** by default, which results in the algorithm automatically selecting to use a minimum number of SVD components (up to a maximum of 100) in the clustering algorithm. One way to think about resolution is how much information you are giving up to reduce the dimensionality of the weighted term-by-document matrix. Low resolution results in more loss of information than high resolution. On the other hand, high resolution means little reduction in dimensionality. As the famous economist Milton Friedman said, "There is no free lunch." In text analytics problems, you should start with the default setting, and then be prepared to experiment with the maximum number of SVD dimensions and resolution. In the Text Mining node, you can control the number of clusters either by using the **Maximum** cluster option (the default) with **Low** SVD resolution or by using the exact number of clusters.

**Display 6.5: Cluster Technique Property Settings**

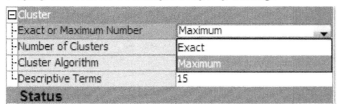

Using the default settings (**Low** SVD resolution and **Maximum** cluster) results in 13 clusters with the SAS Global Forum text corpora as shown in Display 6.6.

**Display 6.6 Cluster Identified with Default Properties**

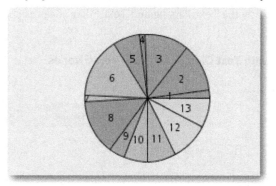

By default, SAS Text Miner uses 15 descriptive terms that best describe each cluster as shown in Display 6.7.

**Display 6.7: Descriptive Terms for Clusters Identified with Default Settings**

| Cluster ID | Descriptive Terms | Frequency | Percentage |
|---|---|---|---|
| 1 | +cdisc +sdtm cdisc sdtm standards +clinical +study +define +final +format +m... | 19 | 2% |
| 2 | +macro coders +file +run +end +statement +variable variables +code options ... | 124 | 12% |
| 3 | coders missing variables +statement +variable +end values statements +step +... | 113 | 11% |
| 4 | preloadfmt 'proc format' levels formats creating +class missing +label proc +pr... | 13 | 1% |
| 5 | customers +customer +business models +model +experience companies +stud... | 71 | 7% |
| 6 | +server +access +business applications users customers +user +file +customer ... | 154 | 15% |
| 7 | 'proc format' preloadfmt proc +final +report +define +procedure statements m... | 16 | 2% |
| 8 | +model models statistics +study +plot +analysis +procedure levels +clinical +cl... | 141 | 14% |
| 9 | +access users options +procedure statements proc best +user references code... | 42 | 4% |
| 10 | +graph +plot graphs +label +style statements +statement creating options +pr... | 56 | 6% |
| 11 | +style +graph 'proc format' graphs formats +class +define creating +file +form... | 77 | 8% |
| 12 | +business +experience +customer standards best customers +access levels co... | 99 | 10% |
| 13 | +customer +model models +plot +analysis +graph customers +business missin... | 75 | 8% |

If SVD resolution is set to **High**, then only five clusters are generated as shown in Display 6.8.

**Display 6.8: Five Clusters and Descriptive Terms Identified with High SVD Resolution Setting**

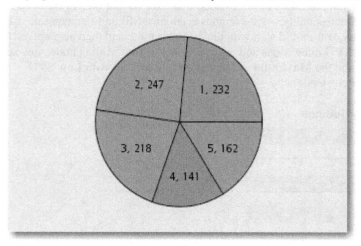

| Clusters | | | |
|---|---|---|---|
| Cluster ID | Descriptive Terms | Frequency | Percentage |
| 1 | +file users +user +process +business creating +work +create +step +run +author +code +ti... | 232 | 23% |
| 2 | +model statistics +analysis results variables +value values +variable +different +number +int... | 247 | 25% |
| 3 | +variable variables +end +proc +step +output +code +run values +set +value +first +numb... | 218 | 22% |
| 4 | creating +proc +file +output values +code +run +variable +create +value +end +set variabl... | 141 | 14% |
| 5 | +business companies best 'product names' 'respective companies' +brand +'other brand' re... | 162 | 16% |

The number of clusters generated is usually different, not just based on SVD resolution, but also based on the type of algorithm used. This is expected in unsupervised classification techniques (such as clustering) and underscores the importance of a user's involvement and judgment that is needed to select the best solution. For the SAS Global Forum corpora, the use of the default setting (low SVD resolution) with hierarchical clustering results in a large number of clusters as shown in Display 6.9.

**Display 6.9: Hierarchical Clustering Results with Low SVD Resolution**

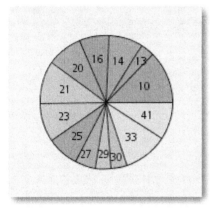

If SVD resolution is set to high, then the number of clusters is reduced to five as shown in Display 6.10. Although the number of clusters in this case happens to be the same as the result of the EM algorithm, note that the frequencies of documents in each cluster and their descriptive terms are quite different.

**Display 6.10: Hierarchical Clustering Results with High SVD Resolution**

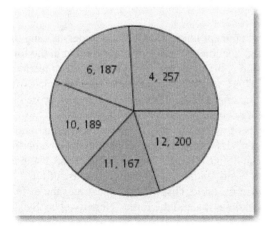

| Clusters | | | |
|----------|----------|-----------|-----------|
| Cluster ID | Descriptive Terms | Frequency | Percentage | Coor |
| 4 | +client +function variables +variable +outcome +e-mail +'data set' +... | 257 | 26% | |
| 6 | kirk +ca lafler +manager +button +illustration +'contact information'... | 187 | 19% | |
| 10 | title1 +ca +ability importing systems +browser rating +difference +co... | 189 | 19% | |
| 11 | +architecture nodes +'lag function' 'grid nodes' scalability +manager... | 167 | 17% | |
| 12 | +buck +advisory kirk lafler services +configuration +architecture +ba... | 200 | 20% | |

Once you run any clustering on the text corpora, the exported data from the Text Cluster node will contain the SVD values. For example, the SVD values for the SAS Global Forum text corpora for the EM algorithm with low SVD resolution are shown in Display 6.11.

**Display 6.11: Text Cluster Node Output Data Set with SVD Values**

EMWS1.TextCluster_TRAIN

| _document_ | TextCluster_SVD1 | TextCluster_SVD2 | TextCluster_SVD3 | TextCluster_SVD4 | TextCluster_SVD5 | TextCluster_SVD6 | TextCluster_SVD7 | TextC |
|------------|------------------|------------------|------------------|------------------|------------------|------------------|------------------|-------|
| 1 | 0.376695 | -0.1442 | 0.274585 | -0.08277 | 0.247017 | -0.04416 | 0.09357 | -0 |
| 2 | 0.502792 | -0.24184 | 0.285088 | 0.010199 | 0.326627 | 0.003744 | -0.04611 | 0.0 |
| 3 | 0.516182 | -0.1758 | 0.052119 | 0.00505 | 0.060874 | 0.002968 | -0.204 | -0 |
| 4 | 0.583391 | -0.23164 | 0.079653 | 0.255741 | 0.177013 | -0.0411 | -0.02976 | 0.2 |
| 5 | 0.580974 | -0.17455 | 0.102683 | 0.173662 | 0.11691 | -0.03183 | -0.05013 | 0.1 |
| 6 | 0.585897 | -0.02892 | -0.13313 | 0.100569 | 0.108039 | 0.076198 | -0.08252 | 0.0 |
| 7 | 0.74512 | -0.0728 | 0.25876 | 0.103738 | 0.068555 | 0.013378 | 0.050091 | 0.0 |
| 8 | 0.549167 | -0.24421 | -0.24803 | -0.14906 | -0.0058 | 0.179597 | -.000516 | 0.0 |
| 9 | 0.58608 | -0.19252 | -0.02157 | 0.016686 | 0.04562 | 0.10382 | -0.05864 | 0.0 |
| 10 | 0.468532 | -0.16417 | 0.024764 | 0.045774 | 0.14972 | 0.066833 | -0.01527 | 0.1 |
| 11 | 0.5907 | -0.11748 | 0.190739 | 0.077337 | -0.0779 | -0.00403 | -0.08665 | 0.0 |
| 12 | 0.493092 | -0.17751 | 0.198721 | -0.03693 | 0.20912 | 0.006333 | 0.072294 | 0.1 |
| 13 | 0.516848 | -0.19514 | 0.256727 | -0.10436 | 0.215215 | 0.039343 | 0.042102 | |
| 14 | 0.561488 | -0.31836 | -0.35651 | -0.16146 | -0.00536 | 0.276953 | 0.143488 | 0.1 |
| 15 | 0.688094 | -0.23573 | 0.154592 | 0.143659 | 0.100665 | -0.03489 | -0.12902 | 0.1 |
| 16 | 0.53501 | -0.25136 | 0.253247 | -0.16684 | 0.317818 | 0.00234 | -0.03092 | -0 |
| 17 | 0.533991 | -0.20267 | 0.107817 | 0.242661 | 0.106509 | -0.05262 | -0.03755 | 0.1 |
| 18 | 0.445089 | -0.10917 | -0.01793 | 0.307929 | -0.03057 | -0.06709 | -0.00473 | 0.1 |

Many other clustering options (including SOM, $k$-means, hierarchical methods such as centroid, average, and so on) are available in SAS Enterprise Miner either through stand-alone nodes (such as SOM/Kohonen or $k$-means) or through the use of a SAS Code node (such as average, centroid, and so on, as the hierarchical clustering option). The exported SVD values from the Text Cluster node can easily be used by analysts to explore other clustering options.

## Topic Extraction

In Chapter 5, you learned about and appreciated the benefits of concept links. They help in understanding the relationships between terms based on their co-occurrence in the documents. They are represented in the form of a hub-and-spoke structure, where the strength of the relationship is indicated by the width of the connecting line. The same concept can be extended to understand how various documents are associated to a topic. A topic is a collection of terms that define a theme or an idea. Every document in the corpus can be given a score that represents the strength of association for a topic. A document can contain zero, one, or many topics. The objective of creating a list of topics is to establish combinations of words that are of interest in the analysis. For example, a car manufacturer might want to explore all product reviews to see negative comments on a particular car feature, for example, the navigation system. The analyst can define a topic, "Negative Nav Sys," using the terms "GPS," "navigation," "bad," "malfunctioning," "useless," and "faulty." You can get a sense of the possible negative terms for defining the topic by browsing the terms table. Display 6.12 illustrates the difference between concept links and text topics by considering the term navigation and the topic "Negative Nav Sys."

**Display 6.12: Similarity between a Concept Link and a Text Topic**

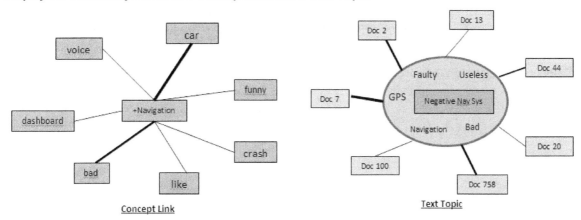

Concept Link                    Text Topic

In this example, "Negative Nav Sys" shows a strong association with Document 7, Document 2, and Document 758. The importance of a particular term within a topic is defined by a weight assigned to each term in that topic. A term can be part of multiple topics and therefore gets multiple weights.

## Text Topic Node

The Text Topic node in SAS Text Miner discovers topics from text. Fundamentally, the node analyzes document contents and summarizes the collection by identifying topics. The Text Cluster node generates clusters where each document can belong to only one cluster. In a text topic analysis, each document can belong to many topics. The Text Topic node enables an analyst to create topics of interest using groups of terms identified in text parsing. The node can be configured to identify single-term topics or multi-term topics in the data. Topic extraction is a computer-intensive task because the node uses rotated SVD in the background to capture information from a sparse term-by-document matrix. We have discussed in detail the mechanics of SVD in earlier sections of this chapter. Here is the SVD example discussed earlier in this chapter. Suppose you have the following set of four text documents:

D1: I love iPad.
D2: iPad is great for kids.
D3: Kids love to play soccer.
D4: I play soccer at OSU.

With two eigenvalues retained, coordinates of the terms represented in the SVD space are shown.

**Table 6.3: Coordinates of Each Term for the Two SVD Dimensions from the Example**

| ID | Type | SVD1 | SVD2 |
|---|---|---|---|
| I | Term | 0.99 | 0.225 |
| Love | Term | 0.876 | -0.105 |
| iPad | Term | 0.739 | -1.078 |
| is | Term | 1.06 | -0.307 |
| Great | Term | 0.405 | -0.806 |
| Kids | Term | 0.949 | -0.637 |
| Play | Term | 1.197 | 0.667 |
| Soccer | Term | 1.197 | 0.667 |
| OSU | Term | 0.656 | 0.498 |

Plotting the terms in the SVD space (as shown in Display 6.13) shows that the terms "iPad," "kids," and "love" are close, and the terms occurring only once in the text like "great" and "OSU," are far from the other terms. The terms "iPad," "kids," and "love" can be combined to form a topic. Similarly, the terms "love," "is," and "I" can be combined to form another topic.

**Display 6.13: Plot of SVD Coordinates from Table 6.3**

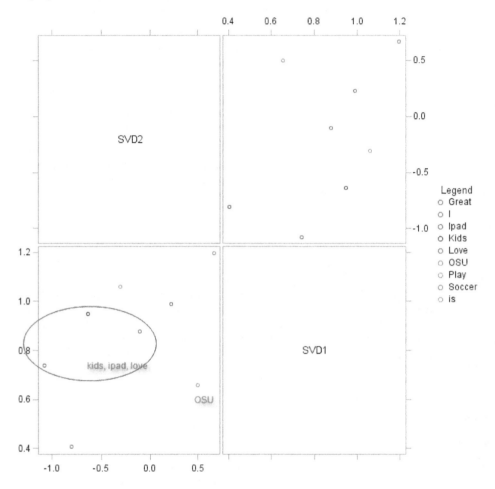

The Text Topic node must be preceded by either the Text Parsing or Text Filter node. When the Text Topic node is connected to a Text Filter node, the term weighting properties set in the Text Filter node are used by the Text Topic node. Otherwise, default settings (such as **Log** for frequency weighting and **Entropy** for term weighting) are used for calculating weights. The properties of the node should be decided carefully based on the size of the document collection.

By default, the node doesn't generate any single-term topics. You can specify this using the **Number of Single-term Topics** setting in the Properties panel. You cannot specify more than 1,000 or more than the number of terms imported into the Text Topic node, whichever is smaller. If you want to represent each extracted term in text parsing as a topic, then create that many single-term topics. SVDs are not required for creating single-term topics. They are generated based on the weights calculated for the terms. However, as shown in the previous example, SVDs are essential for creating multi-term topics. The node, by default, creates 25 multi-term topics, where each topic is essentially an SVD dimension. You can always modify this number using the property setting **Number of Multi-term Topics**.

The requested number should not exceed one of the following:

- 1,000
- the number of documents minus 6
- the number of terms that are imported into the Text Topic node minus 6

Fewer topics are generated than requested in instances where SVD results lead to duplicate topics. To avoid convergence issues, always specify a number that is smaller and not too close to the number of documents and the number of terms that are imported into the Text Topic node.

Consider the SAS Global Forum corpus for topics using the Text Topic node. Connect a Text Topic node to the Text Filter node. In the Properties panel (as shown in Display 6.14), set the **Number of Single-term Topics** property to **10**, and run the node. The **Correlated Topics** property specifies whether topics must be uncorrelated or if they can be correlated.

**Display 6.14: The Text Topic Node Train Properties Panel**

| Train | |
|---|---|
| Variables | ... |
| User Topics | ... |
| Term Topics | |
| Number of Single-term Topics | 10 |
| Learned Topics | |
| Number of Multi-term Topics | 25 |
| Correlated Topics | No |

Display 6.15 shows the Topics window from the results window of the Text Topic node. The 10 single-term topics extracted, like "**zip**," "**act**," "**customer**," etc., are at the top of the list, with document cutoff and term cutoff values as **0.001**. Multi-term topics like topic 11 talk about macros and macro variables. Topic 14, representing software implementations, contains the terms "**server**," "**deployment**," "**metadata**," and "**configuration**." You can see each topic's multi-term topic containing thousands of terms. Only the top five weighted terms for each topic are shown in the **Topic** column.

**Display 6.15: Topics Table from Text Topic Node Results**

| Category | Topic ID | Document Cutoff | Term Cutoff | Topic | Number of Terms | # Docs |
|---|---|---|---|---|---|---|
| Single | 1 | 0.001 | 0.001 +zip | | 1 | 223 |
| Single | 2 | 0.001 | 0.001 +act | | 1 | 146 |
| Single | 3 | 0.001 | 0.001 +customer | | 1 | 271 |
| Single | 4 | 0.001 | 0.001 +factor | | 1 | 252 |
| Single | 5 | 0.001 | 0.001 +key | | 1 | 455 |
| Single | 6 | 0.001 | 0.001 +search | | 1 | 237 |
| Single | 7 | 0.001 | 0.001 +role | | 1 | 186 |
| Single | 8 | 0.001 | 0.001 +element | | 1 | 303 |
| Single | 9 | 0.001 | 0.001 +match | | 1 | 306 |
| Single | 10 | 0.001 | 0.001 +style | | 1 | 250 |
| Multiple | 11 | 0.996 | 0.040 +macro,+macro variable,+file,+variable.cornersas | 1142 | 156 |
| Multiple | 12 | 0.882 | 0.038 +style,+ods,+template,+destination,+html | 1161 | 103 |
| Multiple | 13 | 0.797 | 0.037 +estimate,+model,+hazard.analysisas.statistics | 1183 | 101 |
| Multiple | 14 | 0.860 | 0.037 +server,+configuration,+deployment,+metadata,+tier | 1110 | 126 |
| Multiple | 15 | 0.903 | 0.038 +customer,+skill,+business,+train,+organization | 1397 | 152 |
| Multiple | 16 | 0.822 | 0.036 +model,+estimate,+variance,+model,+effect | 1337 | 121 |
| Multiple | 17 | 0.836 | 0.034 +job,+database,+server,+load,+performance | 1221 | 130 |
| Multiple | 18 | 0.637 | 0.032 +plot,+graph,+graphics,+axis,sgplot | 955 | 85 |

To understand the results of the Text Topic node, you need to understand the difference between term topic weight and document topic weight.

*Term topic weight:* Each term is assigned a weight corresponding to each topic. If there are 25 topics extracted, there will be 25 term topic weights calculated for a single term. Because each topic is an SVD dimension, the term topic weights for a term are nothing but the coordinates of the term in the SVD space. In Table 6.1, the coordinates 0.876 and -0.105 corresponding to the term "love" are the term topic weights corresponding to the two SVD dimensions.

*Document topic weight:* Similarly, every document in the collection is assigned a weight corresponding to each topic. If there are 25 topics extracted, there will be 25 document topic weights calculated for a single document. The document topic weight of a document toward a topic is the normalized sum of tf-idf weightings for each term in the document multiplied by their term topic weights.

Term topic weights and document topic weights are used to calculate cutoff scores for each multi-term topic.

- *Term cutoff:* This is the threshold score that determines whether a term belongs to a topic. This is equal to the mean + 1 standard deviation of all term topic weights for that topic. Any term with an absolute term topic weight greater than this cutoff is assigned to this topic.

- *Document cutoff:* This is the threshold score that determines whether a document belongs to a topic. This is equal to the mean + 1 standard deviation of all document topic weights for that topic. Any document with an absolute document topic weight greater than this cutoff is assigned to this topic.

A better understanding of these weights and cutoffs can be achieved using the interactive topic viewer. An analyst is given complete control to modify these weights to ensure that certain terms are always assigned to or eliminated from a particular topic via the Interactive Topic Viewer.

## Interactive Topic Viewer

The Text Topic node provides a facility to interactively adjust the topics. Click the ellipsis button next to **Topic Viewer** in the Properties panel. Display 6.16 shows the Interactive Topic Viewer. This window is divided into three sections: **Topics**, **Terms**, and **Documents**. The contents of the **Terms** and **Documents** sections update based on the topic selected in the **Topics** section. To select a specific topic, right-click on the last column of that particular topic, and select **Select current Topic**.

**Display 6.16: Interactive Topic Viewer**

In the **Topics** section, you can modify the values in only the **Topic**, **Term Cutoff**, and **Document Cutoff** columns. For example, consider the topic "+macro,+macro variable,+file,+variable,cornersas." You can rename this topic "Macros and Macro variables" by just replacing the text in the **Topic** field. A **Term Cutoff** score of

0.04 is calculated for this topic. From the **Terms** section, all terms with an absolute topic weight value greater than or equal to 0.04 are assigned to this topic. (See Display 6.17.)

**Display 6.17: Topic Cutoff for the Topic "+macro,+macro variable,+file,+variable,cornersas"**

| Topic | Category | Term Cutoff | Document Cutoff |
|---|---|---|---|
| Calcium | User | 0.001 | 0.001 |
| Macros and Macro variables | User | 0.04 | 0.996 |
| +zip | Single | 0.001 | 0.001 |
| +act | Single | 0.001 | 0.001 |
| +customer | Single | 0.001 | 0.001 |
| +factor | Single | 0.001 | 0.001 |
| +key | Single | 0.001 | Change term cutoff, |
| +search | Single | 0.001 | document cutoff here |
| +role | Single | 0.001 | |

Terms

| Topic Weight | + | Term | Role | # Docs | Freq |
|---|---|---|---|---|---|
| 0.04 | | architecturesas | Miscellaneous Proper Noun | 10 | 21 |
| 0.04 | + | catt | | 10 | 37 |
| 0.04 | + | obscount | | 7 | 24 |
| 0.04 | + | vartype | | 6 | 18 |
| 0 | + | trademark | | 966 | 2974 |
| 0 | Change term topic | | | 966 | 1354 |
| 0 | weight here | | Location | 964 | 2191 |
| 0 | | | | 958 | 1942 |

You can add more terms to this topic by adjusting the cutoff values. For example, to add the term "interface" that has a term topic weight of 0.039, you can change the term cutoff value for this topic in the Topics section to 0.039. By decreasing the term cutoff value, you are effectively increasing the number of eligible terms for the topic. This also impacts the weights of the documents. This change results in all terms with a term topic weight of 0.039 being added to this topic, which is not wanted. However, you can directly adjust the term topic weight value for a specific term in the Terms table. In Display 6.18, the topic weight value for the term "interface" is changed to 0.04.

After the changes, the **Category** column of the topic **Macros and Macro variables** shows **User** in the **Topics** section because the topic name is modified by the user. Whenever any changes are made to the cutoff values, click **Recalculate** in the top right corner of the Interactive Topic Viewer for the changes to be applied.

**Display 6.18: Interactive Topic Viewer Showing User-Edited Content**

Similarly, you can eliminate specific terms from a topic by adjusting the term topic weight to a value outside the term cutoff threshold. Similar operations can be performed for including or excluding documents assigned to a topic that are shown in the Documents section. You can investigate the extracted terms by exploring the actual text from the Documents section.

## User-Defined Topics

Another great functionality of the Text Topic node is the ability to define custom topics of interest. You can add custom topics directly using the interface or you can import them as a SAS data set. Click the ellipsis button next to the **User Topics** property from the node's Properties panel. As shown in Display 6.19, you can see the topic **Macros and Macro variables** in this table because this topic has been modified by the user. Scroll to the bottom of the window to add a custom topic named **Calcium** and its associated terms as shown in Display 6.19. Enter a value for the weight for each term. Weight is a relative value between 0 and 1 given to each **Role** and **Term** pair that indicates the importance of the term to the topic. A value of 1 indicates high importance, and a value of 0 indicates low importance.

**Display 6.19: Creating User-Defined Topics**

EMWS1.TextTopic2_INITTOPICS

| Topic | Term | Role | Weight |
|---|---|---|---|
| Macros and Macro variables | command line | NOUN_GROUP | 0.048 |
| Macros and Macro variables | architecturesas |  | 0.04 |
| Macros and Macro variables | nob |  | 0.127 |
| Macros and Macro variables | global forum | NOUN GROUP | 0.117 |
| Macros and Macro variables | file name | NOUN_GROUP | 0.066 |
| Macros and Macro variables | macro variable | NOUN_GROUP | 0.429 |
| Macros and Macro variables | code |  | 0.324 |
| Macros and Macro variables | run |  | 0.274 |
| Macros and Macro variables | index |  | 0.201 |
| Macros and Macro variables | order |  | 0.143 |
| Macros and Macro variables | guide |  | 0.091 |
| Macros and Macro variables | email address | NOUN_GROUP | 0.054 |
| Macros and Macro variables | p.o. | TITLE | 0.042 |
| Calcium | calcium |  | 1.0 |
| Calcium | health |  | 0.8 |
| Calcium | hormone |  | 0.8 |
| Calcium | glucose |  | 0.8 |

Import   OK   Cancel

Run the Text Topic node with the user-defined topics. View the Interactive Topic Viewer after the completion of the run. Display 6.20 shows the results from the Interactive Topic Viewer after the node is run with user-defined topics. The user-defined topics are at the top of the **Topics** table. There are 176 documents that contain the topic **Calcium**. You can enhance this topic by adding more terms and adjusting the cutoffs.

**Display 6.20: Interactive Topic Viewer with User-Defined Topics**

Topics

| Topic | Category | Term Cutoff | Document Cutoff | Number of Terms | # Docs |
|---|---|---|---|---|---|
| Calcium | User | 0.001 | 0.001 | 5 | 176 |
| Macros and Macro variables | User | 0.04 | 0.996 | 1339 | 198 |
| +zip | Single | 0.001 | 0.001 | 1 | 223 |
| +act | Single | 0.001 | 0.001 | 1 | 146 |
| +customer | Single | 0.001 | 0.001 | 1 | 271 |
| +factor | Single | 0.001 | 0.001 | 1 | 252 |
| +key | Single | 0.001 | 0.001 | 1 | 455 |
| +search | Single | 0.001 | 0.001 | 1 | 237 |
| +role | Single | 0.001 | 0.001 | 1 | 186 |

Terms

| Topic Weight | + | Term | Role | # Docs | Freq |
|---|---|---|---|---|---|
| 1 | + | calcium | | 4 | 108 |
| 0.8 | + | health | | 168 | 1109 |
| 0.8 | | health | Miscellaneous Proper Noun | 15 | 25 |
| 0.8 | + | glucose | | 4 | 40 |
| 0.8 | + | hormone | | 4 | 6 |
| 0 | + | information | | 976 | 7536 |
| 0 | + | product | | 968 | 4023 |

The output data set of the Text Topic node contains new variables that represent topics created by the node. There are two variables created for each topic: one is the document cutoff score and the other is a binary variable indicating topic assignment. The binary variable has a value of 1 if the document topic weight is greater than or equal to the document cutoff score. It has a value of 0 otherwise. In our example, 27 topic binary variables are created. These include two user-defined topics, 10 single-term topics, and 15 multi-term topics. Display 6.21 shows the new topic binary variables in the output data set and the existing text variables. In the presence of a target variable, the topic binary variables lend themselves as valuable input variables for performing structured data mining analysis.

**Display 6.21: Topic Binary Variables Created in the Text Topic Node Output Data Set**

| Obs # | Variable N... | Label | Type |
|---|---|---|---|
| 1 | EXTENSION | | CLASS |
| 2 | FILTERED | | CLASS |
| 3 | LANGUAGE | | CLASS |
| 4 | NAME | | CLASS |
| 5 | TEXT | | CLASS |
| 6 | uri | | CLASS |
| 7 | ACCESSED | | VAR |
| 8 | CREATED | | VAR |
| 9 | FILTEREDSIZE | | VAR |
| 10 | MODIFIED | | VAR |
| 11 | OMITTED | | VAR |
| 12 | SIZE | | VAR |
| 13 | TRUNCATED | | VAR |
| 14 | TextTopic2_1 | _1_0_ Calcium | VAR |
| 15 | TextTopic2_10 | _1_0_ +contrast | VAR |
| 16 | TextTopic2_11 | _1_0_ +match | VAR |
| 17 | TextTopic2_12 | _1_0_ +style | VAR |
| 18 | TextTopic2_13 | _1_0_ +sex,+format,+proc,+age,+sort | VAR |
| 19 | TextTopic2_14 | _1_0_ +style,+ods,+template,+destination,+html | VAR |
| 20 | TextTopic2_15 | _1_0_ +server,+configuration,+metadata,+deploym... | VAR |
| 21 | TextTopic2_16 | _1_0_ +estimate,+model,+hazard,analysissas,statisti... | VAR |
| 22 | TextTopic2_17 | _1_0_ +customer,+social,+medium,+site,+business | VAR |
| 23 | TextTopic2_18 | _1_0_ +estimate,+model,+variance,+model,+residual | VAR |
| 24 | TextTopic2_19 | _1_0_ +studio,+warehouse,+integration,+bi,+busin... | VAR |

## Scoring

In the last five chapters, you learned how to perform text mining analysis using SAS Text Miner. Like any typical data mining process, the task that follows modeling is scoring. A model is trained using development data, and then it is applied to customer records in a production or live system to calculate a score for each record. An appropriate business action is performed based on the calculated score. Similarly, a text mining process can be scored on new textual data. In the context of text mining, scoring means assigning a document to a cluster or to a topic. The Score node in SAS Enterprise Miner is used for this purpose. A Score node can be connected to either a Text Cluster node or Text Topic node. If the data for scoring resides on a different database system, you can take the score code generated by any of these nodes (Text Cluster or Text Topic or Score node) and run it on the other database system for scoring. These nodes only generate SAS code for scoring. The scoring process is demonstrated using an example in Case Study 9.

## Summary

In this chapter, we discussed some of the key technical concepts in text mining analysis used for identifying clusters or topics in the textual corpus. We explained the mechanics of the different clustering algorithms available in SAS Text Miner. We demonstrated the topic extraction functionality in SAS Text Miner. Remember that text mining is an iterative and exploration-oriented analysis. There is no one specific clustering technique or topic extraction setting that yields the best results. Hence, it is always suggested to explore different property settings and techniques. Often, an analyst is required to go back in analysis and modify property selections in the Text Parsing and Text Filter nodes. These changes can alter the term-by-document matrix considerably. Further, these changes have significant impact on the clusters and the topics extracted. Cluster analysis and topic extraction suffer from the curse of dimensionality in the term-by-document matrix. SVD is used to reduce the dimensionality in the term-by-document matrix. Although a higher number of SVDs can summarize the data better, they are not wanted due to computing limitations and the higher risk of fitting the noise. SAS Text Miner provides a user with the ability to control the number of SVDs to use in the analysis.

# References

Bradley, P. S., Fayyad, U., & Reina, C. 1998. "Scaling Clustering Algorithms to Large Databases." *KDD 1998: Proceedings of the Fourth ACM SIGKDD International Conference on Knowledge Discovery and Data Mining,* 9-15.

Cattell, R. B. 1966. "The Scree Test for the Number of Factors." *Multivariate Behavioral Research.* 1(2): 245-276.

Esposito Vinzi, V., Chin, W. W., Henseler, J., & Wang, H. (Eds.) 2010. *Handbook of Partial Least Squares: Concepts, Methods, and Applications*. New York: Springer. (Springer Handbooks of Computational Statistics).

Everitt, B. S., Landau, S., & Leese, M. 2001. *Cluster Analysis*. Arnold, London.

Kohonen, T., 1995. "Self-Organizing Maps." *Springer Series in Information Sciences.* Vol. 30. Berlin: Springer-Verlag.

Landauer, T. K., Foltz, P. W., & Laham, D. 1998. "An Introduction to Latent Semantic Analysis." *Discourse Processes.* 25(2-3): 259-284.

Manning, C. D., & Schütze, H. 1999. *Foundations of Statistical Natural Language Processing*. Cambridge: MIT Press.

Radovanović, M., & Ivanović, M. 2008. "Text Mining: Approaches and Applications." *Novi Sad Journal of Mathematics.* 38(3): 227-234.

Sullivan, D. 2001. *Document Warehousing and Text Mining: Techniques for Improving Business Operations, Marketing, and Sales.* John Wiley & Sons, Inc.

Text Analytics Using SAS® Text Miner: Course Notes. SAS Institute Inc., Cary, NC. Course information:

https://support.sas.com/edu/schedules.html?ctry=us&id=1224

# Chapter 7 Content Management

## Introduction

In Chapter 2, we discussed how to extract content from a variety of data sources such as websites, blogs, feeds, local files, etc. In this chapter, we focus on how to organize and manage the data that we collect based on its content. Why is content management so important? Suppose that hundreds of paper documents or files are scattered across your desk. You have encountered a task that requires you to refer to a particular document, and you have absolutely no clue where that document is located in the heap. It means that you have to manually sift through all of the documents to find what you need. If you had chosen to organize all of your files in separate folders and labeled them accordingly, you could have saved a lot of time searching. The same goes for digital documents stored on your laptop or PC. You are always better off creating a folder structure that makes logical sense to you for storing your documents. This process is also called document classification. Document classification is often synonymously referred to as text classification or content categorization, indicating that we consider only the textual data in document content to perform classification. Objects such as images, graphics, or charts are ignored.

Many of us classify documents at our workplace. We typically maintain a manageable number of folders (less than 100), and the total number of documents generally does not exceed more than a few thousand. Assuming that you are the only person accessing your machine, this should be fairly easy to manage. Extend this case to a large enterprise where the total number of documents is a few hundreds of thousands or even in the millions. How about a folder structure with more than a hundred folders and subfolders? What if there were an incremental flow of new documents added to the shared repository each day? In a typical enterprise, there are many knowledge workers contributing to the development of shared document repositories at any given time. This compounds the problem of managing and organizing document content in those repositories. At this point, you would want to automate the process of document classification. Organizing documents on this scale requires a lot of thought, planning, and work. It might appear to be a simple task that you can do all by yourself, which might be true if you are the only person benefiting from it. However, in reality, many users need to benefit from this preclassified repository. As a result, the logical arrangement of the folder structure should be both meaningful and intuitive. This is the reason document classification has grown from being a simple administrative task to an amalgamation of both art and science.

In the context of document classification, the folder structure is called *taxonomy* and the folders and subfolders are called *categories* and *subcategories*, respectively. In large enterprises, it is not efficient to navigate back and forth through a huge folder structure to find documents of interest. It is worth mentioning that you can choose not to create a physical folder structure to organize your documents. All of your documents can be stored in one single storage location on a server. Instead of moving documents to folders, you can opt to simply tag all of your documents with the information of the categories and subcategories to which they belong. These tags form the documents' metadata, and they are useful in the document search and retrieval process. The process of tagging documents is technically termed in literature as *content tagging or document tagging*.

If documents are stored in your company's repository for your employees, but they are unable to find them, it can derail the progress of your workforce. Research studies show that the cost at an employee's workforce of searching for a document and not finding it, in addition to the time and effort involved in creating a new document, add up to approximately 12 million United States dollars per 1,000 employees on average each year. Sometimes, the cost of not being able to find information can prove catastrophic. In a real-world example, a large pharmaceutical company failed to find a patent filed by a competitor that was related to a similar drug that it envisioned to develop. As a result, five years of precious time and money went straight into the gutter!

## Applications

Organizations are increasingly showing interest in using text classification techniques for many business applications. Many media firms deploy document classification techniques to automatically categorize online news articles into various topics such as politics, business, science, sports, entertainment, etc. Spam filtering, automatic classification, and routing e-mails are well-known applications of text classification. Some websites monitor their users for the content that they are reading. These websites provide automatic suggestions with links to similar or related content. Many universities set up their online digital libraries and use abstracts or descriptions of articles, journals, and e-books to classify them into one or more categories. Automatic authorship attribution and identification of a document's genre are some other classic cases of document classification techniques put into practice. In market research, open-ended responses or comments on a customer survey are automatically assigned predefined survey codes for a qualitative study of data using text classification algorithms. An interesting but complex problem, automatic essay grading revealed how text classification programs can be developed to score student-written essays as closely as human graders. Needless to say, scientific advances in machine learning, natural language processing, and statistical techniques laid strong foundations for the evolution of sophisticated text classification methods.

## Content Categorization

Document classification (or content categorization as we refer to it for the remainder of this chapter) can be performed using SAS Content Categorization Studio. This tool offers two features, *categorization* and *concept extraction*, which are covered in great detail in this chapter. Content categorization (as the name implies) helps classify documents into a set of known subject areas or categories. Concept extraction helps extract important pieces of information from the documents called *concepts*. A concept can be the name of a person, place, organization, or event. A much more sophisticated feature, known as contextual extraction, is available if you obtain the Enterprise version of SAS Content Categorization Studio or install SAS Contextual Extraction Studio with SAS Content Categorization Studio. This feature incorporates the combined power of categories and concepts to extract complex factual information. The name of the president of a country, the date on which an article was published, etc., are examples of complex facts. The distinguishing features of contextual extraction compared to concept extraction are discussed later in this chapter. Throughout this chapter, we discuss the features and functionalities of SAS Content Categorization Studio based on version 12.1. Screenshots of features are subject to slight changes in later versions.

In Chapter 4, we showed you how standard entities such as date, currency, address, etc., can be extracted using SAS Text Miner. However, SAS Text Miner cannot extract custom entities and complex facts from documents.

For example, Display 7.1 shows a SAS Global Forum paper abstract in which SAS Add-In for Microsoft Office (a complex fact) represents the name of a SAS software product. To extract the names of software products used in the paper abstracts, you must write concept definition rules to find patterns in the documents with SAS as the prefix. In Display 7.1, the terms highlighted in red (intelligence and dashboard) are closely related to the field of Business Intelligence. As a result, they are helpful in the content categorization process.

Before working in SAS Content Categorization Studio, the subject matter expert familiar with the document content should define the taxonomy structure and categories/sub-categories into which the documents can be classified. A well designed taxonomy will help the end-users to easily narrow down to the specific folder in which they are likely to find a document of their interest. SAS Content Categorization Studio helps to easily build such taxonomies to organize information (content categorization) and identify facts (concept extraction). For these purposes, SAS Content Categorization Studio uses several natural language processing and advanced linguistic techniques. In SAS Content Categorization Studio, you can develop rules for categories and concepts simultaneously, but in separate branches within the taxonomy.

**Display 7.1: Screenshot of a SAS Global Forum Paper Abstract**

> Business users rely on e-mail for communication and collaborative decision making. Have you ever wondered if you could use your Microsoft Outlook environment beyond a simple communication tool and get added value that keeps you more informed about your business? The integration of SAS ® into Microsoft Outlook provides a new outlook into your business. This integration of SAS ® Business Intelligence and Microsoft Outlook increases informed decision making by giving business users intuitive access to reports, stored processes, and dashboards. Business users obtain contextual information from SAS Business Intelligence while performing daily tasks in Microsoft Outlook. This paper highlights the key capabilities provided by seamless integration of SAS with Microsoft Outlook and the application of these capabilities available with SAS ® Add-In for Microsoft Office.

The entire process of taxonomy generation and deployment in real time is shown in Display 7.2.

**Display 7.2: High-Level Process Flow Diagram for Using SAS Content Categorization Studio and SAS Content Categorization Server**

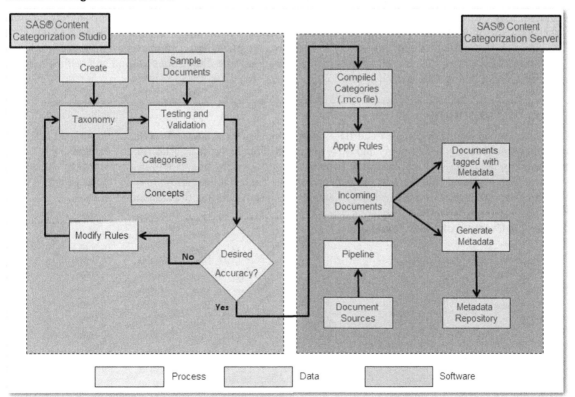

The developed taxonomy can be tested against a set of sample documents preclassified by category to verify the accuracy of classification. Taxonomy can be modified and tested again iteratively until the appropriate accuracy is achieved. Model accuracy is generally assessed by the means of two measures, recall and precision. In a given category, recall indicates the percentage of documents that are correctly classified. Precision indicates the percentage of documents that are correctly classified out of all of the documents classified in the current category. However, these two measures often affect one another, making it difficult to assess model performance. A third measure, traditionally known as F-measure (or F-score), can be used to test model accuracy. F-measure is the harmonic mean of precision and recall. It can be calculated using the following formula:

$$\text{F-measure} = 2 * (p * r)/(p + r) \qquad\qquad \text{(Equation 7.1)}$$

In this formula, $p$ and $r$ are the precision and recall measures of the model. The value of an F-measure ranges between 0 and 1. A higher value of the F-measure indicates a better classification model.

For the purpose of testing category and concept rules, you should collect a significant number of documents (around 10 to 20) representing each category in the taxonomy. It is important to include documents that shouldn't match a category. This helps the taxonomist to efficiently write rules so that documents belonging to one category are less likely to be assigned to another category. For example, documents containing the word "plant" in the context of industrial equipment shouldn't be matched with the term "neem plant," which is a botanical term.

Once the taxonomy is finalized, you can generate the compiled files representing the rules of the categories (.mco) and concepts (.concepts). These files can be used in SAS Content Categorization Server to automatically apply rules to documents feeding from various data sources in real time. The output of this process is metadata information, such as document category and the facts found in the document. SAS Content Categorization Studio supports approximately 30 international languages and many document formats such as HTML, HTM, plain text, XML, RTF, SGML, and XLS. SAS Content Categorization Studio does not support PDF documents or the Microsoft Word format. Hence, you need to convert them to plain text documents using SAS Document Conversion. SAS Content Categorization Studio does not recognize SAS data sets as a data source (unlike SAS Text Miner). Hence, you should use raw documents in recognizable file formats to train and test the models built using SAS Content Categorization Studio.

## Types of Taxonomy

Taxonomy can be simple or complex and flat or hierarchical in its structure. In a flat taxonomy, all categories are at the same level in the taxonomy and there are no children or subcategories for any category. For example, the taxonomy in Display 7.3 represents the structure of various departments within a typical firm. All categories appear at the same level, and there are no dependencies—it is a flat taxonomy. In the project, you need to add a language such as English, and then enable a categorizer to start building the taxonomy.

**Display 7.3: Example of a Flat Taxonomy**

If one or more categories in the taxonomy contains a child or subcategory, it is a hierarchical taxonomy. The child categories can have children, leading to a nested category structure. Display 7.4 shows a partial screenshot of a sample hierarchical taxonomy in which the parent-child relationship between categories and their subcategories can be clearly understood.

**Display 7.4: Example of a Hierarchical Taxonomy**

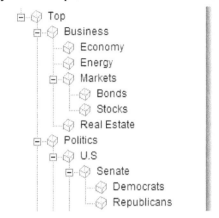

The categories **Business** and **Politics** are at the same level, and they have children or subcategories. The category **U.S** is the child of its parent **Politics**, **Senate** is the child of its parent **U.S**, and both **Democrats** and **Republicans** are children of their parent **Senate**.

## Categorization

Categorization is the process in which documents are assigned to pre-identified categories or topic areas. Although the primary objective of the categorization process is to identify the topic area to which a document pertains, it is important to prevent classifying documents to incorrect categories. There are several methods of performing content categorization in SAS Content Categorization Studio. Display 7.5 shows these methods and how they differ from each other. Each method has its own pros and cons compared to the other methods.

**Display 7.5: Hierarchy of Categorization Methods in SAS Content Categorization Studio**

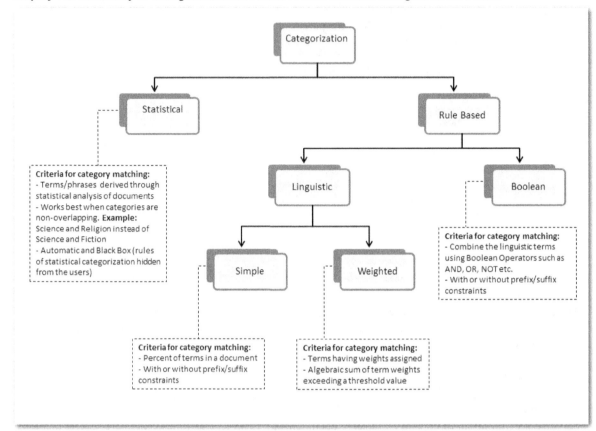

## Training and Testing

Before diving deep into the details of these individual methods, you should understand what are training and testing documents. If you are using a set of documents to train or build the model, the documents are called the training set. If you are using a set of documents to test an already developed or built model, the documents are called the testing set. Both training and testing document sets are required for developing statistical models in SAS Content Categorization Studio. You should organize these documents in separate folder structures reflecting the hierarchy of the taxonomy. If you do not have the folder structures already created, you can create them automatically. To do this, provide the physical path for the testing directory and training directory on the **Data** tab. (See Display 7.6.) Click **Propagate** with the **Create Folders** option checked in the **Propagate Options**. Physical folders reflecting the hierarchy of the taxonomy are created on the storage drive. You can place the relevant documents in the appropriate folders for training or testing.

**Display 7.6: Testing and Training Paths for a Category**

Rule-based categorizers do not require training documents, so you can skip setting the training path. If you like to place documents from all categories and subcategories in only one folder, then check **Identical Path** in **Propagate Options**. Conversely, you can create a taxonomy using an existing folder structure on your physical drive. Right-click **English**, and select **Create Categorizer from Directories**.

**Display 7.7: Defining a Categorizer from an Existing Folder Structure**

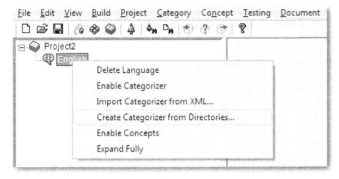

---

## Statistical Categorizer

A statistical categorizer helps users automatically build a model for categorization without having to write any rules. It requires a set of documents pre-identified for each category in the taxonomy to train the model. When training the model for a particular category, documents from all categories in the taxonomy are considered for analysis. A statistical categorizer tries to find uniquely identifiable terms that describe a category while making sure they do not match any other categories. As a result, any changes made to the training documents in any category impact the rules for other categories also. In this case, you need to rebuild the model. Once the model is trained, you can test it on other document to verify its accuracy. The underlying rules of a statistical categorizer cannot be viewed. This is why it is also called a black box model. The statistical categorizer performs well when the number of categories is limited and the categories are significantly different from each other with regard to their content. For example, if the categories are sports, media, events, and entertainment, the statistical categorizer might not be efficient in categorizing documents precisely. The biggest advantage of a statistical model is that it is the easiest to develop, requiring just a set of well-collected training documents. However, due to the difficulty of matching concepts on the basis of a statistical measure, it is difficult to achieve a high level of accuracy, which is a major drawback for the statistical categorizer.

Here is a brief demonstration on how to build a statistical categorizer-based model. For the statistical categorizer, you need a training set of documents to build the model. For this purpose, we have extracted 466 SAS Global Forum paper abstracts published in the past three years from five different sections. **Stats** (Statistics and Data Analysis), **DataMining** (Data Mining and Predictive Modeling), **Reports** (Reporting and Information Visualization), **BusInt** (Business Intelligence), and **SysArch** (Systems Architecture) are the five section categories representing the 466 paper abstracts. These abstracts are split into Test and Train groups. Each group folder contains subfolders representing the five section-based categories and respective paper abstracts in raw files. SAS Content Categorization Studio requires input textual comments in .xml or .txt format. If you have your textual data as a SAS data set, you need to create a unique .xml or .txt file for each textual comment in your data set. Refer to the section "Appendix" in this chapter for SAS code that creates each observation in the SAS data set as a separate text file.

Create a new project in SAS Content Categorization Studio, enable the statistical categorizer, and create these five categories with the same names. Click the **Top** node in the categorizer, set the **Training Path** on the **Data** tab, and point it to the **Train** folder. Deselect any **Propagate Options**, and click **Propagate** next to **Training Path**. Now, the training paths of all categories automatically point to the specific folders containing the paper abstracts in text files. Similarly, set the **Testing Path** in the **Top** node, and propagate the testing paths to all of the categories. After both the testing and training paths are set up, select **Build ▶ Build Statistical Categorizer** to build the statistical model. After the successful completion of the build, you will see the Build Successful window. Select the **DATAMINING** category, and click the **Testing** tab to see all of the files available to test the statistical model. Click **Test** to see which documents have failed and which have passed the test.

**Display 7.8: Test Results for DataMining Category in the Statistical Categorizer**

StatCat
  English
    Categorizer
      Top
        BUSINT
        DATAMINING
        SYSARCH
        REPORTS
        STATS

◉ Test files for this category
○ Test all files everywhere

TEST

C:\Data\Sgf_sectionwise\Test\DATAMINING

| Test File | Result |
|---|---|
| 165-2011.pdf.txt | FAIL |
| 163-2011.pdf.txt | FAIL |
| 159-2011.pdf.txt | FAIL |
| 154-2008.pdf.txt | FAIL |
| 153-2011.pdf.txt | FAIL |
| 164-2011.pdf.txt | PASS |
| 162-2011.pdf.txt | PASS |
| 161-2011.pdf.txt | PASS |
| 160-2011.pdf.txt | PASS |
| 158-2011.pdf.txt | PASS |
| 157-2011.pdf.txt | PASS |
| 156-2011.pdf.txt | PASS |
| 155-2011.pdf.txt | PASS |
| 155-2008.pdf.txt | PASS |
| 154-2011.pdf.txt | PASS |

◇ Taxonomy | ◈ Dependencies | ⦃⦄ Rules | ◈ Testing | Data | ⬚ Document

## Automatic Rule Generation

Automatic rule generation is a powerful feature offered in SAS Content Categorization Studio to automatically generate a list of terms that describe a category based on a training set of documents. The rules derived using this process are called "unqualified linguistic rules." This process helps as a precursor to rule-based categorization. You can export the automatic rules generated into a rule-based categorizer, and then modify them for each category. For example, suppose that you have a need to automatically generate rules for the category Inventions. Place all of the documents related to this category in a single folder on your system, and specify the path to that folder as the **Training Path** on the **Data** tab for the **Inventions** category. Once the training path is set, click **Generate Rules Automatically** in the **Category** menu. Display 7.9 shows the two steps involved in this process.

**Display 7.9: Automatic Rule Generation in SAS Content Categorization Studio**

File  Edit  View  Build  Project  Category  Concept  Testing  Document  Server  Help

Project
  English
    Categorizer
      Top
        Business
        Politics
        Entertainment
        Sports
        Technology
        Inventions

Add Category
Delete Category
Delete All Selected Categories
Rename Category
Import Category from Repository...
Create Directory Tree...
Generate Subcategories...
Generate Rules Automatically
Export All Generated Rules
Clear Generated Rules

Created: December 06, 2012
Modified:

◉ Completed
○ Pending
○ Test Disabled

2. Click here to automatically generate rules for this category

Related Links
Testing Path

1. Set the training path for documents in this category

Training Path
C:\Taxonomy1\Inventions

... Propagate
... Propagate

Propagate Options
☐ Identical Path
☐ Create Folders

◇ Taxonomy | ◈ Dependencies | ⦃⦄ Rules | ◈ Testing | Data | ⬚ Document

Rule terms and phrases that might describe this category are derived based on the statistical analysis of the training documents. (See Display 7.10.) These rule terms and phrases can be exported as rules in the first step to build a rule-based categorizer. Generally, automatically generated rules do not work well if they are used in the rule-based categorizer without any changes. Add a qualifier, assign weights, or use Boolean operators to refine the rules and build a better performing model.

By default, the automatic rule generation feature uses the frequent phrase extraction algorithm to derive the most frequent words or phrases in the training documents. You can change the option to use the maximum entropy classifiers algorithm, which generates words or phrases that are most meaningful for the selected category and are different from the other categories. Select **Project ▶ Settings**, and click the **Rule Generation** tab to change the algorithm. Using this feature, you can either generate weighted linguistic rules or Boolean rules automatically. This feature is efficient when you have training documents for all other categories or subcategories within the taxonomy.

**Display 7.10: Example Showing Terms Derived by the Automatic Rule Generation Feature**

## Rule-Based Categorizer

A rule-based categorizer enables you to write your own rules for each category or subcategory in the taxonomy. Unlike the statistical categorizer, you can control the rules for each category individually independent of other categories. In fact, you can start writing rules for a category without the need for training documents. However, you need to have sample documents for each category in the taxonomy for testing the rule-based categorizer. As shown in Display 7.5, there are fundamentally two types of rule-based methods, linguistic and Boolean.

### Linguistic

In the linguistic approach, rules use terms (word or phrases) that might uniquely represent a category in the taxonomy. In a simple linguistic method, a document is assigned to a category when a certain percentage of the rules is found in the document. For example, consider the category Inventions in Display 7.10, which contains 20 linguistic rules. If you set 40% as the matching criteria percentage, then a document is assigned to this category only when at least eight of the 20 rules are found in that document. This threshold percentage for linguistic rules is called the **Match Ratio**, and it can be set for each category individually on the **Data** tab (s shown in Display 7.11). By default, the match ratio for any category is 10%. A higher percentage might make the rules very rigid in assigning a document to a category. A lower percentage indicates that the rules are too liberal, which can cause a document to be assigned to more than one category. It is important to find an optimal value through thorough testing.

**Display 7.11: Setting for Match Ratio on the Data Tab of a Category**

Rule terms can be modified using qualifiers. A qualifier is a special symbol that, when syntactically applied to a linguistic rule, performs a specific function (most of which is related to natural language processing). For example, a linguistic rule named "position," when suffixed with a qualifier @V and changed to position@V, expands the term "position" to include all of its verb forms in the category rules. Thus, the following terms are included as part of the rules automatically generated during the execution of the classification algorithm:

position

positioned

positioning

positions

In a weighted linguistic approach, rules are assigned different weights. A document is assigned to a particular category if the algebraic sum of the term weights exceeds a threshold value set for that category. You must follow a specific syntax to write these rules, and these rules cannot be mixed with either Boolean or simple linguistic rules within a category. The first line in this type of category should begin with two underscores and the keyword "THRESHOLD," followed by a comma and a threshold value.

\_\_THRESHOLD,*value*

Also, each of the following lines should have a linguistic rule and a weight separated by a comma without any spaces. For example, linguistic rules in the subcategory **Football** are assigned different weights. (See Display 7.12.)

**Display 7.12: Example of Weighted Linguistic Rules for a Category**

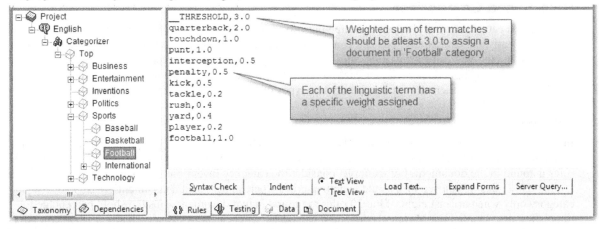

A threshold value of 3.0 is assigned for this category in the first line of the rules. The terms "quarterback," "touchdown," and "football" are assigned higher weights in this category because they are closely affiliated to football. "Kick," "penalty," "tackle," and "yard" are assigned lower weights because these terms are not specific to football and can also be associated with another sport such as soccer. Select **Build ▶ Build Rulebased Categorizer** to compile the rules before testing them on sample documents. If there are any syntax errors in the rules written, the categorizer fails to compile the rules, and errors are displayed at the bottom of the window. You need to fix the errors before building the categorizer. When you test the rules against a sample document, the relevancy score for the category is calculated as follows:

Weighted Linguistic Rule Relevancy Score $= \sum_{i=1}^{N} n_i \cdot W_i$

In this formula, $N$ is the number of weighted linguistic rules defined in the category, $n$ is the frequency of occurrence for the $i^{th}$ rule in the document, and $W$ is the weight of the $i^{th}$ rule defined in the category.

A document is assigned to this category if the calculated relevancy score is higher than the threshold value set for this category. Using the previous formula, the relevancy score is calculated for a sample text document. You can see the rule matches highlighted in red. (See Display 7.13.) The following equation shows how the relevancy score is generated for this document when tested against the Football category.

*1\*0.2(player) + 1\*2.0(quarterback) + 1\*0.4(rush) + 1\*1.0 (punt) = 3.6*

**Display 7.13: Example Showing How the Relevancy Score Is Calculated for the Weighted Linguistic Rule-Based Category**

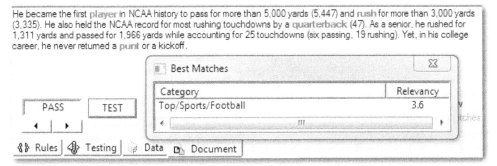

## Boolean

A Boolean rule-based categorizer enables you to join linguistic terms using Boolean operators such as AND, OR, NOT, etc. This method provides the ability to write precise rules to disambiguate between the same words that might occur in different contexts. The result of a Boolean rule expression is always either true or false, indicating whether it has found a match in the document. This method is the most powerful of all categorization techniques available in SAS Content Categorization Studio. It often yields better results compared to all other methods. Boolean rules are the hardest to develop because they require the most sophisticated human effort.

You can create symbolic links to refer to a category in the Boolean rules of another category. This mechanism proves efficient when you have a subcategory that should include the Boolean rules defined in its parent category. You can incorporate concept definitions in Boolean rules if they are defined within the same taxonomy. Similar to a linguistic rule, a Boolean rule allows for the modification of terms using qualifiers. Boolean rules are effective in finding matches in a structured document such as XML, where text data can be embedded inside tags. You can explicitly mention in the rules exactly where to look for the rule terms, such as in the body or title of the file.

Boolean rules follow a specific syntax and structure. They always start with a Boolean operator, followed by one or more arguments. You can either choose to write rules manually in the prescribed syntax or to use the point-and-click interface to add the operators and arguments without having to remember the syntax. To use the point-and-click interface, change the rules window mode from **Text View** to **Tree View** on the **Rules** tab. (See Display 7.14.)

**Display 7.14: Text View and Tree View Options on the Rules Tab**

The structure of Boolean rules facilitates the conditional matching of words or phrases. Rules can be defined to match categories with the occurrence of a term in the presence or absence of one or more terms. There are several types of Boolean operators available in SAS Content Categorization Studio. The biggest advantage of Boolean rule-based categorization is disambiguation. The same word can mean different things when applied in different contexts. For example, consider the word "poker." Poker is a well-known card game in which each player places bets, claiming to hold high-ranking cards. However, "poker" could mean a metal rod used to stir a fire in a fireplace. Linguistic rules are not capable of handling this scenario. Boolean rules can help you disambiguate the contexts when these types of words appear in documents. To distinguish between the two contexts, you can write Boolean rules in the categorizer to look for closely related words in the proximity of the word "poker" as a game. (See Display 7.15.)

**Display 7.15: Example of a Boolean Rule-Based Category**

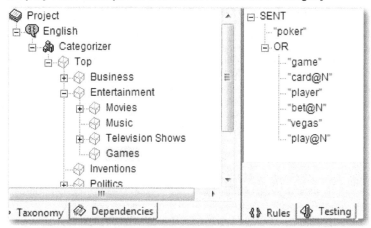

This rule indicates a category match if the word "poker" appears in the same sentence with at least one of the following words: "game," "card," "cards," "player," "bet," "bets," "Vegas," "play," or "plays." The qualifying suffix @N indicates noun stemming. Hence, all noun forms of the qualified word are included in the rule. Click **Expand forms** on the **Rules** tab, and select **Text View** to see the completely resolved rule.

```
(SENT,"poker",(OR,"game",(OR,"card","cards"),"player",(OR,"bet","bets"),"vega
s",(OR,"play","plays")))
```

## Comparison of Statistical versus Rule-Based Categorizers

Now that you have a better understanding of statistical and rule-based categorizers, let's review the distinguishing features between these two methods. Table 7.1 lists the important differences between these methods at a high level to help you understand which method to use if you need to develop a categorization model at your workplace.

**Table 7.1: Differences between Statistical and Rule-Based Categorizers**

| Statistical Categorizer | Rule-Based Categorizer |
|---|---|
| • Fast and easy to develop. | • Demands significant amount of time and human effort to build. |
| • Requires a set of training documents to build model and a set of testing documents for validating model performance. | • No need for training documents. However, a sample set of documents for each category helps validate the rules. |
| • Often less accurate. | • Usually yields higher model accuracy compared to statistical models. |

| Statistical Categorizer | Rule-Based Categorizer |
| --- | --- |
| • Hard to interpret because you cannot view the underlying rules of a statistical model. | • Rules are available and easy to interpret. |
| • Works well when document content is very similar within a category and vastly different between categories. | • Boolean rules can help disambiguate contexts in places where the same concept term with different meanings can appear in the document. This approach makes the categorization process efficient in handling categories that might contain overlapping content. |
| • Cannot develop models for categories separately as you add them to the taxonomy. The statistical categorizer uses all documents in the training set of all categories simultaneously when building a model. | • Can develop models for each category separately, independent of other categories. |

## Determining Category Membership

Earlier in this chapter, we discussed how match ratio plays an important role in assigning categories to documents based on linguistic rules. There could be a scenario where a document might be qualified for more than one category based on the match ratio. In this case, the best matching category is determined using another criterion, relevancy type. There are three choices of relevancy type in SAS Content Categorization Studio: operator-based (default), frequency-based, and zone-based.

A frequency-based relevancy type is based on the number of rule matches in the document. The category for which the document gets the highest score is the best match. For example, consider two linguistic rule-based subcategories. They are **Bollywood** (eight terms and a match ratio of 25%) and **Hollywood** (six terms and a match ratio of 50%) in the category **Movies** (as shown in Display 7.16). The minimum number of terms required for a match to Bollywood is 2 (25% * 8), and 3 (50% * 6) for Hollywood. If a document surpasses the minimum number for both of these categories, the relevancy score helps in breaking the tie to determine which category needs to be assigned to the document.

**Display 7.16: Examples of Simple Linguistic Rule-Based Categories**

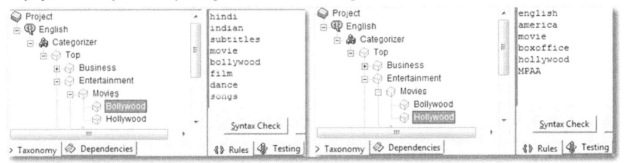

A sample movie review passed the test for both of these categories based on the match ratio. (See Display 7.17.) However, it is assigned to the category Bollywood based on the relevancy score. It is important to choose the frequency-based relevancy type by selecting **Project ▶ Settings** for this to work.

**Display 7.17: Example Showing Role of Match Ratio in Assigning a Category**

For the zone-based relevancy type, the entire document is divided into three equal segments. A document is scanned for any rule matches in all three segments. If the first segment has a match with any of the category rules, it gets a high weight. If a match occurs in the third segment, it gets a low weight. The rationale behind this approach is that whenever rule terms representing a category are found in the beginning sentences of a document, the probability that the document actually belongs to the category is high. If the rule terms are found in the last part of the document, there is a very low chance that the document belongs to the category. In this case, the rule terms might have occurred in the document in reference to some other document or in a different context altogether. This relevancy type might not be the most suitable for all types of data, but it generally works well for categorizing journals, news articles, etc.

The operator-based relevancy type boosts the weight of Boolean rules that have greater coverage of the document content. Although the operator-based relevancy type is most suitable for Boolean rules, it also works for linguistic rules. There are a few other factors such as relevancy cutoff, relevancy bias, and category bias that can influence relevancy calculations. Relevancy cutoff is the minimum threshold that the relevancy score must meet in order for a document to be matched against a category. This cutoff can be specified for all categories in the project settings or specifically for each category separately on the **Data** tab. By default, its value is 0. If a document passes for a match based on the match ratio, but fails to meet the relevancy cutoff, it is branded as conditionally passed. Relevancy bias helps in boosting the relevancy of one category over other categories in the case where the occurrence of certain terms in a document can provide enough credibility to unequivocally decide its category membership. The default value for relevancy bias is 1. Similarly, category bias boosts the relevancy of one or more categories of an entire taxonomy to help in the information retrieval process to return the most relevant information. The primary difference between relevancy bias and category bias is that the former relies on the key terms of a category, and the latter gives more credibility to the categories of a particular taxonomy compared to the categories of other taxonomies.

New Relevancy = (Relevancy Score * Relevancy Bias) + (Default Category Bias * Category Bias)

The default category bias in the project settings is 0. Hence, if you want to use category bias for categories to influence the relevancy score, you need to ensure that the default category bias in the project settings is changed to a nonzero value.

## Concept Extraction

Concept extraction involves the process of identifying and extracting valuable bits of information (or concepts) from text document collections. Concepts can be interesting pieces of information such as the names of entities. Commonly known entities of interest are people, places, events, organizations, etc. Similar to categories, concepts are defined in SAS Content Categorization Studio in a taxonomy structure that can be either flat (concepts only) or hierarchical (concepts and subconcepts). Concept definitions in SAS Content Categorization Studio can be broadly classified into two types: classifier and grammar.

## Classifiers

Classifiers are either literal strings or regular expressions. A literal string is a word or phrase representing the name of an entity. Classifier definitions should contain a match string followed by a mandatory comma. It can have a return string after the comma as follows: below.

```
match_string_literal,<return_string>
```

For example, a concept is defined with a series of literals indicating some key SAS products. (See Display 7.18.) A few of these classifier definitions do not have return strings. Classifier concept definitions work only when you select **Classifier** on the **Definition** tab.

**Display 7.18: Example of Classifier Concept Definition**

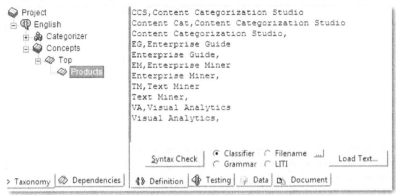

When tested against a sample text file containing information about some of these products, three classifier definitions are detected and highlighted in red (as shown in Display 7.19). Two of the matched strings, "enterprise miner" and "eg," are not exactly defined in the same case in the concept definition. Matching is possible only when you select **Case Insensitive Matching** on the Data tab for this concept. You can choose to make this the default option for the entire project. In that case, this setting applies to all concepts defined in this taxonomy. The compiled concepts binary file can be used in SAS Content Categorization Server to retrieve the return strings in the results when a match is found in the documents.

**Display 7.19: Example Showing Concept Matching with Case Insensitive Matching**

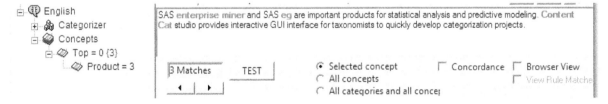

You can use the built-in disambiguation operators within the concept definitions to match concepts identified in a suitable context of interest. __TGIF is the disambiguation operator that returns a match only if the Boolean rule followed by this operator is true. __TGUNLESS works exactly the opposite way. It returns a match only if the Boolean rule followed by this operator is false. The syntax structure is in the following format:

match_string_literal,X:{Y}:<return_string>

In this format, X is either __TGIF or __TGUNLESS and Y is any Boolean rule in the syntax required by SAS Content Categorization Studio.

For example, you can disambiguate between the terms **poker game** and **poker tool** in the concept definition for the classifier concept **Game**. (See Display 7.20.) Boolean rules can include terms such as **card**, **game**, **play**, and **deck** to co-occur with **poker** in the document to return a match.

**Display 7.20: Example Showing Use of Disambiguation Operator in Classifier Concept**

## Regular Expressions

Regular expressions are used when you know the possible patterns in which you might be able to find the concepts. Examples could be phone number, SSN, website address, e-mail address, etc. The rule-writing syntax is similar to the regular expressions that you write in any scripting language such as Perl, Java, or C#. If you are not familiar with regular expressions, refer to a good online guide such as http://www.regular-expressions.info/reference.html. You can find detailed notes on how to write regular expressions in *the SAS Content Categorization Studio 12.1: User's Guide.*

Concept definitions using regular expressions should have the keyword __REGEX__ in its first line for the compiler to understand. Here are some example regular expressions to match several formats of a date concept that you commonly see in documents. You can use the special character # at the beginning of a line to indicate it is a comment, not a rule.

```
__REGEX__

# Example: January 10, 2004,

January\s*[0-9,]*\s*20[0-1]{1}[0-9]{1},

# Example: Jan. 10, 2004,

Jan[\.]*\s+[0-9,]*\s*20[0-1]{1}[0-9]{1},

# Example: 10, Jan 2004,

[0-9,]*\s*Jan[a-zA-Z]*\s*20[0-1]{1}[0-9]{1},

# Example: Jan. 09,

Jan[\.]*\s+[0-9]*[0-9][,]*\s*,

# Example: Jan. 9th, 2004,

Jan[\.]*\s*[0-9]+[ndrhst]{2}[\,\.]*\s*20[0-1]{1}[0-9]{1},

# Example: January. 9th, 2004,

January[\.]*\s*[0-9]+[ndrhst]{2}[\,\.]*\s*20[0-1]{1}[0-9]{1},

# Example: 01-10-2009,

[0-1]{0,1}[0-2]{1}\-[0-3]{0,1}[0-9]{1}\-[1-2]{1}[0-9]{1}[0-9]{1}[0-9]{1},
[0-1]{0,1}[0-2]{1}\/[0-3]{0,1}[0-9]{1}\/[1-2]{1}[0-9]{1}[0-9]{1}[0-9]{1},
[0-3]{0,1}[0-9]{1}\-[0-1]{0,1}[0-2]{1}\-[1-2]{1}[0-9]{1}[0-9]{1}[0-9]{1},
[0-3]{0,1}[0-9]{1}\/[0-1]{0,1}[0-2]{1}\/[1-2]{1}[0-9]{1}[0-9]{1}[0-9]{1},

# Example: yyyy-mm-dd,
```

```
20[0-1]{1}[0-9]{1}\-[01]{0,1}[0-9]{2}\-[0123]{0,1}[0-9]{1},
20[0-1]{1}[0-9]{1}\-[0123]{0,1}[0-9]{2}\-[01]{0,1}[0-9]{1},
20[0-1]{1}[0-9]{1}\/[01]{0,1}[0-9]{2}\/[0123]{0,1}[0-9]{1},
20[0-1]{1}[0-9]{1}\/[0123]{0,1}[0-9]{2}\/[01]{0,1}[0-9]{1},

# Example: in 2009,

in\s*20[0-1]{1}[0-9]{1},
```

Once you compile these concepts, test the rules to verify whether there were any logical mistakes that you made when writing the rules. If the rules are precise, matches found during the testing process are highlighted in red. (See Display 7.21.) It is highly recommended that you include a wide range of concept formats in the test document section and thoroughly test them. By doing this, over a period of time, you can end up with a very good compilation of rules that cover most formats of a concept.

**Display 7.21: Example Showing Test Results of Concepts Written Using Regular Expressions**

## Grammar

Using grammar concept definitions, you can specify parts of speech or use wildcard search rules to match any term. Select **Grammar** on the **Definitions** tab to write grammar concept definitions. The wildcard search symbols #w and #cap identify any words and words that begin with an uppercase letter, respectively. For example, some SAS products such as SAS Enterprise Guide, SAS Enterprise Miner, etc., can be identified by writing a match rule like the following:

#ROOT=*SASConcept

*SASConcept=SAS #cap #cap

All grammar rules should have the format #ROOT=*<ConceptName> in its first line. Part-of-speech tags can be used to capture words or phrases used with a specific sequence in a sentence. For example, the following concept definition is capable of identifying a match when the sentence contains a phrase such as "exceptionally brilliant" or "charmingly intelligent" that follows the specific sequence of an adverb followed by an adjective (:A):

#ROOT=*Concept1

*Concept1 = :Adv :A

Symbol **:Adv** represents an adverb and **:A** an adjective. Similarly, **:digit** represents a numeric value between 0 and 9. **:Det** represents a noun-modifying determinant such as "the," "a," or "an." **:PN** represents a proper noun, and so on. There are many symbols to use in concept definitions depending on the requirement. For a complete list of part-of-speech tags or match symbols, see the *SAS Content Categorization Studio 12.1: User's Guide.*

## Contextual Extraction

Contextual extraction, also known as a LITI (Language Interpretation/Text Interpretation) definition, offers many advanced linguistic capabilities that are not available with concept extraction methods. It offers many more features combining the power of categorization and concept extraction to extract complex facts. Contextual extraction capabilities are available to use only if you have SAS Contextual Extraction Studio licensed and installed with SAS Content Categorization Studio. It is always better to use a LITI definition instead of a regular concept definition when you have SAS Contextual Extraction Studio. You can start writing LITI definitions after you select **LITI** on the **Definitions** tab. There are many types of LITI definitions and each of them offers a variety of capabilities.

## CLASSIFIER Definition

The CLASSIFIER definition is similar to the classifier concept definition that we discussed earlier. It identifies word or phrase matches in a document. LITI definitions provide additional features such as coreferencing, XML field-specific matching, and exporting matches to other concepts. The basic structure of a CLASSIFIER LITI definition is CLASSIFIER:match_string_literal<,*return_string*>. (See Display 7.22.)

**Display 7.22: Example of a LITI Classifier Definition**

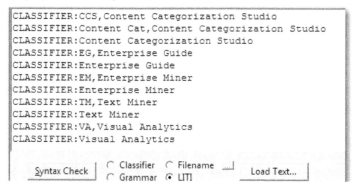

```
CLASSIFIER:CCS,Content Categorization Studio
CLASSIFIER:Content Cat,Content Categorization Studio
CLASSIFIER:Content Categorization Studio
CLASSIFIER:EG,Enterprise Guide
CLASSIFIER:Enterprise Guide
CLASSIFIER:EM,Enterprise Miner
CLASSIFIER:Enterprise Miner
CLASSIFIER:TM,Text Miner
CLASSIFIER:Text Miner
CLASSIFIER:VA,Visual Analytics
CLASSIFIER:Visual Analytics
```

Each definition should be prefixed with CLASSIFIER: to indicate that it is a classifier LITI rule. Both the comma and the return string are optional in a LITI classifier definition. If the document is an XML file, you can specify the exact fields in the file to limit matching. Specify the fields by selecting **Project ▶ Settings** on the **Misc** tab (as shown in Display 7.23). In this case, matching is limited to the **Body** and **Title** fields of the XML document by default. XML tags **Author** and **Date** are ignored from matching.

**Display 7.23: Project Settings to Ignore Specific Fields in an XML File**

| XML Default Field: | XML Tags to Ignore: |
| --- | --- |
| Body,Title | Author,Date |

If you are looking for matches to the string literal "SAS" in the Body field of the XML file, the LITI classifier definition looks similar to CLASSIFIER:_body:SAS,SAS Institute. "SAS Institute" is the return string when a match for "SAS" is found. Similar to the classifier concept definition, the return string information is useful only when the compiled LITI file (.li file) is applied to documents using SAS Content Categorization Server. The export function is useful for exporting matches to other concepts when matches are found for the classifier string defined in the concept in which the export rule is defined. For example, you can export "Text Analytics" and "SAS" to two concepts, "Domain" and "Company" respectively, only after you find a match in the document for "Content Categorization Studio." As a first step, define the concepts "Domain" and "Company" with CONCEPT: expDom and CONCEPT: expComp, respectively. Both expDom and expComp are the acronyms for the Domain and Company concepts, respectively, that are used to export matches from other rules when the export function is defined. Define a new concept named **Product** to export the concept matches for **Company** and **Product** (as shown in Display 7.24).

**Display 7.24: Example Showing LITI Classifier Definition Using the Export Function**

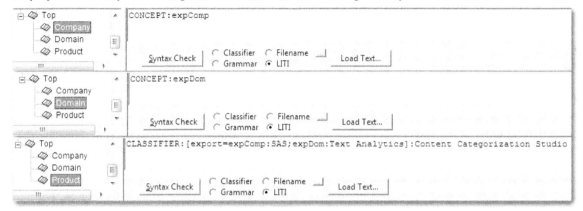

When the match string **Content Categorization Studio** is found in the document, the other matches, **SAS** and **Text Analytics** are exported to the **Company** and **Domain** concepts, respectively. (See Display 7.25.) Using this method, you can export more than one concept. Matches highlighted in red belong to the category **Product**. Matches highlighted in blue are exported to the two other concepts **Company** and **Domain** based on the export function used in the LITI classifier definition.

**Display 7.25: Results of LITI Classifier Definition Used with Export Function**

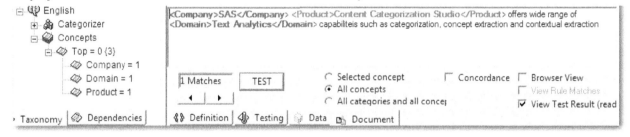

LITI classifier definitions enable you to coreference terms within a document. For example, terms such as **He** and **him** in a document might be referring to a single person. In this case, you can write a LITI definition such as CLASSIFIER:[coref=He,him]:Barack Obama. When tested on a document containing these terms, all of the three matches were returned when **Barack Obama** was found as a match (as shown in Display 7.26).

**Display 7.26: Example Showing LITI Classifier Definition Using Coreferencing Operator**

Barack Obama got elected for the second time as the president of United States. He addressed a large gathering shortly after the results were announced. Througout his speech, audience kept cheering him.

| 3 Matches | TEST |
|---|---|
| ◀ ▶ | |

- ⊙ Selected concept
- ○ All concepts
- ○ All categories and all concep

☐ Concordance  ☐ Browser View
☐ View Rule Matches

## Regular Expressions

Regular expressions in LITI definitions are similar to those used in classifier concept definitions. The syntax slightly differs in a LITI definition with the keyword "REGEX:" prefixed for each regular expression written on every line. In LITI, regular expressions are used for pattern matching after the document is broken into individual tokens or terms. Hence, a pattern embedded within another term is not returned as a match in a LITI definition. In a classifier concept definition, regular expressions work in the document before terms are tokenized. Hence, any embedded patterns are returned as a possible match.

**Example:** REGEX: \d+lbs

**Match Example:** I weigh 185lbs approximately.

**Non-Match Example:** You should be 185lbsor more.

In the case of a non-match, you can see that **185lbs** is embedded inside the token **185lbsor**. This is a match if the classifier concept definition is used instead of a LITI definition.

## CONCEPT Definition

A CONCEPT definition in LITI provides the combined capabilities of classifier and grammar concept extraction features. The CONCEPT definition enables you to reference concepts within other concepts and provide priorities in the case of overlapping concept matches. Overlapping matches are possible when the terms or phrases in a document are matched with definitions in more than one concept. For example, "Columbus" could be a reference to Christopher Columbus, the European explorer and navigator who sailed to America in the fifteenth century. However, it could also be a reference to one of the many cities in the United States with the same name (for example, Columbus in the state of Ohio). There can be scenarios where references to both Columbus, the city, and Columbus, the person, can coexist in a single document. In this case, you might need to give priority to guide concept matches whenever there is an overlap. To deal with this case, you can create concepts like the following:

**PERSON**

```
CLASSIFIER:[coref=he,His]:Columbus
CONCEPT:PRIORITY=20:Christopher Columbus
CONCEPT:PRIORITY=20:Columbus's :V
CONCEPT:PRIORITY=20:Columbus's :Npl
```

**CITY**

```
CONCEPT:States StateShrt
CONCEPT:PRIORITY=20:Columbus, StateShrt
CONCEPT:PRIORITY=20:city of Columbus
CONCEPT:PRIORITY=20:Columbus city
CONCEPT:PRIORITY=20:Columbus in States
CONCEPT:PRIORITY=20:Columbus, States
```

**States**

```
CLASSIFIER:Ohio
CLASSIFIER:Georgia
CLASSIFIER:Arizona
CLASSIFIER:Mississippi
CLASSIFIER:Montana
CLASSIFIER:Michigan
CLASSIFIER:Kansas
CLASSIFIER:Kentucky
CLASSIFIER:Indiana
CLASSIFIER:Illinois
```

**StateShrt**

```
CLASSIFIER:OH
CLASSIFIER:GA
CLASSIFIER:AR
```

```
CLASSIFIER:MI
CLASSIFIER:MT
CLASSIFIER:MS
CLASSIFIER:KS
CLASSIFIER:KY
CLASSIFIER:IN
CLASSIFIER:IL
```

LITI classifier rules **States** and **StateShrt** are defined and used as references in the **CITY** concept to qualify this rule for any of the listed states. Classifier concept **Columbus** is coreferenced to terms **he** and **His** in the PERSON concept. Part-of-speech tags for verb **:V** and plural noun **:Npl** are used to identify pattern matches in sentences referring to Columbus as a person. If Columbus is mentioned in the document as a city, higher priority is given to match it with the CITY concept. **PRIORITY=20** in the concept definitions boosts the priority (default value for PRIORITY is 10) to match with the CITY concept. Display 7.27 shows how the PRIORITY option works with the concept definitions to find the best matches within a document.

**Display 7.27: Example Showing Test Results for LITI Concept Definition Using the PRIORITY Option**

<PERSON>Christopher Columbus</PERSON> was not the first European to reach Americas. <PERSON>he </PERSON> was preceded by the Norse expedition led by Leif Ericson in 11th Century. <PERSON>His</PERSON> voyages led to the first lasting contact with Americas. <CITY>Columbus, Ohio</CITY> and <CITY>Columbus, GA </CITY> got their names in the remembrance of this great voyager.

## C_CONCEPT Definition

The C_CONCEPT definition has all of the capabilities of a CONCEPT definition with the added ability to match concepts with a context and detect partial matches. The C_CONCEPT definition uses a context marker _c to identify the words or phrases indicating the context in which it should find matches. For example, if you want to highlight or match **SAS** in the context of the company **SAS Institute Inc.**, but not SAS Scandinavian Airlines, you can write a C_CONCEPT rule like the following:

```
C_CONCEPT: _c{SAS}> Institute Inc.
```

The greater than symbol > after the context term returns partial matches for all occurrences of the term **SAS** in a document if the complete rule is found at least once in the document.

**Display 7.28: Example Showing Context-Based LITI Concept Definition**

SAS Institute Inc. recently celebrated its 35th anniversary at their world headquarters in Cary, NC. SAS is a global market leader in Business Analytics.

## CONCEPT_RULE Definition

A CONCEPT_RULE definition can be interpreted as having all of the features in the C_CONCEPT definition, plus the ability to use Boolean operators in the definition. The CONCEPT_RULE definition mandatorily requires a word or a phrase with a context marker _c and at least one Boolean operator to work. The syntax for Boolean operators is similar to what we use in the Boolean rule-based categorizer. For example, you can find matches only when certain keywords appear in the Title and Body fields of an XML document. The following simple example shows a concept rule designed to find matches only when the keywords **Abstract** and **SAS®** appear within 200 words and in the specified order. However, the context is to find cases where **SAS®** is followed by any word or name of the enterprise software with version number.

```
CONCEPT_RULE:(ORDDIST_200,"Abstract", (OR,"_c{SAS® _cap _cap :digit}",
"_c{SAS® _w}"))
```

Using this rule, you can find matches where SAS software is mentioned in the abstract of a technical paper. The context marker _c helps highlight the keyword **SAS®** and the rest of the pattern when there is a match. **SAS®**

**programs** and **SAS® Enterprise Miner 6.1** are highlighted and returned as matches when found in the abstract of a paper (as shown in Display 7.29).

**Display 7.29: Example Showing Usage of a LITI Concept Rule Definition**

Abstract

Segmentation is the process of dividing a market into groups so that members within the groups are very similar with respect to their needs, preferences, and behaviors but members between groups are very dissimilar. Marketers often use clustering to find segments of respondents in data collected via surveys. However, such data often exhibits response styles of respondents. For example, if some respondents use only the extreme ends of scales for answering questions in a survey, the clustering method will identify that group as a unique segment, which cannot be used for segmentation.
In this paper, we first discuss the different data transformation methods that are commonly used before clustering. We then apply these different transformations to survey data collected from 959 customers of a business-to-business company. Both hierarchical and k-means clustering are then applied to the transformed data. Our results show that double-standardization performs better than other transformations in eliminating groups that identify response styles. We show how double-standardization can be achieved on any data using SAS® programs and SAS® Enterprise Miner 6.1.

## NO_BREAK Definition

The NO_BREAK definition enables you to prevent partial matches where the context is not suitable for a match. For example, you can write a definition to match the name of a person while making sure it doesn't return a match when the same name is used to mention a landmark, place, or city. The following example shows how you can prevent a match return when the keywords **Boone Pickens** are followed by a noun indicating a non-match:

```
CLASSIFIER:Boone Pickens
NO_BREAK: _c{Boone Pickens :N}
```

**Boone Pickens** is returned as a match when detected as a person, but not when followed by the noun **stadium** (as shown in Display 7.30). Instead of using part-of-speech tags in the NO_BREAK definition, you can reference a CLASSIFIER LITI concept, listing all possible keywords to prevent partial matches.

**Display 7.30: Example Showing Usage of NO_BREAK LITI Definition**

Boone Pickens was born in Holdenville, Oklahoma in the year 1928. He is an American business tycoon and chairman of BP Capital Management. Boone Pickens Stadium at Oklahoma State University in Stillwater, Oklahoma received its name after him for his financial contribution to help develop the stadium.

## REMOVE_ITEM

It helps to prevent a concept match when there is an overlapping match found by a match key. This requires the Boolean operator ALIGNED and a context marker to specify the concept. For example, Wall Street is a famous landmark in New York where the New York Stock Exchange and many financial giants are located. You can indicate **Wall Street** as one of the landmark locations in New York and avoid a match for the **StreetName** concept when it is found in the document. These concepts are defined as follows:

**Landmark**

```
CLASSIFIER:Wall Street
CLASSIFIER:Times Square
CLASSIFIER:Brooklyn Bridge
CLASSIFIER:Union Square
```

**StreetName**

```
CONCEPT: _cap Street
REMOVE_ITEM:(ALIGNED,"_c{StreetName}", "Landmark")
```

In this example, **Wall Street** is recognized as a **Landmark** instead of as a **StreetName** in New York. **Broad Street** is identified as a **StreetName** (as shown in Display 7.31). Though both are valid street names in New York, the **REMOVE_ITEM** definition enabled you to avoid this overlap and give preference to the **Landmark** concept when matched.

**Display 7.31: Example Showing Usage of REMOVE_ITEM LITI Definition**

> Warren Buffet is a business man, investor and financial guru in the United States. He is also one of the wealthiest people in the world. Most of the financial giants on <Landmark>**Wall Street**</Landmark> considers him as the most successful investors of the 20th century. There are also few investment banks that have set up their operations in <StreetName>Broad Street</StreetName>.

## SEQUENCE and PREDICATE_RULE Definitions

These definitions are useful to locate and extract facts from textual data. A fact is a structured relationship between two or more concepts that is not known before analyzing documents. Concepts can be names of people, cities, locations, countries, etc. When concepts occur together in a sentence to form facts, they can be extracted using either SEQUENCE or PREDICATE_RULE definitions. SEQUENCE is useful to extract facts when the individual concepts are in a specific sequence. It requires at least two arguments to assign the returned matches based on the concepts found. For example, you can extract the names of people, their designations, and their affiliated organizations from a document without explicitly mentioning their names in the definition. Here is a list of concepts defined for this purpose:

**Organizations**

```
CLASSIFIER:SAS
CLASSIFIER:Apple
CLASSIFIER:Microsoft
CLASSIFIER:Google
```

**Designations**

```
CLASSIFIER:CEO
CLASSIFIER:Chief Executive Officer
CLASSIFIER:Chairman
CLASSIFIER:VP
CLASSIFIER:Vice President
CLASSIFIER:CTO
CLASSIFIER:Chief Technology Officer
```

**Tags**

```
CLASSIFIER:Inc.
CLASSIFIER:Incorporated
CLASSIFIER:Corporation
CLASSIFIER:Corp.
CLASSIFIER:Private Limited
CLASSIFIER:Pvt Ltd.
```

**Executives**

```
SEQUENCE:(a,b): _a{_cap _cap} is the _b{Designations of Organizations Tags}
SEQUENCE:(a,b): _a{_cap _cap}, _b{Designations of Organizations Tags}
```

In the following example, **Jim Goodnight** is returned as the first argument, and **CEO of SAS Institute Inc.** is returned as the second argument (as shown in Display 7.32). The fact that Jim Goodnight is the CEO of SAS can be extracted from the document without any prior knowledge of the person's name.

**Display 7.32: Example Showing Usage of SEQUENCE LITI Definition**

PREDICATE_RULE is much more powerful than SEQUENCE in finding facts because the order in which the concepts occur is not relevant anymore. It uses Boolean operators to find facts in the context of two or more contexts. PREDICATE_RULE avoids the hassle of writing multiple SEQUENCE definition rules to catch all of the patterns in which a fact can be found in the documents. For the same previous example, you can rewrite the definition using a PREDICATE_RULE instead of writing two SEQUENCE definitions (as shown in Display 7.33):

PREDICATE_RULE: (a,b,x,y): (SENT, "_a{_cap _cap}", "_b{Designations}", "_x{Organizations}", "_y{Tags}")

All the four arguments returned are related to each other, but there is no indication of sequence. However, PREDICATE_RULE can be written to perform a much more powerful factual extraction with the inclusion of more Boolean operators nested within each other.

**Display 7.33: Example Showing Usage of PREDICATE_RULE LITI Definition**

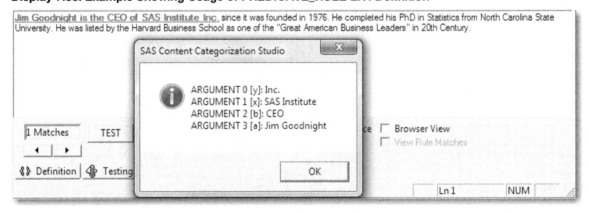

## Scoring Category Rules and Concept Definitions

As you build and test a taxonomy using SAS Content Categorization Studio, it is important that the rules defining the taxonomy are applied to all documents in the organization. Category rules, concept definitions, and LITI rules are compiled to generate the binary files (.mco) concept files (.concepts), and LITI files (.li), respectively. These files can be uploaded to SAS Content Categorization Server to allow for automatic document categorization, concepts, and complex facts extraction in real time. SAS Content Categorization Server uses the SAS Document Conversion server to convert documents in other file formats such as PDF, XML, HTML, Excel, CSV, etc., to plain text files. Using the SAS Content Categorization Server Administration Web Page, you can monitor the statistics and matches for categorization rules and concept matches. This chapter mainly focuses on building models and not on the scoring process. Refer to the *SAS Content Categorization Single User Servers 12.1: Administrator's Guide* for complete information about how to score documents using SAS Content Categorization Server.

## Automatic Generation of Categorization Rules Using SAS Text Miner

Earlier in this chapter, we mentioned how precisely written Boolean rules in a rule-based categorizer can be very efficient in the categorization process. However, this is possible only if the analyst working on building the taxonomy has significant knowledge about the document content. It would be a great benefit for the analyst to have basic blocks for these rules for each category instead of writing them from the scratch. Starting in version 12.1, SAS Text Miner offers a new feature called the Text Rule Builder that can be used to generate Boolean rules using important terms in the data to help predict the categories. A sufficient set of documents separated by category is required in a SAS data set to train the model. To train the model, assign the role Target to the variable containing the pre-identified categories, and assign the role Text to the variable holding the content of the documents. The trained model generates Boolean expressions that represent one or more terms occurring in the presence or absence of one or more terms using the Boolean operators AND, OR, and NOT. These expressions are in the same syntax as the code used in SAS Content Categorization Studio to define categories. Thus, the rules generated by SAS Text Miner can be exported to SAS Content Categorization Studio as preliminary definitions for the Boolean rule-based categorizer. They can be subsequently modified to achieve the appropriate level of accuracy and precision. Users can choose to modify the values of the target category to be required to assist the model training process and to achieve better accuracy. The important value added by this feature is its ability to show the terms explaining the reason why a document, based on its content, falls into a category.

Let's generate Boolean rules that are useful to understand and predict a target category for a set of documents. We are using a new SAS data set named SGFPAPERS_BYSECTION.sas7bdat, which contains 466 SAS Global Forum paper abstracts for the past three years from five randomly chosen sections. This data set is available by selecting **Chapters ▶ Chapter 7** in the data provided with this book. When we discussed the statistical categorizer, we used text files for these abstracts. Continue to use the SAS Enterprise Miner project that you created in Chapter 3 or create a new project. Register the data set by creating a data source. In the Data Source Wizard, step 5, change the roles of the variables as shown in Display 7.34. Variable **name** contains the unique identification number of the paper, **text** contains the abstract of the paper, and **type** contains section names in short form.

**Display 7.34: Roles and Levels of Variables in SAS Data Set SGFPAPERS_BYSECTION**

| Columns: | ☐ Label | | ☐ Mining | | ☑ Basic | | ☐ Statistics | |
|---|---|---|---|---|---|---|---|---|
| Name | Role | Level | Report | Order | Drop | Lower Limit | Upper L |
| name | ID | Nominal | No | | No | | . | |
| text | Text | Nominal | No | | No | | . | |
| type | Target | Nominal | No | | No | | . | |

Create a new diagram named TextBuilderDemo, and drag and drop the registered data source SGFPAPERS_BYSECTION onto the diagram. Drag, drop, and connect the **Data Partition**, **Text Parsing**, **Text Filter**, and **Text Rule Builder** nodes in sequence as shown in Display 7.35. Change the data set allocations for Training, Validation, and Testing to 70.0, 30.0, and 0.0, respectively, in the properties for the Data Partition node. Change the **Term Weight** option to **Mutual Information** in the Text Filter node Properties panel. Right-click the **Text Rule Builder** node, and select **Run** to execute the entire process flow.

**Display 7.35: Text Mining Process Flow to Run Text Rule Builder Node**

Once the entire process flow is completed, click **Results** to examine the output generated by the Text Rule Builder node. In the Output window, scroll down to view the distribution of predictions for all of the categories. You will see that approximately 79% of the total abstracts in the Training data set are correctly predicted by the model (as shown in Display 7.36). Similarly, about 68% are correctly classified in the Validation data set. Based on the misclassification rate, the model has performed reasonably well, although the misclassification rate increased a bit for the Validation data set compared to the Training data set.

**Display 7.36: Classification Results (from Training Data Set) from Text Rule Builder Node Output**

| Target | Outcome | Target Percentage | Outcome Percentage | Frequency Count | Total Percentage |
|---|---|---|---|---|---|
| BUSINT | BUSINT | 78.3784 | 64.4444 | 29 | 8.9783 |
| REPORTS | BUSINT | 10.8108 | 5.7971 | 4 | 1.2384 |
| SYSARCH | BUSINT | 10.8108 | 6.6667 | 4 | 1.2384 |
| BUSINT | DATAMINING | 10.4167 | 11.1111 | 5 | 1.5480 |
| DATAMINING | DATAMINING | 72.9167 | 79.5455 | 35 | 10.8359 |
| STATS | DATAMINING | 8.3333 | 3.8095 | 4 | 1.2384 |
| SYSARCH | DATAMINING | 8.3333 | 6.6667 | 4 | 1.2384 |
| BUSINT | REPORTS | 5.0000 | 6.6667 | 3 | 0.9288 |
| REPORTS | REPORTS | 88.3333 | 76.8116 | 53 | 16.4087 |
| STATS | REPORTS | 3.3333 | 1.9048 | 2 | 0.6192 |
| SYSARCH | REPORTS | 3.3333 | 3.3333 | 2 | 0.6192 |
| BUSINT | STATS | 3.9063 | 11.1111 | 5 | 1.5480 |
| DATAMINING | STATS | 7.0313 | 20.4545 | 9 | 2.7864 |
| REPORTS | STATS | 7.0313 | 13.0435 | 9 | 2.7864 |
| STATS | STATS | 75.7813 | 92.3810 | 97 | 30.0310 |
| SYSARCH | STATS | 6.2500 | 13.3333 | 8 | 2.4768 |
| BUSINT | SYSARCH | 6.0000 | 6.6667 | 3 | 0.9288 |
| REPORTS | SYSARCH | 6.0000 | 4.3478 | 3 | 0.9288 |
| STATS | SYSARCH | 4.0000 | 1.9048 | 2 | 0.6192 |
| SYSARCH | SYSARCH | 84.0000 | 70.0000 | 42 | 13.0031 |

Click the ellipsis button for the content categorization code in the Text Rule Builder node Properties panel to view the rule expressions generated by the node to predict each of the five target categories (as shown in Display 7.37). You might observe that most of the terms are intuitive and make sense for the respective sections. For example, terms such as **graphs**, **reports**, and **maps** are more appropriate for the **Reports** section. Terms such as **procedure** and **analysis** make sense for the statistical analysis section. Copy the rules of each category and build a rule-based categorizer in SAS Content Categorization Studio.

**Display 7.37: Partial Output from Automatic Content Categorization Code Generated by Text Rule Builder Node**

```
F_type =REPORTS ::
(OR
, (AND, (OR, "graphs" , "graph" ))
, (AND, (OR, "reports" , "report" ))
, (AND, (OR, "map" , "maps" ))
, (AND, (OR, "details" , "detail" )))
F_type =STATS ::
(OR
, (AND, (OR, "model" , "models" ))
, (AND, (OR, "procedure" , "procedures" ))
, (AND, (OR, "analyses" , "analysis" )))
```

Let's go back to the SAS Content Categorization Studio project in which you built the statistical categorizer using SAS Global Forum paper abstracts. Copy the automatically generated rules from the Text Rule Builder node for each category into the respective categories in the SAS Content Categorization Studio project. Click **Build ▶ Build Rule based Categorizer** to build a Boolean rule-based categorization model. Select the **BUSINT** (Business Intelligence) category, and click the **Testing** tab to view the files listed for testing the model. Click **Test** to run the model on these files to see how many passed and how many failed. You should see two files that failed the test for this model in the **BUSINT** category. After carefully examining the two files that failed, you discover that they failed the test because none of the terms defining the rules for this category were present in the document. However, the presence of terms such as **portlet**, **SAS portal**, and **cube** in the abstracts might be good candidates to represent the category. Modify the Boolean rules for this category to accommodate the new terms (as shown in Display 7.38). Rebuild the categorizer, and then test the files again.

**Display 7.38: Boolean Rules Modified in SAS Content Categorization Studio**

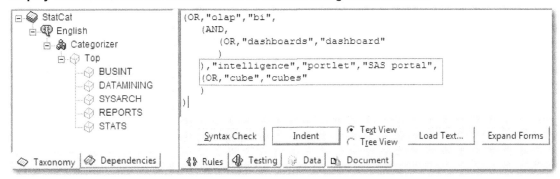

The two files that failed earlier have now passed the test by including specific terms identifiable with the Business Intelligence category in the rules. You can use the Text Rule Builder feature in combination with SAS Content Categorization Studio to build categories with a higher classification accuracy. In general, the following steps should help you as an analyst build the taxonomy for categorization:

1. Identify and define classes or categories for the taxonomy based on your knowledge about the document content. Consult with another subject matter expert if needed.
2. Collect and organize two sets of documents (Training (50 to 100) and Testing (25 to 50)) for each category that you intend to define in the taxonomy.
3. Create a project in SAS Content Categorization Studio, and define these categories.
4. Set up the Training Path and Testing Path for all of the categories.
5. Use the Generate Rules Automatically feature to derive a primitive list of linguistic terms defining the categories.
6. Modify these linguistic terms either with qualifiers or by adding weights to create a simple linguistic categorizer or a weighted linguistic categorizer, respectively.
7. Build a statistical categorizer for all of the categories in the taxonomy using the Training set.
8. Use the Text Rule Builder feature in SAS Text Miner to automatically generate Boolean expression rules for these categories.
9. Copy the generated rules from the Text Rule Builder to SAS Content Categorization Studio for all of the categories.
10. Modify the Boolean rules copied from the Text Rule Builder node after carefully analyzing the testing documents.
11. Make changes to the match ratio, relevancy type, relevancy cutoff, category bias, relevancy bias, and default category bias as required based on the type of categorizer.
12. Thoroughly test all of the categorizers that you have built so far, and modify the rules as necessary to improve the performance.
13. Identify the best categorizer that yields high recall and precision values for classification.
14. In the case of the simple linguistic categorizer, weighted linguistic categorizer, and Boolean rule-based categorizer, you can combine the Testing and Training set of documents and use all of them for testing.
15. Repeat steps 6, 10, 11, and 12 in an iterative fashion to produce reasonably performing models.

## Differences between Text Clustering and Content Categorization

In Chapter 6, we discussed how text clustering in SAS Text Miner helps in grouping documents of similar content. Although content categorization (or document classification) has a similar purpose (i.e., to organize documents of closely related content), it is not the same as text clustering. Let's review the key distinguishing characteristics between the text clustering feature in SAS Text Miner and the content categorization methods in SAS Content Categorization Studio. The differences should help you understand which feature works best for organizing documents at your organization.

**Table 7.2: Differences between Features in SAS Text Miner and SAS Content Categorization Studio**

| Text Clustering (SAS Text Miner) | Content Categorization (SAS Content Categorization Studio) |
|---|---|
| • Unsupervised learning. | • Statistical categorizer – Supervised learning, requires corpus of documents to train model.<br>• Rule-based categorizer – Unsupervised learning, requires users to write rules and test model. |
| • Categories not predefined. | • Categories should be predefined. |
| • Completely based on statistical techniques. | • Statistical categorizer based on statistical techniques.<br>• Rule-based categorizer based on natural language processing and advanced linguistic techniques. |
| • Uses latent semantic analysis and singular value decomposition for dimension reduction. | • Statistical categorizer – Uses complex statistical and algebraic methods.<br>• Rule-based categorizer – Not applicable. |
| • Cannot disambiguate between same words with distinct meanings.<br>**Example:** In the sentence, "Mr. White found guilty in the trial," **White** refers to the name of a person. In the sentence, "He came to the office dressed in white," **white** refers to a color. Both are nouns, but they represent two different entities and shouldn't be treated identically. | • Natural language processing methods provide the ability to custom build rules to disambiguate based on context.<br>**Example:** For the same example, you can write a Boolean rule to check whether a concept term suggesting any color (such white, red, blue, black, green, etc.) is preceded by a title (such as Mr., Ms., Mrs., etc.). If so, you can choose not to assign the Color category for the document. In this way, you can write exclusion rules to disambiguate the context. |
| • Concepts or entities are not exclusively included for context-based document clustering. You cannot conditionally force a document containing one or more entities to fall in a specific cluster. | • You can use concepts or entities in addition to category definitions to perform context-based category matching. You can choose to include concept definitions in category matching using one or more of the Boolean, context, frequency, and sequence operators available. |

# Summary

In this chapter, you learned how to use SAS Content Categorization Studio to create taxonomies; define categories and subcategories; build statistical, linguistic, and rule-based categorizers; write definitions to extract concepts; write LITI rules to extract complex facts; and use the Text Rule Builder feature in SAS Text Miner to automatically generate content categorization code. Organizations can leverage SAS Content Categorization Studio to organize, catalog, and efficiently maintain their document repositories. In the next chapter, we discuss how to perform sentiment analysis on documents containing user opinions about your organization's brand, product, or service in the market. We introduce SAS Sentiment Analysis Studio, a proprietary software product from SAS specifically designed to meet your organizational needs in conducting sentiment analysis.

# Appendix

```
/***********************************************************************
***************/
/**    Use this macro to create .txt files from a SAS data set.
       **/
/**
       **/
/**            The SAS data set should have two variables.
          **/
/**            1) Comment 2) Id
       **/
/**
       **/
/**    The macro will create text files with each observation as a separate
text file.    **/
/**    Each file is named with the "Id" value in the folder C:\Destination.
          **/
/**    Create a folder in C:\ and name it Destination.
             **/
/**    Before running the code, set the name of your input data set to macro
variable DATA,**/
/**    set the text variable name to macro variable COMMENT, and set the Id
variable **/
/**    name to the macro variable ID.
        **/
/**
       **/
/***********************************************************************
***************/

dm "log; clear; output; clear;";

%let DATA= <<input-data>>;
%let COMMENT= <<text variable>>;
%let ID= <<id variable>>;

%macro createtxt(data,comment,id,numvars);
%do i=1 %to &numvars;
    data _null_;
    obsnum=&i;
    length text $2500.;
        set &data. POINT=obsnum;
        file &&name&i;
        text=compbl(strip(tranwrd(&comment.,'"','')));
        put text;
        STOP;
    run;
%end;
%mend;

data _null_;
    set &data. nobs=count;

call symput("name"||left(_n_)," 'C:\Destination\"||&id.||".txt' ");

  if _n_=1 then call symput("numvars", trim(left(put(count, best.))));
run;

%createtxt(&data,&comment,&id,&numvars);
```

# References

Albright, R. 2004. "Taming Text with the SVD."

Available at: ftp://ftp.sas.com/techsup/download/EMiner/TamingTextwiththeSVD.pdf [WebCite Cache].

Antonie, M. L., & Zaiane, O. R. 2002. "Text Document Categorization by Term Association." *ICDM '02: Proceedings of the 2002 IEEE International Conference on Data Mining.* IEEE, 19-26.

IDC. 2009. *Hidden Costs of Information Work: A Progress Report*, Doc #217936.

Jackson, P., & Moulinier, I. 2007. *Natural Language Processing for Online Applications: Text Retrieval, Extraction and Categorization,* Vol. 5. Philadelphia: John Benjamins Publishing.

Joachims, T. 2002. *Learning to Classify Text Using Support Vector Machines: Methods, Theory, and Algorithms.* Boston: Kluwer Academic Publishers.

Kaski, S., Honkela, T., Lagus, K., & Kohonen, T. 1998. "WEBSOM–Self-Organizing Maps of Document Collections." *Neurocomputing.* 21(1-3): 101-117.

Mavroeidis, D., Tsatsaronis, G., Vazirgiannis, M., Theobald, M., & Weikum, G. 2005. "Word Sense Disambiguation for Exploiting Hierarchical Thesauri in Text Classification." *Knowledge Discovery in Databases: PKDD 2005.* . Berlin: Springer-Verlag, 181-192.

Pottenger, W. M., & Yang, T. H. 2001. "Detecting Emerging Concepts in Textual Data Mining." *Computational Information Retrieval.* Philadelphia: Society for Industrial and Applied Mathematics, 89-105.

Reamy, T. 2010. Knowledge Architecture Professional Services. "Enterprise Content Categorization – The Business Strategy for a Semantic Infrastructure [White paper]." Retrieved from

http://www.kapsgroup.com/presentations/ContentCategorization-Business%20Value.pdf

SAS® Content Categorization Single User Servers 12.1: Administrator's Guide. Cary, NC: SAS Institute Inc.

SAS® Content Categorization Studio 12.1: User's Guide. Cary, NC: SAS Institute Inc.

SAS® Enterprise Content Categorization 12.1: User's Guide. Cary, NC: SAS Institute Inc.

Schütze, H. 1998. "Automatic Word Sense Discrimination." *Computational Linguistics.* 24(1): 97-123.

Sebastiani, F. 2002. "Machine Learning in Automated Text Categorization." *ACM Computing Surveys (CSUR).* 34(1): 1-47.

Yang, Y. 1999. "An Evaluation of Statistical Approaches to Text Categorization." *Information Retrieval.* 1(1-2): 69-90.

Yang, Y., & Joachims, T. 2008. "Text Categorization." *Scholarpedia.* 3(5): 4242.

# Chapter 8 Sentiment Analysis

## Introduction

*In this age, in this country, public sentiment is everything. With it, nothing can fail; against it, nothing can succeed. Whoever molds public sentiment goes deeper than he who enacts statutes, or pronounces judicial decisions.*
                                                      Abraham Lincoln, 1858

It is clear from President Lincoln's famous quote that politicians realized the power of public sentiment a long time ago. Although keeping track of the public's opinions and sentiments about topics and events of general interests has been in the domain of political and social scientists, marketers and businesses are typically interested in a consumer's opinions and sentiments that specifically relate to their products, services, and so on. Monitoring what consumers are saying about a company's brands and products and how they are expressing their opinions and sentiments to others has always been important to businesses. The difference now is the scale and scope of a consumer's opinions and expressions and the toolset that is available to marketers compared to what existed just a few years ago. Until the last century, businesses typically used surveys and focus groups from time to time to gauge and track consumer sentiments. With the widespread adoption of the Internet, the proliferation of social media channels (such as Twitter, Facebook, and others), and the abundant opportunity for consumers to express their opinions and sentiments, monitoring sentiment continuously has become more critical. Conceptually, there is a difference between asking people questions about their feelings and sentiments and getting answers versus people expressing their opinions and sentiments freely without any prompting. Although companies still continue to capture a customer's opinions and sentiments via regularly scheduled surveys and focus groups, those methods need to be supplemented by what is freely expressed by consumers on the Internet, particularly in social media.

"Conventional marketing wisdom long held that a dissatisfied customer tells ten people; but in the age of new social media, he or she has the tools to tell millions," says Paul Gillin, author of *The New Influencers: A Marketer's Guide to the New Social Media*. Therefore, it becomes imperative for companies to actively monitor conversations on the web among consumers about a company's brands, products, or services, and to glean information about what positive and negative sentiments are being expressed. This is a challenging task for many reasons, such as too many sites on the Internet, too many formats in which opinions and sentiments are stored and displayed on the web, idiosyncrasies in expressed opinions that are both time and domain dependent, and so on. In this chapter, sentiment analysis is used as an umbrella term that covers the analysis of people's expressed opinions, expressions, emotions, and sentiments about brands, products, services, events, topics, and their attributes (Bing Liu 2012). Sentiment analysis (sometimes referred to as opinion mining in academic literature) is a field of study that originated in computer science, but has gained traction in business and marketing literature. In the past few years, many applications of sentiment analysis have been published using a variety of textual data and business contexts, such as tweets about movies (Castellanos et al. 2011), reviews about movies (Joshi et al. 2010), tweets about events related to retailers (Duraidhayalu et al. 2012; Grover et al. 2013), and reviews about products and services (Liu et al. 2013; Nagarajan et al. 2013; Pantangi et al. 2013; Sarkar et al. 2013).

## Basics of Sentiment Analysis

The basic task involved in sentiment analysis is identifying and quantifying the polarity or valence of sentiments (such as positive, negative, neutral, or mixed) expressed typically in written opinions, expressions, reviews, comments, and so on. Therefore, it involves many of the text analytics steps such as tokenization, sentence identification, part-of-speech tagging, and so on, that have been discussed in previous chapters. But, sentiment analysis has to go beyond the basic steps in text analytics. For example, often tweets, reviews, and postings have different types of sentences (declarative, imperative, and interrogative) that must be identified because they might not be appropriate for sentiment analysis. For example, a declarative sentence (such as "It is an amazing TV.") often states the views of the author and is appropriate for sentiment analysis. An imperative sentence (such as "Do not buy this TV.") can be used to infer sentiments about the product. However, an interrogative sentence (such as "Which is the best TV?") might not be a good candidate for sentiment analysis. Depending on the context, sentences can be non-comparative (where opinion is restricted to one thing) or comparative (where multiple things might be compared). Comparative sentences are more difficult to handle than non-comparative sentences. Just as in text analytics, sentiment analysis requires the analyst to spend considerable time preprocessing text and making a large number of subjective judgments to get meaningful results.

Sentiment analysis starts with determining whether a text contains an opinion (sentiment). If it does contain sentiment, at what granularity level does the sentiment exist? Consider the following example of a review by a customer for a TV:

> The TV is wonderful. Great **size**, great **picture**, easy **interface**. It makes a cute little song when you **boot** it up and when you shut it off. I just want to point out that the 43" does not in fact play videos from the USB .This is really annoying because that was one of the major perks I wanted from a new TV. Looking at the product description now, I realize that the feature list applies to the X758 series as a whole, and that each model's capabilities are listed below. Kind of a dumb oversight on my part, but it's equally stupid to put a description that does not apply on the listing for a very specific model.

Many questions come to mind as you read the review. There are some statements that are clearly subjective and contain an opinion, and other statements are just facts. Considering the opinion statements, is the overall sentiment (in the entire review) about the TV expressed by the customer positive, negative, or mixed? What about the sentiment for attributes of the TV, such as size, picture, and interface? What about the video-playing ability of the TV from the USB? A clear understanding of the level of analysis and the terms used to specify these levels is needed. Liu (2012) defined three granularity levels for any sentiment analysis problem as follows:

*Document level:* At this level, the task is to figure out whether the entire document can be classified as positive or negative. This is possible only if the document involves a single entity (such as the TV in the previous example). If a document involves multiple entities, then it might not be possible to arrive at a document-level sentiment classification.

*Sentence level:* At this level, the task is to classify each sentence in a document as a positive, negative, or mixed sentiment sentence. In the previous example, the first sentence, "The TV is wonderful." expresses positive sentiment. The third statement, "I just want to point out that the 43" does not in fact play videos from the USB."

expresses negative sentiment. Depending on the nature of the text, some sentences might not express any sentiment and therefore cannot be classified.

*Entity (or Object) and Attribute (or Aspect or Feature) level:* An entity is typically the target of the opinion. In the previous example, the target of the review is clearly the TV, which is the entity for the entire review. However, in many sentences, the sentiments reflect the reviewer's opinions about attributes (or aspects or features) of the entity. For example, in the second sentence, ("Great size, great picture, easy interface."), positive sentiment is being expressed for three specific attributes (size, picture, and interface) of the entity, the TV.

Often a hierarchical taxonomy is used to represent an entity and its various attributes. Sentiments can then be applied to each attribute. In this example, the TV is the root entity (object). Each non-root attribute (size, picture, and interface) is a component or subcomponent of the root entity or object. You can express an opinion about the root node or the object (TV) or any components or subcomponents. In SAS Sentiment Analysis Studio, the term "feature" is used instead of attribute and the term "product" is used instead of entity. These terms are used hierarchically in the sense that features are nested in a product. However, in SAS Sentiment Analysis Studio, it is possible to define intermediate entities that frequently occur in the documents so that they can be referenced and used again in the same analysis.

## Challenges in Conducting Sentiment Analysis

Sentiment analysis starts with text data. As a result, all of the typical natural language processing (NLP) problems associated with analyzing text data, such as correctly identifying part-of-speech tags, disambiguating terms and lexicons, correcting spelling errors, and so on, can plague sentiment analysis. The commonly used lexicon-based sentiment analysis approach depends on correctly identifying opinion words that express positive or negative sentiments in a sentence. There are general opinion words whose polarity is always the same, independent of the context. An example is the word "beautiful," which expresses a positive sentiment. These are usually easier to handle. But, there are also context-dependent lexicons in which the polarity of the word depends on the domain or context. For example, the word "small" can be positive or negative depending on the context. The sentence, "The size seems *small*." can be positive for a USB flash drive with 1 TB capacity. But, the same sentence can be interpreted as negative if the context is an LED big-screen TV. In addition to opinion words, there are idiom lexicons—typical expressions such as "costs an arm and a leg" that embody sentiments. In general, the difficulty with correctly identifying sentiments increases as you move from general to context-dependent to idiom lexicons in texts.

Other challenges in conducting sentiment analysis arise from the nature of the text that is being analyzed. For example, tweets are short, and they are typically focused on one topic only. In that sense, they are easier to analyze. But, tweets often contain a lot of special meaning characters, such as RT (retweets), hashtags (#), emoticons (such as smiley faces), that need to be handled carefully. Customer reviews are typically on one entity or object. Therefore, there is less ambiguity in the entity detection task when analyzing reviews. Analysis of discussions, free-flow comments, and blog postings is often the hardest because they typically cover multiple entities, make comparisons instead of expressing direct opinions, use a lot of sarcasm, etc.

## Unsupervised versus Supervised Sentiment Classification

The sentiment classification task can be formulated as a supervised or an unsupervised classification problem, depending on whether there are known examples of documents belonging to positive or negative sentiments. Unsupervised sentiment classification involves the application of a sentiment lexicon of opinion-related positive or negative terms to evaluate text in the document. On the other hand, supervised sentiment classification applies machine-learning algorithms (such as support vector machines (SVMs) and neural networks) to textual feature representations to derive the relationships between features of the text segment and the opinions expressed in the document. In many practical situations, known class examples are created by experts who read the documents or use rules. Then, if a text review's numeric rating is four or more stars, then the review is positive. If no known class examples are possible, then analysts have to use an unsupervised classification of sentiments.

Supervised classification is typically performed at the document level. If enough labeled examples are available, any commonly used classification models can be trained, validated, and tested to check their performances. Published research on the topic of model performance in supervised classification is often based on product review data, which typically has a text review and an overall numeric rating on a scale of one to five stars. Often, a review rating of four to five stars is considered a positive rating, and a review rating of one to two stars is considered a negative rating. Many researchers have shown that the naïve Bayes classifier and (SVMs) perform reasonably well for the supervised classification task (Pang et al. 2008). The main challenge for modelers is to carefully select the inputs from text features such as terms and their frequencies (often weighted or normalized), part-of-speech tags, opinion lexicons (general, context-specific, and idiom), syntactic dependency (from parsing trees), and the handling of negation words (such as "not").

Unsupervised classification is typically performed at the sentence level. There are two types of unsupervised classification methods: lexicon-based and syntactic-pattern based. The lexicon-based approach can be used for sentence- and aspect-level sentiment classification. It uses lexicons (opinion words) and involves identifying which opinion words are related to which attributes in a sentence. The relationships between opinion words and attributes are identified via dependency relationships obtained through parsing. For example, in the sentence, "The picture quality is outstanding," the opinion word "outstanding" and the attribute "picture quality" share the same dependency relationship with the verb "is." Consequently, the sentence is deemed to be expressing a positive sentiment about picture quality. If a clear dependency is not observed between an opinion word and an attribute, then how close an opinion word is to an attribute in a sentence can be used to judge the polarity of the attribute. This process can get very complex, depending on how long the sentence is, how many attributes are being mentioned in the same sentence, whether both positive and negative polarity words are used in the same sentence, whether negation is used, and so on. Once sentiment values are computed for each word-attribute combination, they are typically combined using appropriate normalization or weights to come up with an overall sentiment score. On the other hand, a syntactic pattern-based approach involves defining part-of-speech tags and the keywords AND, NOR, OR, NOT, BUT, etc. Primarily useful in contextual analysis when performing phrase-level analysis, this method can be used to develop a variety of rules for better accuracy. For example, a simple pattern such as <subject> <NOT> <verb> can be used to extract negative phrases like, *"This <feature> does <not> < work> as advertised."* We delve deep into this approach in the rule-based models section later in this chapter.

## SAS Sentiment Analysis Studio Overview

SAS Sentiment Analysis Studio is a comprehensive solution to the multifaceted challenges of analyzing sentiment in input documents. It enables users to automatically evaluate the opinions and feelings expressed about an entity (or object, person, event, or experience), either at the entity level or at the attribute (or feature) level. Generally speaking, SAS Sentiment Analysis Studio can use statistical models (such as naïve Bayes classifiers), rule-based models (using advanced linguistic technologies), or a combination of statistical and rule-based (hybrid) models.

Here is the architecture of SAS Sentiment Analysis Studio:

**Display 8.1: Architecture**

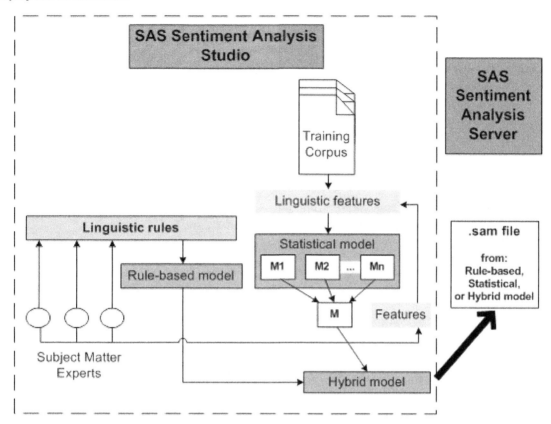

The training corpus typically consists of documents that have known class examples of sentiments (such as positive, negative, neutral, and unclassified). These are often compiled by domain experts. The software first extracts linguistic features from the training corpus. Then, using those linguistic features as inputs and known class examples as targets, the software trains and validates several statistical models and identifies the best model. The linguistic features are used by subject matter experts to write linguistic rules that are used in training and validating rule-based models. Hybrid models are created and tested by combining rule-based and statistical models. Any of the developed models can be used to score new documents.

If you also have a license for SAS Sentiment Analysis Workbench, you can use the software to create visually appealing graphs and charts to track sentiment over time and to customize search or drill down to explore sentiments at a very granular level.

## Statistical Models in SAS Sentiment Analysis Studio

Version 12.1 of SAS Sentiment Analysis is used in this chapter. Some screenshots of the tool might differ slightly in later versions. The default settings for the statistical model (displayed at the start of the project) are shown below. By default, all of the options are selected. These options refer to and use different system files (such as tokenizers, tagger, extraction, and stop words filter). Of these, the stop words filter file is a text (.txt) file that can be easily edited by users to add customized stop words based on the project context. The words in the stop words file will not be used by the statistical models in this project.

**Display 8.2: Statistical Model Settings**

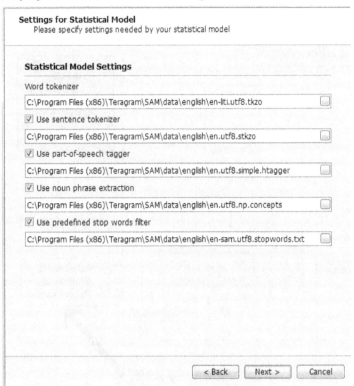

To build statistical models, known class examples of documents must be available. These are typically stored as text (.txt) files in separate folders for each class (such as positive, negative, neutral, or unclassified). At this time, only positive and negative sentiments can be classified via the statistical model. However, it is possible to import the training documents that express ambivalent opinions into the neutral folder, or to import the documents that have not been reviewed by domain experts into the unclassified folder. A typical folder and file structure is shown in Display 8.3.

**Display 8.3: Folder and File Structure**

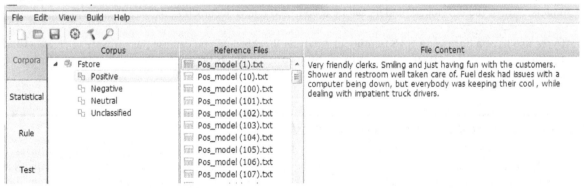

The statistical models are grouped under two broad sets of options: simple and advanced. Regardless of the option that you select, the algorithm, by default, uses 80% of the data for training and 20% for validation. The percentages of training and validation data can be changed by the user based on what's appropriate in the context of the analysis. By default (if the best option is not selected), four statistical models are built using smoothed relative frequency text normalization combined with four versions of a feature-ranking algorithm in the simple model scenario. If the best option is selected, then 16 models are built in the simple model scenario by combining four different text normalization methods (relative frequency, smoothed relative frequency, Okapi

BM25, and pivoted length normalization) with four versions of a feature-ranking algorithm (none, risk ratio, chi-square, information gain).

The idea behind text normalization is that instead of using just the raw frequency of feature words, the algorithm uses a weighted relative index of frequency that takes into account the variability in document lengths and the number of feature words per document. The four different text normalization methods vary in terms of how they correct for document length and the number of feature words per document. The none feature-ranking algorithm uses an approach based on Bayes' theorem to derive the weights of the feature words during the training process. The risk ratio, chi-square, and information gain feature-ranking algorithms modify the weights calculated via the Bayesian model by using different frequency ratios of positive and negative training examples and how the feature words appear or do not appear in them. The exact calculations and formulas for these options are the intellectual property of SAS and are not shared in the product documentation.

The training results from any model are displayed both textually and graphically. To understand the model statistics, consider the 2x2 table of Actual (Positive and Negative) versus Predicted (Positive and Negative) values from the model.

| Predicted | Actual | |
|---|---|---|
| | Positive | Negative |
| Positive | True Positive | False Positive |
| Negative | False Negative | True Negative |

The models are evaluated using three precision estimates calculated on the validation data. Positive precision is calculated as the ratio of true positives to true positives and false positives. Negative precision is calculated as the ratio of true negatives to true negatives and false negatives. Overall precision is calculated as the ratio of true positives and true negatives to total documents. The model with the highest overall precision in the validation data is selected as the best model (as shown in Display 8.4).

**Display 8.4: Simple Model Text Results**

**Display 8.5: Simple Model Graphical Results**

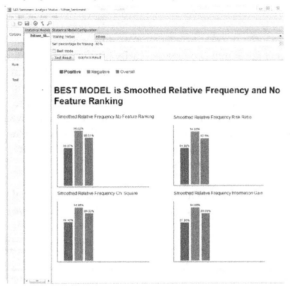

The advanced statistical model enables users to have more control over the model building process. The basic solution uses the Bayes method, but users can choose any one of the four text normalization algorithms (relative frequency, smoothed relative frequency, Okapi BM25, and pivoted length normalization) that was mentioned in the simple models. The Okapi BM25 uses normalization weights derived from a 2-Poisson probabilistic model. The pivoted length normalization adjusts the weights based on the association between document length and probability of relevance from the training collections.

The advanced model interface (as shown in Display 8.6) enables users to use other options such as probability threshold, the use of the contextual extraction file, and the use of run-time stop words. The probability threshold is a cutoff value used to classify a document as positive or negative during the validation step. When SAS Sentiment Analysis Studio has a document, the text is assigned a probability score. The value of this score is between 0 and 1. Any document that has a probability score above the threshold is considered to be positive. A probability score that falls below the threshold is considered to be negative. A document that has a score equal to the threshold is considered to be neutral. The contextual extraction file option enables users to use a noun-phrase file for languages such as Chinese or Japanese. Or, users can use an .li file that contains concepts extracted from SAS Contextual Extraction Studio. The .li file, if specified, is applied to the training documents, and the concepts that are extracted are used as additional features to train the model. The run-time stop words option enables users to specify a customized .txt file that contains additional stop words that are ignored during the analysis.

**Display 8.6: Advanced Model User Interface**

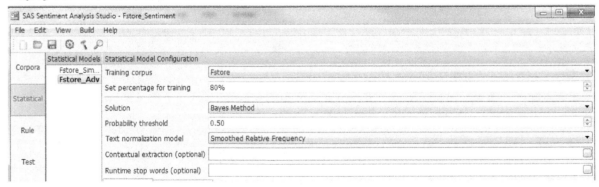

Once a model is created (either simple or advanced), it can be tested on documents that have not been used in training or validating the model. Testing can be done at the directory level (i.e., on a folder containing all of the test documents as separate .txt files) or at the individual document level. If the documents have known class sentiments, then testing can reveal how well the built model performs in classifying unseen documents. If the sentiments of the documents are unknown, then it becomes a scoring situation. Scoring is best achieved using SAS Sentiment Analysis Server. It takes the compiled (trained, validated, and tested) model from SAS Sentiment Analysis Studio and scores new input documents. The output documents from scoring can be passed onto SAS Sentiment Analysis Workbench. The SAS Sentiment Analysis Workbench interface can create graphs and charts to track sentiments over time. It allows users to drill down to get deeper insights into sentiments across different input document sources, categories, and so on. SAS Sentiment Analysis Workbench can score documents. When documents are added to a project in SAS Sentiment Analysis Workbench, they are scored using the model assigned to that project.

The testing interface (as shown in Display 8.7) provides several options to customize the testing process and results. The first option, **Default test type**, allows users to select either a statistical or rule-based model. The **Probability threshold** is a cutoff value used to classify a document as positive or negative. This is the same concept explained earlier in the advanced statistical model options. The **Activated statistical model** allows users to select a specific model from statistical models that might have been trained and validated in the current project. The **Relative weight of positive rules in rule-based model** determines how much importance is assigned to positive rules compared to negative rules in calculating an overall sentiment score for each document. The default value of 100% means both positive and negative rules are equally weighted. If you have built a hybrid model (a combination of a statistical and a rule-based model), the **Weight of statistical model in hybrid model** option enables you to designate the relative importance of the statistical model over the rule-based model.

### Display 8.7: Testing Interface

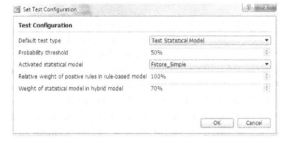

The testing results at the directory (folder) level can be displayed either as text or graphics as shown in Display 8.8. The +, -, and * signs in parentheses to the left of the individual documents in the directory indicate whether the document is classified as having positive, negative, or neutral sentiments during the testing process.

### Display 8.8: Results of Testing at the Directory Level

Users can select and test particular documents to understand the terms that were important (shown in boldface) in figuring out the sentiment score for the document. In addition, highlighting any boldface term in the text displays the likelihood of positive or negative sentiment associated with that term as shown in Display 8.9.

**Display 8.9: Results of Testing at the Document Level**

---

# Rule-Based Models in SAS Sentiment Analysis Studio

The default settings for the rule-based model (displayed at the start of the project) are shown in Display 8.10. Although some of the options are similar to the options in statistical models (such as tokenizers and tagger), there are some new options as well. Some of the new options are needed because of the additional complexity in developing rule-based models that explicitly work on linguistic features in the documents compared to statistical models that are built using numerical data only. The advantage of the rule-based model over the statistical model is its ability to determine sentiment at a more granular level such as product features or attributes.

**Display 8.10: Rule-Based Model Settings**

To develop a good rule-based model, analysts must have some domain expertise and be willing to write, test, and modify different types of rules such as CLASSIFIER, CONCEPT, C_CONCEPT, CONCEPT_RULE, PREDICATE_RULE, and REGEX. Each of the different rule types enable different user capabilities such as the following:

- Match specific words or strings.
- Match any word using _w and _cap markers.
- Reference parts of speech.

- Reference defined entities with _def marker.
- Use Boolean operators such as AND or SENT.
- Use regular expressions to match patterns of characters and digits.

All of these rule types are compatible with the definitions used in SAS Content Categorization Studio. The *SAS Sentiment Analysis Studio: User's Guide* provides detailed instructions and examples of each type of rule. A brief description of the rule types and their typical use is given below.

## CLASSIFIER Rule

This rule consists of a word or string. When a match occurs with this rule, the entire string is highlighted in the input document. This rule is typically used in the definitions of tonal keywords to locate the terms that express sentiment.

## CONCEPT Rule

This rule specifies matched and highlighted terms the same way as the CLASSIFIER rule. Add the _def marker to locate referenced concepts in an input document. Also, specify part-of-speech tags and the _cap and _w markers to locate types of terms.

## C_CONCEPT Rule

This rule is similar to a CONCEPT rule, but it is used to specify a coreference. Use a C_CONCEPT rule to locate matches within a specified context.

## CONCEPT_RULE Rule

Like the C_CONCEPT rule, the _c marker is required for a CONCEPT_RULE. However, a CONCEPT_RULE also requires Boolean operators. Use the _def marker and other markers to specify the matches that are returned by this rule.

## PREDICATE_RULE Rule

This rule is the most complex and exact of the rule types. You can specify arguments that return highlighted matches in a tested document. These matches include all of the strings between the first and last located arguments.

## REGEX Rule

Use this type of rule to locate information that follows a preset pattern such as price, percent, and various word spellings. Do not specify any of the Boolean operators or markers with this rule.

Here is a general approach that is recommended for developing these rules. First, review a sample of documents to understand how products and features are mentioned and to identify how sentiments are expressed at the product or feature level in the documents. Use knowledge gained from this task to identify product features and tonal keywords that are basic building blocks for rule-based models. Build a statistical model (starting with the default options), and use that model as a starting point for identifying keywords in your documents and for importing keyword-based classifier-type rules from the statistical model. Use your judgment and domain expertise to write different types of rules and, more importantly, to test and refine those rules. You can use SAS Text Miner for the preliminary analysis of the textual corpus to identify themes and topics as explained in the next section. Case Study 3 demonstrates the advantage of performing text mining analysis as a precursor to sentiment analysis. Display 8.11 shows the SAS Sentiment Analysis Studio interface with tonal keywords depicting different types of rules and weights, intermediate entities, and products and features for one of the case studies in this book. The weights reflect the relative degree of sentiment expressed. In this example, the word **exceptional** has a higher weight for positive sentiment than the word **favorite**.

**Display 8.11: Rule-Based Model Interface**

| | | | Type | Body | Weight |
|---|---|---|---|---|---|
| | | 1 | CONCEPT | love@ | 0.17 |
| | | 2 | CONCEPT | outstand@ | 0.17 |
| | | 3 | CLASSIFIER | great | 0.16 |
| | | 4 | CLASSIFIER | best | 0.16 |
| | | 5 | CLASSIFIER | cleanest | 0.16 |
| | | 6 | CLASSIFIER | Awesome | 0.15 |
| | | 7 | CLASSIFIER | exceptional | 0.15 |
| | | 8 | CONCEPT | help@ | 0.14 |
| | | 9 | CLASSIFIER | favorite | 0.14 |
| | | 10 | CLASSIFIER | exceptionally | 0.14 |
| | | 11 | CLASSIFIER | wonderful | 0.14 |
| | | 12 | CLASSIFIER | Greatly | 0.14 |
| | | 13 | CLASSIFIER | friendliest | 0.14 |
| | | 14 | CLASSIFIER | brand new | 0.14 |
| | | 15 | CLASSIFIER | enjoy | 0.14 |

To a large extent, these weights need to be decided subjectively by the analyst, and then modified based on the results of rule evaluation and testing. The imported rules from the statistical model are best treated as a starting point only. To start modifying the weights, analysts need some idea about the degree of sentiment in words such as "outstanding" versus "good." If "outstanding" is believed to be more positive than "good," then higher weights are assigned to "outstanding" than "good." Although native English speakers might have some idea about the degree of sentiment in common English words, that knowledge is often inadequate for assigning relative weights to all the keywords in the documents. Often, analysts need to turn to other resources where many English words have been categorized based on their valence and tagged properly. (For a comprehensive listing of free resources for research purposes, see http://www.wjh.harvard.edu/~inquirer/homecat.htm, http://mpqa.cs.pitt.edu/, and http://sentiwordnet.isti.cnr.it/.)

Once the rule-based model is developed and validated, it can be used to test new documents the same way the statistical model does. Test results can be viewed at the directory level (in textual or graphical format) or at the individual document level as shown in Display 8.12. The results are more detailed than those from the statistical model because the sentiments are evaluated at the product level (in this case, **Fstore**) as well as at the attribute or feature level (such as **shower**, **parking**, and so on).

**Display 8.12: Graphical Results of Testing a Rule-Based Model at the Directory Level**

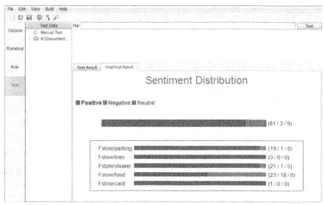

Testing each document separately provides detailed results with respect to rules, products, and features. The matched words in the text are displayed in different colors (such as green for positive, red for negative, blue for products and features), which makes the results more attractive and easier to understand.

**Display 8.13: Results of Testing a Rule-Based Model at the Document Level**

# SAS Text Miner and SAS Sentiment Analysis Studio

SAS Sentiment Analysis Studio is a stand-alone product that is designed to handle all of the tasks needed to run sentiment analysis on a corpus of documents. However, applying SAS Text Miner on the same corpus helps analysts get better insight about the corpus and create better sentiment analysis models. In particular, the Text Parsing node results are very useful to get a handle on all of the terms and concepts in a corpus. Both the Text Topic and Text Cluster node results help in identifying concepts, terms, and features that can be used directly as concepts or as intermediate entities in sentiment analysis. For example, the output from the Text Topic node shown in Display 8.14 shows some of the terms that are good candidates for tonal keywords and concepts in sentiment analysis.

**Display 8.14: Results of Text Topic Node**

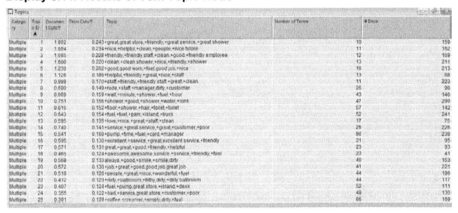

The Text Rule Builder node can be used to build Boolean rules using terms in the corpus for predicting a target variable. If known class examples of positive versus negative documents are used to train and validate the Text Rule Builder node, and a good model can be found, then the Boolean rules from this node can be used to define and modify tonal keywords in the sentiment analysis. For example, the results (as shown in Display 8.15) show a reasonably good model from the application of the Text Rule Builder node in one of the case studies. The overall misclassification rate is about 21.5% in the validation data. More importantly, the model correctly predicts about 79.8% of negative documents and 75.8% of positive documents in the validation data.

**Display 8.15: Classification Results from Text Rule Builder Node**

```
90    Classification Table
91
92    Data Role=TRAIN Target Variable=Type Target Label=' '
93
94                        Target      Outcome    Frequency    Total
95    Target    Outcome   Percentage  Percentage  Count    Percentage
96
97    NEGATIVE  NEGATIVE    83.2558    88.6139      895     53.3055
98    POSITIVE  NEGATIVE    16.7442    26.9058      180     10.7207
99    NEGATIVE  POSITIVE    19.0397    11.3861      115      6.8493
100   POSITIVE  POSITIVE    80.9603    73.0942      489     29.1245
101
102
103   Data Role=VALIDATE Target Variable=Type Target Label=' '
104
105                        Target      Outcome    Frequency    Total
106   Target    Outcome   Percentage  Percentage  Count    Percentage
107
108   NEGATIVE  NEGATIVE    79.8535    85.8268      218     51.6588
109   POSITIVE  NEGATIVE    20.1465    32.7381       55     13.0332
110   NEGATIVE  POSITIVE    24.1611    14.1732       36      8.5308
111   POSITIVE  POSITIVE    75.8389    67.2619      113     26.7773
112
```

The Boolean rules output from the Text Rule Builder node (See Display 8.16) shows terms that might be good candidates for tonal keywords, concepts, and features for use in SAS Sentiment Analysis Studio.

**Display 8.16: Boolean Rules from Text Rule Builder Node**

## Summary

Sentiment analysis plays an important role in understanding a customer's experience with a product or service. In this chapter, you learned how SAS Sentiment Analysis Studio can be used to build sophisticated models to classify a consumer's opinion or feedback toward an entity or attribute of an entity as positive, negative, or neutral. Using known examples of classes of sentiment, a quick sentiment analysis of data can be performed by building a statistical model. A rule-based model that uses advanced linguistic technologies can be built by creating different types of rules: CLASSIFIER, CONCEPT, C_CONCEPT, CONCEPT_RULE, PREDICATE_RULE, and REGEX. Rule-based models provide the enhanced capability to extract sentiment at a more granular level: as a sentence or product feature. A hybrid model can be created by integrating the results from a statistical model into a rule-based model, which can yield better results in many situations. Similarly, a better sentiment analysis model can be built by using the results from the SAS Text Miner Text Rule Builder node. The Boolean rules from the Text Rule Builder node can be a good starting point for developing tonal keywords, concepts, and features for use in SAS Sentiment Analysis Studio. Case Study 3 and Case Study 6 discuss in detail how to use SAS Sentiment Analysis Studio with SAS Text Miner to build better sentiment analysis models.

# References

Castellanos, M. Dayal, U., Hsu, M., Ghosh, R., Dekhil, M., Lu, Y., Zhang, L., and Schreiman, M. 2011. "LCI: A Social Channel Analysis Platform for Live Customer Intelligence". *Proceedings of the 2011 ACM SIGMOD International Conference on Management of Data,* 1049-1058.

Duraidhayalu, H. Garla, S., and Chakraborty, G. 2012. "Analyzing Sentiments in Tweets about Wal-Mart's Gender Discrimination Lawsuit Verdict Using SAS® Text Miner". *Proceedings of the SAS Global Forum 2012 Conference,* 306.

Grover, S., Jacob, J., and Chakraborty, G. 2013. "Analysis of Change in Sentiments towards Chick-fil-A after Dan Cathy's Statement about same sex Marriage Using SAS® Text Miner and SAS® Sentiment Analysis Studio". *Proceedings of the SAS Global Forum 2013 Conference,* 251.

Joshi, M. Das, D., Gimpel, K., and Smith, N. A. 2010. "Movie Reviews and Revenues: An Experiment in Text Regression", *Human Language Technologies: Proceedings of the 2010 Annual Conference of the North American Chapter of the ACL,* 293-296.

Liu, B. 2012. Sentiment Analysis and Opinion Mining (Synthesis Lectures on Human Language Technologies), Morgan & Claypool.

Liu, J., Sarkar, M., and Chakraborty, G. 2013. "Feature-Based Sentiment Analysis on Android App Reviews Using SAS® Text Miner and SAS® Sentiment Analysis Studio". *Proceedings of the SAS Global Forum 2013 Conference,* 250.

Meng, Yanyan. 2012. "Sentiment Analysis: A Study on Product Features", University of Nebraska – Lincoln. http://digitalcommons.unl.edu/cgi/viewcontent.cgi?article=1031&context=businessdiss

Nagarajan, D., Harasudhan, H., and Chakraborty, G. 2013. "Investigating the Impact of Amazon Kindle Fire HD 7" on Amazon.com Consumers Using SAS® Text Miner and SAS® Sentiment Analysis". *Proceedings of the SAS Global Forum 2013 Conference,* 261.

Pang, B., and Lee, L. 2008. "Using Very Simple Statistics for Review Search: An Exploration". *COLING 2008: Companion Volume – Posters and Demonstrations,* 75-78.

Pantangi A., Mandati, S., Ravuri, S., and Chakraborty, G. 2013. "Feature Extraction and Rating of a Smartphone Photosharing Application Using SAS® Sentiment Studio". *Proceedings of the SAS Global Forum 2013 Conference,* 241.

Sarkar, Mantosh, Kumar, and Chakraborty, G. 2013. "Opinion Mining and Geo-Positioning of Textual Feedback from Professional Drivers," *Proceedings of the SAS Global Forum 2013 Conference,* 500.

# Case Studies

From the chapters in this book you have learned so far about text analytics concepts and read about applications of those concepts using the SAS Text Analytics suite. Going with the famous quote from Benjamin Franklin, *"Tell me and I forget. Teach me and I remember. Involve me and I learn,"* as much as understanding concepts and theory via reading of chapters are important, perhaps real case studies with detailed applications of tools are more important from a learning perspective. One of the key objectives of this book is to provide you, the reader, an opportunity to try hands-on applications of most of the concepts and tools discussed in this book.

This book contains nine case studies. Each case study has step-by-step instructions on how to perform text analytics using different SAS tools. We have provided you with access to data for seven case studies. Case Study 8 is a demonstration on the new web-based application that combines the power of SAS® Text Miner and SAS® Content Categorization in a single interface called SAS® Contextual Analysis. The data for this case study is not currently available.

The table below lists the SAS Text Analytics tools used in each of the case studies.

| Case Study | Title | SAS Text Analytics Tools |
|---|---|---|
| 1 | Text mining SUGI/SAS Global Forum paper abstracts to Reveal Trends | *SAS® Text Miner* |
| 2 | Automatic Detection of Section Membership for SAS Conference Paper Abstract Submissions | *SAS® Text Miner,* *SAS® Enterprise Content Categorization* |
| 3 | Features-based Sentiment Analysis of Customer Reviews | *SAS® Text Miner,* *SAS® Sentiment Analysis Studio,* *SAS® Enterprise Content Categorization* |
| 4 | Exploring Injury Data for Root Causal and Association Analysis | *SAS® Text Miner,* *SAS® Enterprise Content Categorization* |
| 5 | Enhancing Predictive Models Using Textual Data | *SAS® Text Miner* |
| 6 | Opinion Mining of Professional Drivers' Feedback | *SAS® Text Miner,* *SAS® Sentiment Analysis Studio* |
| 7 | Information Organization and Access of Enron Emails to Help Investigation | *SAS® Information Retrieval Studio,* *SAS® Text Miner,* *SAS® Enterprise Content Categorization* |
| 8 | Unleashing the Power of Unified Text Analytics to Categorize Call Center Data | *SAS® Contextual Analysis,* |
| 9 | Evaluating Health Provider Service Performance Using Textual Responses | *SAS® Text Miner,* *SAS® Sentiment Analysis Studio* |

# Case Study 1 Text Mining SUGI/SAS Global Forum Paper Abstracts to Reveal Trends

Zubair Shaik
Satish Garla
Goutam Chakraborty

## Introduction

Text Mining is mainly used for information retrieval and text categorization. With the growth of social media, the popularity of text mining as a technique to discover trends in topics has caught up tremendously. Given a set of documents with a time stamp, text mining can be used to identify trends of different topics that exist in the text and how they change over time. In this case study we aim to demonstrate the application of text mining for understanding trends in conference proceedings.

Many professional conferences or forums are held every year across the globe. The total number of papers presented each year at all the conferences is likely in the hundreds of thousands. While there are conferences which focus on only one specific field, there are many conferences which act as a platform for different fields. Unlike academic journals, which are heavily indexed and searchable, most conference papers are not indexed properly. In addition, there are many academic institutions that publish working papers and many consulting firms that publish white papers every year which are also not indexed. While it may be possible to use a structured query to get a listing of papers published about a certain topic from indexed journals/conferences, it is virtually impossible to gain broad based knowledge about the hundreds of topics that are presented or published during a time period and how such topics may have changed over time.

### Data

The data used in this case study are abstracts of all SAS conference papers published each year at SUGI/SAS Global Forum from 1976 to 2011. Initially, we considered three types of textual data for text mining,

- Title of the paper
- Abstract of the paper
- Complete body of the paper

Title of the paper may not be a good input because it is often restricted in length and may not fully reflect the theme of the paper semantically. Considering the complete paper for analysis will likely add a lot of noise because a full paper can include tables, images, references, SAS programs, etc., which are problematic and may not add much value in text mining for topic extraction. We felt analyzing the abstract of a paper to be most appropriate since it captures a detailed objective of the paper and does not contain extraneous items such as tables, images, etc. We are thankful to SAS for making available SUGI/Global Forum proceedings for all of the years the conference was held. We faced many challenges starting from downloading papers in PDF format from the SAS website to preparing a final data set that contains only the abstract for each paper. We used the %TMFILTER macro for preparing SAS data sets from a repository of SAS papers in .pdf and .txt format. We had to make some strategic choices to prepare the data sets. For complete details on the data preparation process refer to 2012 SAS Global Forum paper 'SAS Since 1976: An application of text mining to reveal trends'. In the paper, the authors created four data sets, one for each decade, with the number of observations (i.e., paper abstracts) shown in Display C1.1.

**Display C1.1 Number of paper abstracts in each decade**

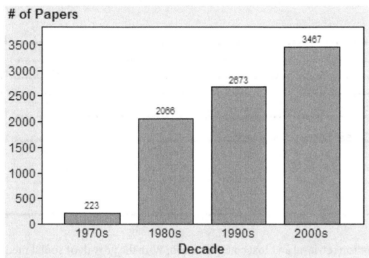

For this case study we start with the same four data sets with only one exception, the latest decade data set also includes papers from the 2012 SAS Global Forum conference.

The four data sets used for this analysis are:

- sas1976_1980
- sas1981_1990
- sas1991_2000
- sas2000_2012

The data sets are available in the folder *Case Studies ▶ Case Study 1* provided with this book. See the website for this book.

The following sections take you through the detailed, step-by-step text mining process from text parsing to cluster generation on each data set separately.

1. Create a SAS Enterprise Miner project and create four data sources using the above data sets. In each data source, set the Role of the variable 'text' to Text and the role of the variable 'name' to ID, as shown below.

**Display C1.2 Variables list window in data source creation**

| Name | Role | Level | Report |
|------|------|-------|--------|
| name | ID | Nominal | No |
| text | Text | Nominal | No |

2. Create a Diagram and drag the data source SAS1976_1980 to the diagram workspace.
3. Connect a Text Parsing node to the data source node (Display C1.3) and Run the node.

**Display C1.3  SAS Enterprise Miner diagram space**

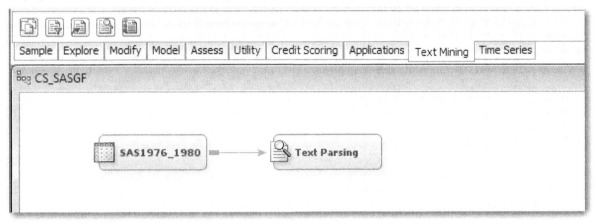

4. After the completion of the run, select Results on the Run status dialog box
5. Maximize the Terms window. This window (Display C1.4) contains list of all terms extracted with term properties such as role, attribute, frequency, Keep/Drop status, parent/child status and the rank assigned to the term.

**Display C1.4 Terms window**

| Term | Role | Attribute | Freq | # Docs | Keep | Parent/Child Status | Parent ID | Rank for Variable numdocs |
|------|------|-----------|------|--------|------|---------------------|-----------|---------------------------|
| + be | Verb | Alpha | 1630 | 217 N | + | | 62 | 1 |
| + sas institute | Company | Entity | 643 | 182 Y | + | | 16559 | 2 |
| data | Noun | Alpha | 765 | 176 Y | | | 77 | 3 |
| + have | Verb | Alpha | 266 | 142 N | + | | 29 | 4 |
| + use | Verb | Alpha | 256 | 135 N | + | | 415 | 5 |
| s | Noun | Alpha | 195 | 111 N | | | 178 | 6 |
| r | Noun | Alpha | 196 | 109 N | | | 676 | 7 |
| e | Noun | Alpha | 192 | 108 N | | | 885 | 8 |
| the | Prop | Alpha | 161 | 107 N | | | 209 | 9 |
| + analysis | Noun | Alpha | 223 | 106 Y | + | | 96 | 10 |
| t | Noun | Alpha | 194 | 97 N | | | 276 | 11 |
| + procedure | Noun | Alpha | 175 | 91 Y | + | | 181 | 12 |
| + not | Adv | Alpha | 145 | 90 N | + | | 33 | 13 |
| + use | Noun | Alpha | 133 | 89 N | + | | 84 | 14 |
| l | Noun | Alpha | 194 | 88 N | | | 912 | 15 |
| + system | Noun | Alpha | 257 | 88 Y | + | | 49 | 15 |
| h | Noun | Alpha | 100 | 82 N | | | 176 | 17 |
| i | Noun | Alpha | 159 | 79 N | | | 358 | 18 |
| two | Num | Alpha | 115 | 77 N | | | 381 | 19 |
| + program | Noun | Alpha | 138 | 75 Y | + | | 45 | 20 |
| statistical | Adj | Alpha | 148 | 72 Y | | | 145 | 21 |
| + university | Noun | Alpha | 112 | 72 Y | + | | 342 | 21 |
| + group | Noun | Alpha | 87 | 71 Y | + | | 102 | 23 |
| i | Noun | Alpha | 134 | 71 N | | | 175 | 23 |

A great deal of understanding about what is happening for the settings being used in the nodes can be made by looking at the Terms table. Once you have a sense of which terms are being kept or dropped and their Role, you can then adjust the settings accordingly. Let us explore how you can make use of the 'Entity' extraction property in the parsing node. This property has a default value of 'None'. In this case study you will not have a need to identify entities since you are analyzing just the abstract of a paper. However, due to limitations in the data preparation task, you will find names of authors, location, company, and address appearing in the text.

Most of these entities are parsed as a proper noun when the entity property is disabled. These terms may be considered as noise in our analysis because our interest is in the trend.

6.  Exclude these terms from analysis by changing the 'Find Entities' property from None to Standard in the properties panel of the text parsing node. Then, click on the ellipsis button next to 'Ignore Types of Entities' and select all terms except Miscellaneous Proper Noun (Prop_misc) as shown in Display C1.5.

**Display C1.5 Text Parsing node property panel**

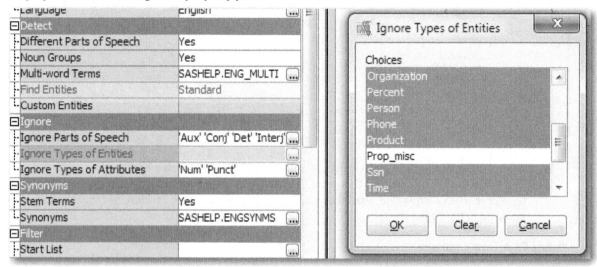

7.  Parts of speech will add no value in our current analysis. Hence, set the Different Parts of Speech property to No. With this setting you will find words such as "the", "an", "also", etc, being included in the analysis. You can drop these terms from the analysis by using the 'Ignore Parts of Speech' property, as shown in Display C1.6.

**Display C1.6 Ignore Parts of Speech property window**

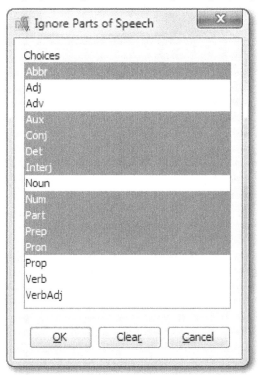

When a synonyms list is already available, you can use them with the Text Parsing node. Otherwise these lists can be custom generated for this data using the Text Filter node.

8. Connect a Text Filter node to the Text Parsing node. Enable the spell-checking property and set the Maximum Number of Documents property value 5 and Run the node.
   After the run completion, the Interactive Filter Viewer window can be accessed to explore and interactively manage the terms. The main objective is to filter the terms not needed for analysis. Generating a Start/Stop list and a synonym list are integral to this task.

The simplest way to filter words is to select a number of terms from the Terms window and uncheck the Keep column. From the interactive filter window, try sorting the Terms table multiple times for each different column to understand groups of related terms. For example, sorting the 'Term' column will help in identifying synonyms that are very close in spelling, such as the terms airline, aviation, aircraft, etc.

9. Sort the Terms table by the KEEP column. You will see all the terms retained in the analysis in the order of document frequency. Clearly you will always see certain terms that exist across almost all the documents that provide no value in discriminating the documents. In this case, high frequency terms like data, paper can be dropped. Other high frequency terms like set, type, ed, carolina, inc, north, introduction, institute can be dropped since they are not useful in this analysis.
10. Select all these terms as shown in Display C1.7, right-click and select 'Toggle KEEP'. This will uncheck the KEEP column for all these terms and they are dropped from analysis.

**Display C1.7 Terms window from Interactive filter viewer**

| TERM | FREQ | # DOCS | KEEP ▼ | WEIGHT | ROLE | ATTRIBUTE |
|---|---|---|---|---|---|---|
| ⊞ data | 770 | 178 | ✓ | 0.095 | | Alpha |
| ⊞ analysis | 214 | 105 | ✓ | 0.173 | | Alpha |
| ⊞ user | 174 | 104 | ✓ | 0.165 | | Alpha |
| ⊞ procedure | 185 | 94 | ✓ | 0.194 | | Alpha |
| ⊞ system | 264 | 92 | ✓ | 0.22 | | Alpha |
| ⊞ paper | 108 | 90 | ✓ | 0.176 | | Alpha |
| ⊞ program | 195 | 88 | ✓ | 0.207 | | Alpha |
| ⊞ statistical | 169 | 80 | ✓ | 0.238 | | Alpha |
| ⊞ university | 125 | 80 | ✓ | 0.218 | | Alpha |
| ⊞ set | 185 | 80 | ✓ | 0.231 | | Alpha |
| ⊞ group | 101 | 76 | ✓ | | Add Term to Search Expression | |
| ⊞ variable | 168 | 75 | ✓ | | Treat as Synonyms | |
| ⊞ number | 134 | 71 | ✓ | | Remove Synonyms | |
| ⊞ problem | 99 | 68 | ✓ | | Toggle KEEP | |
| ⊞ time | 127 | 66 | ✓ | | View Concept Links | |
| north | 69 | 65 | ✓ | | Find | |
| ⊞ information | 121 | 64 | ✓ | | Repeat Find | |
| ⊞ type | 101 | 63 | ✓ | | Clear Selection | |
| ed | 68 | 61 | ✓ | | Print... | |
| ⊞ carolina | 74 | 60 | ✓ | | | |
| ⊞ present | 85 | 60 | ✓ | 0.261 | | Alpha |
| ⊞ include | 75 | 59 | ✓ | 0.26 | | Alpha |
| ⊞ inc | 59 | 58 | ✓ | 0.248 | | Alpha |
| ⊞ design | 118 | 58 | ✓ | 0.293 | | Alpha |
| ⊞ introduction | 58 | 57 | ✓ | 0.251 | | Alpha |
| ⊞ form | 87 | 57 | ✓ | 0.276 | | Alpha |
| ⊞ research | 109 | 56 | ✓ | 0.319 | | Alpha |
| ⊞ observation | 102 | 56 | ✓ | 0.299 | | Alpha |
| ⊞ file | 143 | 55 | ✓ | 0.298 | | Alpha |
| ⊞ institute | 57 | 55 | ✓ | 0.259 | | Alpha |
| ⊞ value | 111 | 55 | ✓ | 0.316 | | Alpha |

11. Now sort the Terms table by the TERM column. Scroll down the table to identify similar terms and create a synonym list. Navigate to the terms starting with letter "b" by hitting the letter "B" on the keyboard. Scroll down the list to the term "bio". You will see a lot of terms with the letters "bio" that can be treated as synonyms. To make all these terms synonyms, select the terms bio, biochemical, biological, biomedical, biometrics, biometrika, biometry, biometries, biostatistics, then right-click and select 'Treat as Synonyms' as shown in Display C1.8.

**Display C1.8 Terms window showing creating synonyms**

| | TERM ▲ | FREQ | # DOCS | KEEP | WEIGHT | ROLE | ATTRIBUTE |
|---|---|---|---|---|---|---|---|
| | bimulation | 1 | 1 | ☐ | 0.0 | Miscellaneous Pro… | Entity |
| | bination | 1 | 1 | ☐ | 0.0 | | Alpha |
| | bine | 1 | 1 | ☐ | 0.0 | | Alpha |
| | bio | 1 | 1 | ☐ | 0.0 | | Alpha |
| | bioassay | 3 | | | | | Alpha |
| | bioav | 1 | | | | | Alpha |
| | biochemical | 1 | | | | | Alpha |
| | bioiissay | 1 | | | | is Pro… | Entity |
| | biol 1lll | 1 | | | | is Pro… | Entity |
| | biological | 5 | | | | | Alpha |
| | biological data | 1 | | | | | Alpha |
| ⊞ | biological experiment | 1 | | | | | Alpha |
| | biome | 2 | | | | | Alpha |
| | biomedical | 2 | | | | | Alpha |
| | biomedical computer | 2 | 2 | ☐ | 0.0 | Noun Group | Alpha |
| | biomeo | 1 | 1 | ☐ | 0.0 | Miscellaneous Pro… | Entity |
| | biometrics | 1 | 1 | ☐ | 0.0 | Miscellaneous Pro… | Entity |
| | biometries | 1 | 1 | ☐ | 0.0 | | Alpha |
| | biometrika | 1 | 1 | ☐ | 0.0 | | Alpha |
| | biometrika article | 1 | 1 | ☐ | 0.0 | Miscellaneous Pro… | Entity |
| | biometry | 1 | 1 | ☐ | 0.0 | Miscellaneous Pro… | Entity |
| | biometry | 2 | 1 | ☐ | 0.0 | | Alpha |
| ⊞ | biostatistics | 5 | 5 | ☑ | 0.701 | | Alpha |
| | biostatistks senic | 1 | 1 | ☐ | 0.0 | Miscellaneous Pro | Entity |

Context menu overlaying table:

- Add Term to Search Expression
- Treat as Synonyms
- Remove Synonyms
- Toggle KEEP
- View Concept Links
- Find
- Repeat Find
- Clear Selection
- Print...

12. In the following dialog (see Display C1.9), select "bio" as the keyword to represent the group. Hence when these terms were represented individually, all the terms were dropped from the analysis due to their low document frequency (less than 5 as in the setting). With the creation of synonym group, the term "bio" will now be considered in the analysis since the document frequency meets the cut-off. You will have to re-run the node to see this term included in your analysis.

**Display C1.9 Window to select parent synonym term**

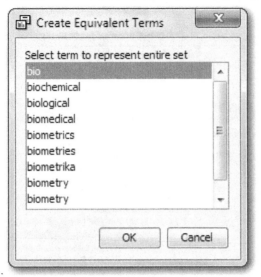

Similarly, browse through the complete list of terms to keep the dropped terms and to drop the kept terms wherever appropriate for the analysis. This is the most important and time-consuming task of text mining analysis. This task is purely subjective and completely depends on the problem and the analyst's level of domain expertise.

13. In the end, save the custom synonyms by clicking on File ▶ Export Synonyms. Select the Library where you intend to save your work and give a meaningful name to the data set as shown in Display C1.10. Close the interactive filter window and Save all your changes when prompted.

**Display C1.10 Creating a custom synonym list**

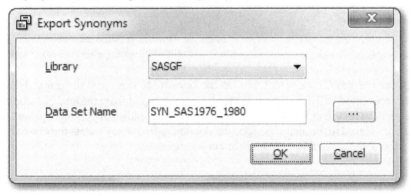

As far as the case study is concerned, it is highly difficult to list all of the operations performed at this stage. Hence we provide you with a custom synonym list that can be used with the Text Filter Node.

14. Connect a Text Cluster node to the Text Filter node. Set the Maximum number of the clusters property to 20 and run the node.
The Results window (see Display C1.11) of the Text Cluster node displays 11 clusters generated by the Expectation–Maximization (EM) method. It is worth trying other methods like Hierarchical clustering with different SVD resolution levels before you conclude on a final result. For EM techniques with Low SVD resolution settings we clearly see a reasonable separation of the documents. As shown in display C1.11 Cluster 8 includes papers on the GLM procedure, linear regression; cluster 2 includes papers on experiments, effects, treatment; cluster 4 contains papers on graphs, plots, line options, etc.

**Display C1.11 Cluster results for 1970s data**

| Cluster ID | Descriptive Terms | Frequency | Percentage |
|---|---|---|---|
| 1 | students +patient +first +tool +center +language +management programs records reports +college patients +tape +aid feature... | 23 | 11% |
| 2 | +block treatments +treatment blocks experiments patients ana repeated +patient measurements +proe +trial usual +drug appl... | 8 | 4% |
| 3 | +virginia west +package packages areas +state +university +code files +easy led +language +sample records +single ... | 26 | 12% |
| 4 | +plotter plotting plots +graphical +plot +proe +line +option +distribution +produce +sample periods +create +second +test ... | 11 | 5% |
| 5 | pp +annual +third +'third annual conference' +conference +international +proc +group 'sas institute' users sas +regression +... | 33 | 15% |
| 6 | processing +base +survey 'data management' +summary +management systems areas ment +national capabilities +pro +proce... | 31 | 14% |
| 7 | days hours numbe +rat +person led tute periods tasks +task +period +project fixed +tool +area ... | 4 | 2% |
| 8 | +linear +vector equations effects +model squares glm +effect models +general +variance +matrix +main +dependent paramet... | 35 | 16% |
| 9 | assumptions nature distributed +college +constant +technique +regression +apply +assume basic +assumption +independent ... | 6 | 3% |
| 10 | +entry +discussion +record +input clinical +trial +system +'data base' +report +important packages +work +base +file +inform... | 20 | 9% |
| 11 | +tape +'data set' +macro means +read +option +order +output +mean +process +code statements +statement +table records ... | 22 | 10% |

What you have seen up until now is a typical and basic process flow in text mining analysis. By performing these steps, you gain a fair understanding of the text. Since the objective of this case study is to perform trend analysis, carry out the same set of steps with the remaining three data sets, as shown in Display C1.12, representing the latter three decades and compare the results.

**Display C1.12 Diagram with process flows for all four decades data**

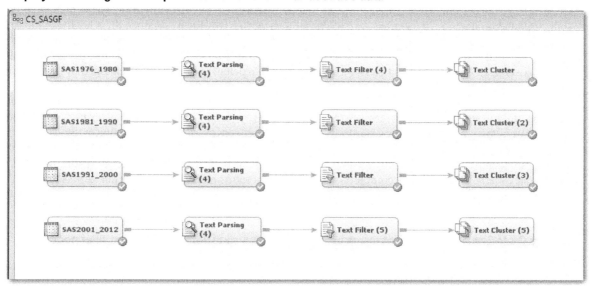

## Results

As discussed in Chapter 5, "*Data Transformation*," a quick and easy route to understanding the textual data is by looking at the high frequency terms in the corpus. The results window of the Text Parsing node or the Text Filter node can be used to look at the high frequency words. Display C1.13 shows the top occurring terms in each of the four data sets from the Text Filter Node results (only terms that are kept in the analysis are shown here).

**Display C1.13 Top frequency terms in each of the four data sets from text filter node**

| Term | # Docs | Term | # Docs | Term | # Docs | Term | # Docs |
|---|---|---|---|---|---|---|---|
| + analysis | 124 | + system | 1135 | + paper | 1610 | + paper | 2768 |
| + user | 104 | + paper | 981 | + system | 1406 | + information | 1301 |
| + statistical | 100 | + procedure | 959 | + user | 1017 | + proc | 1182 |
| + procedure | 94 | + program | 947 | + program | 1004 | + application | 1180 |
| + system | 92 | + user | 835 | + software | 939 | + process | 1168 |
| + paper | 90 | + analysis | 763 | + application | 929 | + create | 1147 |
| + program | 88 | + software | 676 | + process | 821 | + system | 1135 |
| + university | 80 | + information | 660 | + information | 817 | + user | 1118 |
| + group | 76 | + application | 660 | + create | 804 | + program | 1056 |
| + variable | 75 | + variable | 641 | + include | 770 | + include | 988 |
| + number | 71 | + process | 633 | + time | 720 | + analysis | 963 |
| + problem | 68 | + time | 631 | + code | 714 | + code | 937 |
| + time | 66 | + base | 583 | + variable | 699 | + time | 906 |
| + information | 64 | + develop | 570 | + procedure | 666 | + report | 864 |
| + present | 60 | + present | 563 | + present | 663 | + present | 850 |

(1970s) (1980s) (1990s) (2000s)

Clearly, there are certain words that occur in all four decades like, system, information, paper, present, program, time, user and procedure. In fact one can exclude all these terms from analysis since these are the common terms that can be easily expected to be present in SAS Conference proceedings. Terms like these definitely do not help in trend detection. However, you will find terms like application, process, and code occurring in all the decades except 1970s. This shows that not a lot of application oriented papers were published in 1970s, rather the papers were more about explaining techniques available in SAS at that time.

Another way to identify important terms in the text corpus is by looking at the important terms as identified by the term weighting methods. In this case study we used default settings for term weighting methods. It is always recommended to try different settings and explore the results. With default settings, the important terms for four decades are identified as shown below. You can find these in the text filter node results. Just sort the terms file by weight. Clearly, you will see a completely different set of terms identified as important across the four data sets. The weight is calculated using the entropy (default) technique that uses document frequency along with the overall frequency of the terms.

**Display C1.14 The most important terms in each of the four data sets from the text filter node**

| Term | Weight ▼ | Term | Weight ▼ | Term | Weight ▼ | Term | Weight ▼ |
|---|---|---|---|---|---|---|---|
| + forecast | 0.804 | + dominant | 0.886 | + newsletter | 0.885 | + credential | 0.851 |
| + demand | 0.791 | + zap | 0.870 | + ab | 0.863 | + google maps | 0.847 |
| + faculty | 0.762 | + oracle | 0.863 | + cache | 0.853 | windows server | 0.842 |
| + public | 0.752 | + promotion | 0.859 | + gdwd | 0.849 | sustainability | 0.842 |
| + block | 0.750 | + dental | 0.853 | + folder | 0.844 | remote compute ... | 0.842 |
| growth | 0.749 | + foster | 0.848 | + coder | 0.844 | + recurrent | 0.839 |
| + trend | 0.748 | + converter | 0.846 | + meta-analysis | 0.844 | + table definition | 0.836 |
| regional | 0.743 | + quantile | 0.839 | + ope | 0.837 | + community care | 0.836 |
| confidence | 0.743 | + career | 0.839 | + container | 0.837 | + bubble | 0.836 |
| + disease | 0.743 | addressable | 0.839 | polynomial | 0.836 | + pivot table | 0.836 |

(1970s) (1980s) (1990s) (2000s)

# Trends

Trends of topics across data sets can be tracked by looking at the frequency of words related to a topic in the four data sets. Another way to explore trends is by understanding the context in which a particular term is frequently used. This can be accomplished by looking at the concept links for a term. You can view the concept links from the interactive filter window (to activate the interactive viewer, in the properties panel for the text filter node, click on the ellipsis button next to Filter Viewer). In the terms window select the term of interest, right-click and select View concept links. Display C1.15 and C1.16 show the concept links for the term "forecast" in the 1980s and 2000s.

**Display C1.15 Concept links for the term "forecast" in the 1980s data**

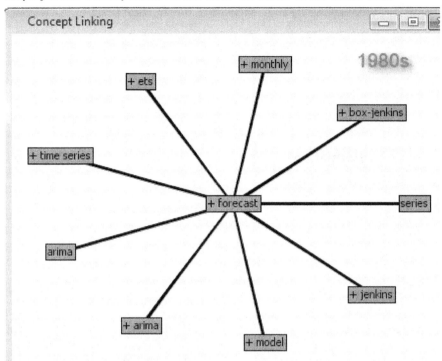

**Display C1.16 Concept links for the term "forecast" in the 2000s data**

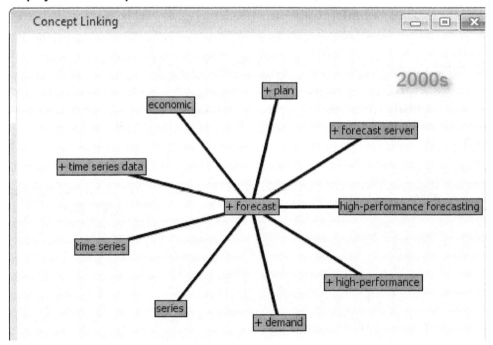

In the 1980s the papers that talked about "forecasting" discussed the technique that is clear from the associated terms like arima, Box-Jenkins, time series, etc., whereas in the 2000s the papers on "forecasting" talked about demand, economic, high-performance forecasting, etc. Similarly, if you observe the term "information", as shown in Display C1.17, you will understand the context in which the term is heavily used in the conference papers across the four decades.

**Display C1.17 Associated terms for the term "information" from all four data sets**

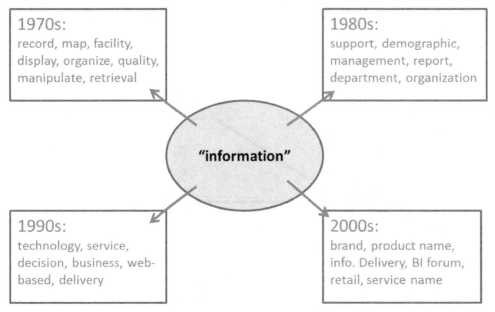

1970s:
record, map, facility, display, organize, quality, manipulate, retrieval

1980s:
support, demographic, management, report, department, organization

"information"

1990s:
technology, service, decision, business, web-based, delivery

2000s:
brand, product name, info. Delivery, BI forum, retail, service name

Trends in industry can be explored by creating synonyms using terms related to a specific industry or simply counting the frequencies of various terms that represent a particular industry. The occurrences of industry-specific terms across the four decades can be used to understand the trend for that industry. A frequency value of "n" for a term means that particular term was mentioned in the abstracts of "n" distinct conference papers. The frequency value here is not the overall occurrence of the term, but the number of documents that the term occurs. This value can be obtained from #Docs column from the Terms window of the Text Parsing or the Text Filter node results.

For example, open the results window of the Text Filter node from the 2000s process flow. In the Terms window sort the table by Term column in A to Z order by clicking on the column header. Scroll down to the term "agriculture." As shown in Display C1.18, the term "agriculture" appears in 12 documents (#Docs). You can easily copy and paste the whole Terms table to a spreadsheet and calculate the percentage of occurrence for each term. In 2000s there are a total of 3,887 terms and, therefore, the percent of occurrence of the term "agriculture" is about 0.31 as shown in Display C1.19.

**Display C1.18 Terms window**

| Term ▲ | # Docs | Role | Attribute | Status |
|---|---|---|---|---|
| + agree | 12 | | Alpha | Keep |
| + agreement | 20 | | Alpha | Keep |
| + agriculture | 12 | | Alpha | Keep |
| + aid | 52 | | Alpha | Keep |
| + aim | 83 | | Alpha | Keep |
| + airline | 14 | | Alpha | Keep |
| + akaike | 7 | | Alpha | Keep |

Display C1.19 shows the representation count for the terms "agriculture" and "clinical/pharmaceutical" across the decades. The values in the chart represent the percentage of the papers with these topics compared to the total conference proceedings for the decade. It is clearly evident that the number of papers on agriculture decreased greatly with the years.

**Display C1.19 Representation trend for the topic "Agriculture" and "Clinical/ Pharmaceutical"**

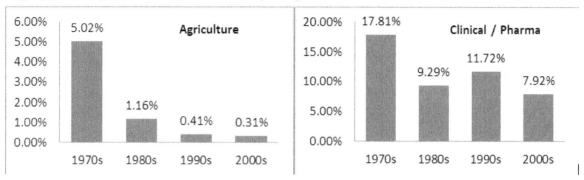

Similarly, you can identify the trend for other industries such as manufacturing, education, health, entertainment, etc. Insights on the trends at a very high level can be gained by looking at the clusters generated for each data set. The theme of each cluster can be understood from the descriptive terms reported by SAS Text Miner. You can request SAS Text Miner to report as many terms as needed in order to describe a cluster. In this analysis we used eight terms to describe a cluster. Earlier in Display C1.17 you saw the cluster results for the 1970s. Display C1.20 shows the clusters generated for the 1980s.

**Display C1.20 Cluster results for 1980s data**

| Cluster ID | Descriptive Terms | | Percentage |
|---|---|---|---|
| 1 | +code +macro +statement macros statements +step +input +compiler | ... | 6% |
| 2 | +clinical +drug +patient +study +trial trials studies +research | ... | 4% |
| 3 | +hypothesis +test +treatment +variance effects squares tests +linear | ... | 5% |
| 4 | +population +probability +random +sample +size distributions samples sampling | ... | 3% |
| 5 | +access +database +library +version +base +system +management files | ... | 6% |
| 6 | +graph +output +statement +variable proc +format values +produce | ... | 13% |
| 7 | +county +geographic +map mapping maps +graph +area areas | ... | 2% |
| 8 | +company +cost +department +management systems +area +major +process | ... | 8% |
| 9 | +'time series' +forecast forecasting forecasts models series +time +estimation | ... | 2% |
| 10 | +compiler c +ed r +language +tion +library +code | ... | 3% |
| 11 | +estimate +estimation +linear +regression estimates least models squares | ... | 6% |
| 12 | +command +display +interactive +menu +screen +user +file +program | ... | 10% |
| 13 | +covariance +matrix +multivariate +linear estimates +estimation +probability models | ... | 4% |
| 14 | +care +health +hospital +medical +patient hospitals +major +information | ... | 2% |
| 15 | +quality +research +process +statistical +study developed studies +control | ... | 7% |
| 16 | 'repeated measures' +experimental measures repeated subjects +multivariate +subject +covariance | ... | 2% |
| 17 | +computer +support users systems processing +software +system +user | ... | 17% |

In contrast to statistical techniques and experimentation papers that you have seen in clusters from 1970s, you now see more papers on computer processing (cluster 17), time series forecasting (cluster 9), C-programming (cluster 10), database management (cluster 5) etc., Though you can make a high-level guess on the kind of representation using clusters, a deep understanding on the trend can only be achieved by looking at the terms via concept links and frequencies. The clusters from the latter decades highlight papers on the internet, web-application development, retail, manufacturing, social media etc.

## Summary

In this case study you applied text mining to figure out trends in research topics related to various industries in SAS/SUGI conference papers. This case study is intended to showcase the capabilities in SAS Text Miner to track trends of topics in text corpus from different periods of time. The concept links feature of SAS Text Miner is a great tool to discover the context in which a particular term/topic is frequently used. Generating a synonym list on the go makes the text mining analysis more objective. This analysis can be easily extended to find trends in research topics related to different methods or technology. A similar approach can also be used to analyze the many conference proceedings corpus that is available with various organizations across the globe.

## Instructions for Accessing the Case Study Project

Along with this book we provide you with both the data used and the SAS Enterprise Miner project created for this case study. The case study includes most of the steps performed in this analysis. However, it is highly impossible to include every single operation performed in text filtering stage. Hence if you start with the data and follow the steps in the case study your results may not match the ones in the case study. Nevertheless, you can use the project to view the exact results presented in the case study. The following instructions will guide you on how to access the project.

1. In the data provided with this book go to the folder Case Studies ▶ Case Study 1. You will find all the data sources needed for this case study and a folder **SASGF_Case**. The folder is the SAS Enterprise Miner project.
2. Open SAS® Enterprise Miner and click **New Project.**
3. As shown below enter project name as **SASGF_Case** and enter the path to the folder and Click **Next**

4. You will see the below prompt window. Click **Yes.**

If you do not see this prompt, then you may have used a different name/path and then this project may not work correctly.

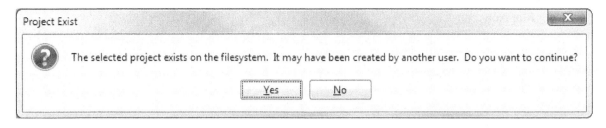

5. Click **Next** on the following windows and finally click **Finish.**
6. The project panel should contain four data sets and one diagram as shown below.

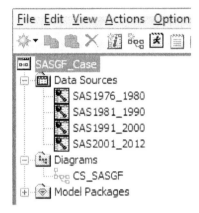

7. Before you run the process flow, you will have to create a library that points to these data sets.
8. Go to **Project Start code** from the property panel.

9. Modify the **LIBNAME** statement with the path pointing to the data folder as shown below and click

**Run Now.**

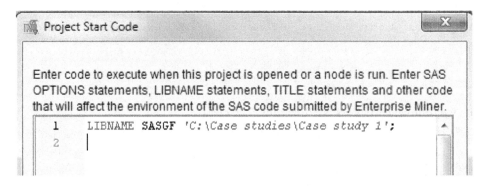

**NOTE:** If you are using a different LIBREF instead of "SASGF", you will have to delete the existing data sources and create each data source again using the new LIBREF. You will also have to update the Exported Synonyms property of the Text Filter node to point to the respective decade's synonyms data set. Hence we suggest using **SASGF** as the library name for this case study.

10. From the **Log** tab on Project Start Code window verify that the library was created successfully. Then click **Ok.**
11. Open the Diagram and run the Text Cluster node for each process as shown below.

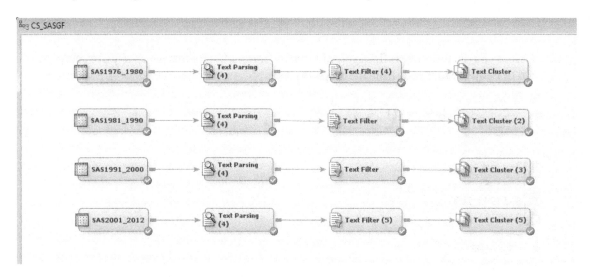

You can now view the results for each node as reported in the case study.

# Case Study 2 Automatic Detection of Section Membership for SAS Conference Paper Abstract Submissions

Murali Pagolu
Goutam Chakraborty

## Introduction

Every year since its inception, SAS® and users of SAS® products have been actively making significant knowledge contributions to the SAS® user community using SAS® regional conferences and SAS® Global Forum as a platform. So far, thousands of papers have been published in these conference proceedings under various topic sections. Each year the number of contributions increase compared to its previous year and this trend is likely to continue in future. The array of SAS products, industry solutions and its user base is growing across the globe. You can anticipate a growing library of SAS® conference papers serving as free online educational material showcasing several innovative applications of SAS® put into practice. What you see in the online proceedings are papers published after they were accepted through careful selection and scrutiny from a big pool of submissions every year.

Several scholars, efficient and experienced professionals from industry are handpicked, appointed as leaders/heads of individual sections based on their area of expertise, experience and knowledge. These leaders are bestowed with a huge task of reading through all the paper abstracts submitted and select those that qualify to be presented at the conference and later published in the conference proceedings. The number of paper abstracts submitted to SAS® Global Forum 2013 was rumored to be somewhere between 600 and 650. However, only about 90 of them were finally accepted. Paper acceptance criteria may depend on a lot of factors. Some of these factors include:

- Type of submission (internal – submitted as a submission by SAS® employee or external – submitted by an external user of SAS®)
- Choice and relevance of the topic to the current section.
- Displaying theoretical accuracy and writing skills in the content.
- Showcasing a possible solution for a recurring problem in an industry or technology.
- Providing a business application using trending SAS® product(s) or technology.
- Discussing an innovative idea or technique.
- Preference pre-set by the section leaders and conference organizers in anticipation of attendees' background and interests.
- Range of competitive topics covered by other authors in their submissions for that section.

Though this list may not be exhaustive and accurate, one can determine that many of these factors play important roles in deciding the fate of an abstract submission. Except for the range of competitive abstracts submitted in that section by other authors, authors can make their best possible efforts to work on all other factors to increase the chance of their submission being selected. Once authors have finished working on their abstract(s), the most important step that lies in their hands is to choose the appropriate section to submit their abstract. Some sections are so popular that they are often inundated with submissions, creating a tough challenge for the evaluators to make their decisions in the selection process. Experienced authors may find it easy to narrow down to their top section choices (2 or 3) in which they may fit well according to the section description and the abstract topic. In such cases, their submission, though rejected in one section, may be accepted in other section due to one or more of the many reasons we discussed earlier in selection process. For example, a paper abstract discussing the usage of a unique segmentation method to distinctly identify several customer groups for better marketing and sales strategy may be applicable for both the 'Customer Intelligence' and 'Data Mining' sections. A custom-written SAS macro to address a data integration issue may qualify for both the 'Data Integration' and 'Coder's Corner' sections. Hence, it is critical for an author to determine the most appropriate sections for submission.

## Objective

In this case study, we attempt to address the issue of determining the section membership of a paper abstract submission based on its content. For this purpose, we use SAS® Text Miner and SAS® Content Categorization Studio to develop a rule-based categorizer. This taxonomy should serve as an application to automatically score and identify the most relevant and appropriate conference section in which an abstract should be submitted for a better chance of acceptance.

## Data

For this case study, we collected SAS® paper abstracts from SAS® Global Forum online proceedings http://support.sas.com/events/sasglobalforum/previous/online.html. We downloaded 466 papers from 2008 to 2012 encompassing 5 sections: 'Business Intelligence', 'Reports', 'Data Mining', 'Statistical Analysis' and 'Systems Architecture'. Using the %TMFILTER macro, we converted these papers from the PDF file format to plain text files and parsed the content to retain only the abstracts in them. We also created a SAS data set **'SGFpapers_sectionwise.sas7bdat'** to hold these abstracts, file names and names of the sections to which they belong under three different columns. We use this data set in SAS® Text Miner to automatically build Boolean rules and use them in building rule-based categorization models in SAS® Content Categorization Studio. In addition to this data set, we also provided plain text files containing these abstracts in individual folders. These are used to test the categorization models in SAS® Content Categorization Studio. The data needed for this case study is available in the data folder *Case Studies* ▶ *Case Study 2* provided with this book. See the website for this book. Copy the folder to a location (say for example, C:\Data) on your laptop or PC hard drive or use data directly from the given folder.

**Note:** We tried our best to provide you with the step-by-step instructions in this case study for you to follow. However, steps such as text parsing and text filtering involves extensive analysis of results and careful modification of keep terms, parts of speech tags, entities, stop list, synonym list, etc. Thus, it may not be possible for us to exclusively specify or mention all of the changes we make to the text mining nodes in this case study. An analyst would typically perform a series of iterative modifications and re-run the process flow in a text analysis exercise such as this until a desired result is achieved. Hence, we would like to explicitly mention to you that following these instructions may not exactly yield the same results for you. If you would like to see exactly the same results, we suggest you to open the EM project provided to you that includes the EM diagram and process flow used for this case study. You may follow same instructions provided at the end of Case Study 1 to open the SAS Enterprise Miner project 'SGF_CS_2' available in the *Case Studies* ▶ *Case Study 2* folder.

## Step-by-Step Instructions

1. Create a new project in SAS® Enterprise Miner and name it **'SGF_CS_2'**. Create a new diagram and name it **'Build_Rules'**. Create a library pointing to the location where the data reside using the project start code or File ▶ New ▶ Library menu.
2. Add the SAS data set **'SGFpapers_bysection.sas7bdat'** to the project and assign the roles of the variables as shown in Display C2.1. Variable 'type' is assigned the role 'Target' and will be used to build a category prediction model using the Text Builder feature in SAS® Text Miner.

**Display C2.1: Data Source Wizard for assigning roles to variables**

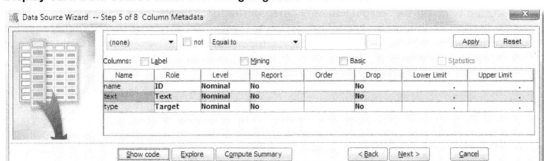

3. After the data set is added, drag the data source into the diagram as an **'Input Data'** node. Now, connect a **'Data Partition'** node and change the 'Data Set Allocations' property for Training, Validation, and Test data sets to 70.0, 30.0 and 0.0, respectively. Run the data partition node to see the result as shown in Display C2.2. You will see that the 'Stats' category contains more abstracts than compared to other sections. This is not intentional but in general more papers are published in 'Statistical Analysis' section every year hence the difference in counts.

**Display C2.2: Distribution of abstracts by section name in Train, Validation and Test groups**

```
Summary Statistics for Class Targets

Data=DATA

              Numeric    Formatted    Frequency
Variable      Value      Value        Count      Percent    Label

   type         .        BusInt         65       13.9485
   type         .        DataMining     64       13.7339
   type         .        Reports       100       21.4592
   type         .        Stats         150       32.1888
   type         .        SysArch        87       18.6695


Data=TRAIN

              Numeric    Formatted    Frequency
Variable      Value      Value        Count      Percent    Label

   type         .        BusInt         45       13.9319
   type         .        DataMining     44       13.6223
   type         .        Reports        69       21.3622
   type         .        Stats         105       32.5077
   type         .        SysArch        60       18.5759


Data=VALIDATE

              Numeric    Formatted    Frequency
Variable      Value      Value        Count      Percent    Label

   type         .        BusInt         20       13.9860
   type         .        DataMining     20       13.9860
   type         .        Reports        31       21.6783
   type         .        Stats          45       31.4685
   type         .        SysArch        27       18.8811
```

4. Connect a 'Text Parsing' node to the data partition node and run it with the default property settings. Display C2.3 shows a partial screenshot of the text parsing node results displaying a list of terms found after the abstracts are parsed. You can see the terms with Attribute type 'Abbr' (Abbreviation) are not

so frequently occurring in the data set. Also, there are generic terms such as data, paper, include, information etc. with the 'keep' status – 'Y' in the terms list. It means these terms are kept or retained in the process flow for the next node/feature to use. Terms with 'keep' status – 'N' in the terms list are thus excluded from further analysis. Similarly, terms such as 'miner', and 'enterprise miner' represent the same thing but appear as different terms in the abstracts. Hence, synonyms should be added to the list whenever possible to reduce the size of terms list.

**Display C2.3: Partial output of Text parsing results showing term list**

| Term | Role | Attribute ▲ | Freq | # Docs | Keep | Parent/Child Status | Parent ID | Rank for Variable numdocs |
|---|---|---|---|---|---|---|---|---|
| + app. ...Abbr | Abbr | | 1 | 1Y | | + | 10612 | 2896 |
| et al. ...Abbr | Abbr | | 1 | 1Y | | | 8164 | 2896 |
| mr. ...Abbr | Abbr | | 5 | 1N | | | 3517 | 2896 |
| + be ...Verb | Alpha | | 1121 | 294N | | + | 35 | 1 |
| + use ...Verb | Alpha | | 417 | 218N | | + | 578 | 3 |
| data ...Noun | Alpha | | 565 | 203Y | | | 117 | 4 |
| + paper ...Noun | Alpha | | 238 | 179Y | | + | 213 | 5 |
| + have ...Verb | Alpha | | 190 | 125N | | + | 62 | 6 |
| + provide ...Verb | Alpha | | 151 | 112N | | + | 233 | 7 |
| + include ...Verb | Alpha | | 138 | 100Y | | + | 308 | 8 |
| + analysis ...Noun | Alpha | | 186 | 95Y | | + | 51 | 9 |
| + not ...Adv | Alpha | | 125 | 93N | | + | 19 | 10 |
| + create ...Verb | Alpha | | 149 | 88Y | | + | 900 | 11 |
| information ...Noun | Alpha | | 129 | 84Y | | | 1355 | 12 |

5.  Based on our analysis of the text parsing results in the previous step (4), you can make the following changes to text parsing node properties and re-run the node.
    - Add 'Abbr' and 'Num' to the 'Ignore Parts of Speech' property.
    - Add 'Abbr' to 'Ignore Types of Attributes' property.

6.  Connect a 'Text Filter' node to the text parsing node. In the text filter node property panel, change the 'Term Weight' weightings property to 'Mutual Information' and run the node. The categorical variable is defined with a role of 'Target' in the data source; hence, this is the most appropriate term weight to use. Let the other properties set to default.

7.  Now that the text filtering node is run at least once, click on the ellipsis button next to 'Filter Viewer' under the 'Results' property. It opens an 'Interactive Filter Viewer' providing the list of terms, total frequency of occurrence in the corpus, number of documents in which they occur at least once, keep flag, term weight, role and attribute. As discussed in step (4), you may choose to modify the keep flag to either drop or add certain terms based on your intuition and frequency of occurrence of terms to arrive at a better classification model for the section category. If there are terms which need to be closely investigated, you may choose to use that term to search and find the abstracts data containing that term. For example, the term 'model' can mean either a statistical model or a predictive model. Hence, right click on that term in the Terms list and click 'Add Term to Search Expression.'
Click **Apply** next to the Search window to find all the documents containing this term. Display C2.4 shows the document search result for the term 'model'. All the words 'modeling', 'models', 'modeled' stemmed to the root word 'model' can be found highlighted in bold from the 'TESTFILTER_SNIPPET' column. You can also see the 'Type' of the abstract to which these documents belong in the same table. You will see that the documents containing the term 'model' are fairly distributed between the topic sections 'Data Mining' and 'Stats'.

**Display C2.4: Searching for terms in documents using Interactive Filter Viewer**

8. This case study does not require a great deal of modification to the terms list. Optionally, you can start creating a synonym list based on the closely related terms. You can highlight those terms that represent the same thing, right click and select 'Treat as Synonyms.' For example, terms 'miner' and 'miner$^{TM}$' can be treated as synonyms for SAS Enterprise Miner (See Display C2.5).

**Display C2.5: Build synonyms list choosing terms meaning the same**

| TERM ▲ | FREQ | # DOCS | KEEP | WEIGHT | ROLE | ATTRIBUTE |
|---|---|---|---|---|---|---|
| ⊞ mine procedure | 1 | 1 | ☐ | 0.0 | Noun Group | Alpha |
| ⊞ mine technique | 1 | 1 | ☐ | 0.0 | Noun Group | Alpha |
| miner | 13 | 12 | ☑ | 0.741 | Prop | Alpha |
| miner | 20 | 11 | ☑ | 0.824 | Noun | Alpha |
| minertm | 1 | 1 | ☐ | 0.0 | Prop | Alpha |
| miner™ | 10 | 10 | ☑ | 0.711 | Prop | Mixed |
| miner™ softwar... | 1 | 1 | ☐ | 0. | Add Term to Search Expression |
| minimal | 2 | 2 | ☐ | 0. | Treat as Synonyms |
| minimal effort | 1 | 1 | ☐ | 0. | Remove Synonyms |
| minimal mainten... | 1 | 1 | ☐ | 0. | Keep Terms |
| minimalistic | 1 | 1 | ☐ | 0. | Drop Terms |
| minimalistic appr... | 1 | 1 | ☐ | 0. | View Concept Links |
| ⊞ minimize | 3 | 3 | ☐ | 0. | Find |
| minimum | 3 | 3 | ☐ | 0. | Repeat Find |
| minimum | 3 | 2 | ☐ | 0. | Clear Selection |
| ⊞ minimum admini... | 1 | 1 | ☐ | 0. | Print... |
| ⊞ minimum value | 1 | 1 | ☐ | 0. | |

Choose one of these highlighted terms to be used as the equivalent term to represent all of these synonymous terms in the next pop-up window (See Display C2.6). Similarly, in this case study you may also treat terms such as 'mine', 'data mine', 'data mining', 'mining' as synonyms with 'mining' as the equivalent term representing all these terms. Hence, in that case, you can select all these terms at once and use the 'Treat as Synonyms' option to create the synonym list. It is important to export the synonyms list that you have created by clicking on File ▶ Export Synonyms. Give a name for the data set and store it in the library that you have created from the project start up code (default). Close the 'Interactive Filter Viewer' window and click 'Ok' on the prompt window to save results.

**Display C2.6: Choose the term to represent the entire data set**

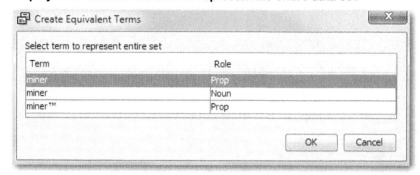

9.  Drag a 'Text Rule builder' node into the diagram, connect it to the text filtering node and run using the default properties. Once the node run is complete, click on Results to view the output. In the Fit Statistics, you will find the misclassification rate to be approximately 20% for the training data and 32% for the validation data (See Display C2.7). This is a very good model given that SAS® Text Miner has automatically built rules to classify abstracts using the training corpus. Close the results window.

**Display C2.7: Fit Statistics results from Text Rule Builder node**

| Target | Target Label | Fit Statistics | Statistics Label | Train | Validation | Test |
|---|---|---|---|---|---|---|
| type | | _ASE_ | Average Squared Error | 0.052155 | 0.069666 | |
| type | | _DIV_ | Divisor for ASE | 1615 | 715 | |
| type | | _MAX_ | Maximum Absolute Error | 0.999985 | 0.996873 | |
| type | | _NOBS_ | Sum of Frequencies | 323 | 143 | |
| type | | _RASE_ | Root Average Squared Error | 0.228374 | 0.263944 | |
| type | | _SSE_ | Sum of Squared Errors | 84.22993 | 49.81141 | |
| type | | _DISF_ | Frequency of Classified Cases | 323 | 143 | |
| type | | _MISC_ | Misclassification Rate | 0.20743 | 0.328671 | |
| type | | _WRONG_ | Number of Wrong Classifications | 67 | 47 | |

10. If you click on the ellipsis button next to the 'Content Categorization Code' under the 'Score' property, you will find the rule expressions automatically built by the text rule builder node (See Display C2.8). These rules are in the same syntax as that of 'SAS Content Categorization Studio' and can be used for building a Boolean rule-based categorizer for all those section categories.

**Display C2.8: Automatic Content Categorization Code generated by the text rule builder node**

```
Content Categorization Code

   F_type =DATAMINING ::
   (OR
   , "mining"
   , "miner"
   , "miner"
   , (AND, (OR, "credits" , "credit" ))
   , (AND, (OR, "costs" , "cost" )))
   F_type =BUSINT ::
   (OR
   , "olap"
   , "bi"
   , (AND, (OR, "dashboards" , "dashboard" ))
   , "intelligence" )
   F_type =SYSARCH ::
   (OR
   , (AND, (OR, "server" , "servers" ))
   , (AND, (OR, "configuration" , "configurations" ))
   , "metadata"
   , "storage"
   , "performance"
   , "operational" )
   F_type =REPORTS ::
   (OR
   , (AND, (OR, "graphs" , "graph" ))
   , (AND, (OR, "reports" , "report" ))
   , (AND, (OR, "map" , "maps" ))
   , (AND, (OR, "details" , "detail" )))
   F_type =STATS ::
   (OR
   , (AND, (OR, "model" , "models" ))
   , (AND, (OR, "procedure" , "procedures" ))
   , (AND, (OR, "analyses" , "analysis" )))
```

11. Launch SAS® Content Categorization Studio, create a new project and name it SGF_Cat_CS_2. Right click on 'SGF_Cat_CS_2' and click 'Add Language'. Select 'English' as the language and click Ok.

12. Right click on 'English' and select the option 'Create Categorizer from Directories'. Browse to the location on your PC or Laptop where you have copied the folder **Case Study 2** and its subfolders from the data provided with this book. Navigate to the 'Top' subfolder contained within the **'SGF_SECTIONWISE'** folder and click Ok to create the categories based on the folder structure (See Display C2.9). Categories are created as BI – Business Intelligence, DM – Data Mining, REPORTS – Visualization and Reporting, STAT – Statistical Analysis and SYSARCH – Systems Architecture.

**Display C2.9: Categories created from an existing folder structure**

13. Select any of these section categories and click on the 'Data' tab. You will find the training path automatically populated for each of these categories since you created them using the existing folder structure instead of creating them manually.
14. Change the Training and Testing Paths of the categories and point them to the designated 'Train' and 'Test' folders (See Table C2.1) to prepare for building a Statistical Categorization model.

**Table C2.1: Testing and Training Paths by section category for a statistical model**

| Category | Training Path | Testing Path |
|---|---|---|
| BI | C:\Data\SGF_SECTIONWISE\Train\BUSINT | C:\Data\SGF_SECTIONWISE\Test\BUSINT |
| DM | C:\Data\SGF_SECTIONWISE\Train\DATAMINING | C:\Data\SGF_SECTIONWISE\Test\DATAMINI |
| REPORTS | C:\Data\SGF_SECTIONWISE\Train\REPORTS | C:\Data\SGF_SECTIONWISE\Test\REPORTS |
| STAT | C:\Data\SGF_SECTIONWISE\Train\STAT | C:\Data\SGF_SECTIONWISE\Test\STAT |
| SYSARCH | C:\Data\SGF_SECTIONWISE\Train\SYSARCH | C:\Data\SGF_SECTIONWISE\Test\SYSARCH |

15. Now that the training and testing paths are set for all the categories, click on Build ▶ Build Statistical Categorizer to generate a statistical model. Once you receive a message 'Build Successful', click Ok and go to the Testing tab on any of the categories, for example DM (Data Mining). You will find a list of files populated from the 'Test' folder of the category ready to be tested against the statistical model that you just built.
16. Click 'Test' and view the results to find out how many of those files have failed the test and how many passed the test (See Display C2.10). As you can observe, there are a few files that failed the test but there are some that passed. Double-click on any of the listed files to open the actual abstract contained within the file. However, the statistical categorizer is a black box model which is why you cannot see the rules working behind the scenes for the categorization process. There is not much you can do to better the performance of a statistical model other than increasing the size of the training corpus for each of these categories. Statistical models largely depend on the quality of training documents by which they are truly separated by each category with respect to another category.

**Display C2.10: Test results for DM (Data Mining) category using statistical model**

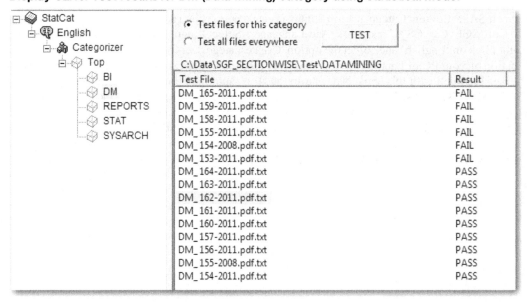

Click 'Testing ▶ Full Test Report' to generate precision and recall scores specific to each category (Display C2.11). If you look at the recall values (In-Cat% column), you can clearly observe a very low score (6%) for BI, medium score (48%) for REPORTS, reasonable scores for DM, SYSARCH and very good score (81%) for

STAT categories. This is a basic model based on the statistical analysis of the training corpus that you can build very quickly using Content Categorization Studio.

**Display C2.11: Full Test Report results of all categories using Statistical model**

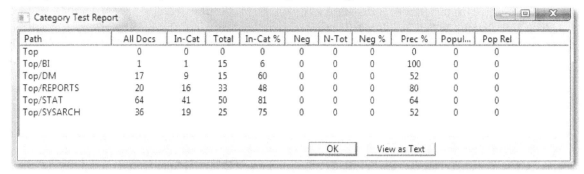

| Path | All Docs | In-Cat | Total | In-Cat % | Neg | N-Tot | Neg % | Prec % | Popul... | Pop Rel |
|---|---|---|---|---|---|---|---|---|---|---|
| Top | 0 | 0 | 0 | 0 | 0 | 0 | 0 | 0 | 0 | 0 |
| Top/BI | 1 | 1 | 15 | 6 | 0 | 0 | 0 | 100 | 0 | 0 |
| Top/DM | 17 | 9 | 15 | 60 | 0 | 0 | 0 | 52 | 0 | 0 |
| Top/REPORTS | 20 | 16 | 33 | 48 | 0 | 0 | 0 | 80 | 0 | 0 |
| Top/STAT | 64 | 41 | 50 | 81 | 0 | 0 | 0 | 64 | 0 | 0 |
| Top/SYSARCH | 36 | 19 | 25 | 75 | 0 | 0 | 0 | 52 | 0 | 0 |

17. Now you have a base model (statistical categorizer model) in Content Categorization Studio to compare against the Boolean rule-based model that you can build using the content categorization code automatically generated from the text rule builder node in SAS® Text Miner. You do not require setting training data to build rule based categorization models. Hence, you may now change your testing paths for all the categories as shown in Table C2.2 and keep the training paths blank.

**Table C2.2: Testing Paths by section category for a rule-based model**

| Category | Testing Path |
|---|---|
| BI | C:\Data\SGF_SECTIONWISE\Top\BUSINT |
| DM | C:\Data\SGF_SECTIONWISE\Top\DATAMINING |
| REPORTS | C:\Data\SGF_SECTIONWISE\Top\REPORTS |
| STAT | C:\Data\SGF_SECTIONWISE\Top\STAT |
| SYSARCH | C:\Data\SGF_SECTIONWISE\Top\SYSARCH |

18. Copy the content categorization code that you have previously generated using the text rule builder node and paste them under the 'Rules' tab for each of the categories. Click on 'Syntax Check' button each time that you copy and paste those rules for every category (See Display C2.12).

**Display C2.12: Syntax check of content categorization code in the Rules tab**

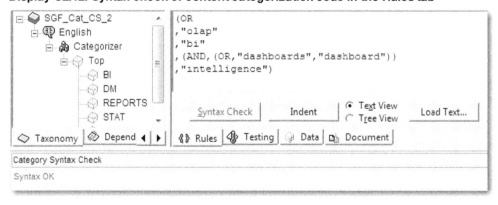

Go to Build ▶ Build Rulebased Categorizer to build a Boolean rule-based categorization model using the rule expressions that you have imported from SAS® Text Miner. You will receive a message 'Build Successful' once you were able to successfully build a Boolean Rulebased Categorization model.

19. Since you have set the testing paths for each of the 5 categories, you may click on any category and go to 'Testing' tab to view the test files. Click 'Test' to test the files based on the Boolean Rulebased categorization model you have built. You will find the test results (pass or fail) and relevancy scores for each of the test files that passed the test.

20. Click 'Testing ▶ Full Test Report' to generate a full test report on the model performance with recall and precision scores (Display C2.13). In general, you will observe that more files pass the test in this model compared to the statistical model that you built previously. This is because Boolean rule based models are flexible so you can write your own rules based on linguistic terms and incorporate Boolean operators for improved accuracy. In this case, you have just exported the automatic rules generated from Text Rule Builder node into SAS® Content Categorization Studio and used them 'as is.' However, after careful examination of the test documents and using the domain knowledge of individual section categories these rules can be further modified to improve accuracy.

**Display C2.13: Full Test Report results of all categories using Boolean Rule based model**

| Path | All Docs | In-Cat | Total | In-Cat % | Neg | N-Tot | Neg % | Prec % | Popula... | Pop Rel |
|------|----------|--------|-------|----------|-----|-------|-------|--------|-----------|---------|
| Top | 0 | 0 | 0 | 0 | 0 | 0 | 0 | 0 | 0 | 0 |
| Top/BI | 107 | 70 | 82 | 85 | 0 | 0 | 0 | 65 | 0 | 0 |
| Top/DM | 98 | 62 | 79 | 78 | 0 | 0 | 0 | 63 | 0 | 0 |
| Top/REPORTS | 145 | 76 | 100 | 75 | 0 | 0 | 0 | 52 | 0 | 0 |
| Top/STAT | 279 | 152 | 174 | 87 | 0 | 0 | 0 | 54 | 0 | 0 |
| Top/SYSARCH | 160 | 80 | 99 | 80 | 0 | 0 | 0 | 50 | 0 | 0 |

21. Click on the 'BI' category and go to the testing tab and carefully examine the files that have failed the test. You will find many terms that are unique to this section category that were not picked up by the text rule builder node during the automatic rule generation process. Terms such as 'information map(s)', 'web report studio', 'information delivery portal' and 'KPI' can specifically identify the topic "business intelligence (BI)" because these are the names of products and features used in SAS Enterprise Business Intelligence suite. Whenever these are identified in the documents, you can conveniently relate them to the BI category. Modify the rules in this category as shown in the Display C2.14. As you can observe, terms such as cube, data, aggregation and table(s) are also added to the rules bound by Boolean operators to ensure more variety of patterns captured.

**Display C2.14: Modified Boolean rules for Business Intelligence (BI) category**

```
(OR
,"olap"
,"bi"
, (AND, (OR,"dashboards","dashboard"))
,"intelligence"
,"KPI"
,"business intelligence"
, (AND,"information", (OR,"map","maps"))
,"web report studio"
,"information delivery portal"
, (SENT,"cube","data")
, (AND,"Aggregation", (OR,"table","tables"))
)
```

Similarly, rules can be modified for the data mining category to include terms specifically related to predictive modeling (regression, decision trees and neural networks), clustering, model comparison (receiver operating characteristic) and so on. Terms representing products or features such as enterprise miner, credit scoring, model manager, etc., related to the data mining field are generally found useful in modifying the rules to match the category (Display C2.15). It is also important to remember that this rule modification is an iterative process requiring careful understanding of terms that can lead to category matching.

**Display C2.15: Modified Boolean rules for the Data Mining (DM) category**

```
(OR
, (OR, "mining", "data mining")
, (AND, (OR, "enterprise", "text"), "miner")
, "logistic regression"
, "sentiment analysis"
, "content categorization"
, "credit scoring"
, "weight of evidence"
, "cluster analysis"
, (OR, "rate-making", "rate making")
, (AND, (OR, "regression", "neural network", "neural networks", "decision tree", "decision trees"),
                 (OR, "model", "models", "modeling"))
, (AND, "predictive", (OR, "model", "models", "modeling", "classification"))
, (AND, "model", (OR, "manager", "management"))
, (OR, "segmentation", "clustering", "segments", "clusters")
, (OR, "AUC", "area under curve", "receiver operating characteristic", "ROC")
)
```

22. Continue to analyze the terms which may represent the products, features or capabilities that better define a particular category and test them well before moving on to the next category. Once all the category rules are modified, rebuild the Boolean rule-based categorizer model again and generate the Full test report (See Display C2.16) to compare its performance against other models that you have built so far. Observe that the model accuracy has increased overall compared to using either the default automatic rules generated from the text rule builder node in SAS® Text Miner or the Statistical categorizer model.

**Display C2.16: Full Test Report using Boolean rule based model with modified rules**

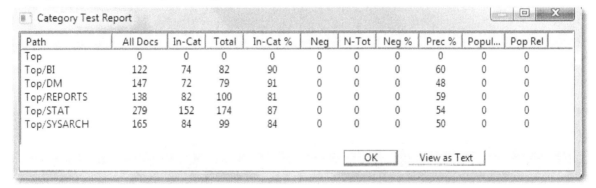

| Path | All Docs | In-Cat | Total | In-Cat % | Neg | N-Tot | Neg % | Prec % | Popul... | Pop Rel |
|------|----------|--------|-------|----------|-----|-------|-------|--------|----------|---------|
| Top | 0 | 0 | 0 | 0 | 0 | 0 | 0 | 0 | 0 | 0 |
| Top/BI | 122 | 74 | 82 | 90 | 0 | 0 | 0 | 60 | 0 | 0 |
| Top/DM | 147 | 72 | 79 | 91 | 0 | 0 | 0 | 48 | 0 | 0 |
| Top/REPORTS | 138 | 82 | 100 | 81 | 0 | 0 | 0 | 59 | 0 | 0 |
| Top/STAT | 279 | 152 | 174 | 87 | 0 | 0 | 0 | 54 | 0 | 0 |
| Top/SYSARCH | 165 | 84 | 99 | 84 | 0 | 0 | 0 | 50 | 0 | 0 |

23. We have provided you the categorization project that contains the modified Boolean rules for each of the categories. Open the project 'SGF_Cat_CS_2' locally in your environment using SAS® Content Categorization Studio and verify the modified Boolean rules. These are the rules that were modified from their original form (when imported from SAS® Text Miner) to improve the accuracy of the models. However, this is a very subjective job and the style and approach of modifying these rules can vary from analyst to analyst.

24. It is important to remember that usually categorization models are not 100% in their predictive ability. Hence, even if you write rules of high precision and quality it can only improve the performance to a certain extent after which it may degrade with the addition of more terms and/or rules there by losing its generality. We suggest that you to practice rule writing and ensuring that those rules are neither too broad nor too specific.

## Summary

- SAS® Content Categorization Studio is an easy-to-use point and click interface used in quickly building models for automatic text categorization process.

- Statistical categorizer utilizes a set of documents from each category in the taxonomy to train the model. However, in terms of model performance statistical categorizer often performs below par.

- Boolean rule-based categorizer works well when the rule terms and Boolean operators are carefully chosen to categorize documents. You can iteratively build the model while testing the rules on a set of documents. It has an additional advantage that you don't need a separate set of documents to train the model.

- The Text rule builder node in SAS® Text Miner is a powerful feature useful to generate preliminary Boolean rule expressions that can be exported to SAS® Content Categorization Studio. It requires a set of documents separated by category to train the model and generate rules.

# Case Study 3 Features-based Sentiment Analysis of Customer Reviews

Jiawen Liu
Mantosh Sarkar
Goutam Chakraborty

## Introduction

Sentiment analysis, (also called sentiment or opinion mining), is a popular technique for summarizing and analyzing consumers' textual reviews about products and services. Feature-based sentiment analysis is used to reveal customers' sentiments not just at the overall product/brand level but also at the product/service specific feature level. There are two major approaches for performing sentiment analysis in SAS® Sentiment Analysis Studio: Statistical model-based approaches and Natural Language Processing (NLP) based approaches. In this case study we demonstrate both of these approaches.

With the rapid evolution and growth of mobile devices, mobile products and services have become "must haves" in our lives. Every day new mobile products and services, often with different features or attributes, are entering the market. Consumers often write online reviews of these products and services and make comments not just about their overall sentiment about such products and services but also about which feature or attribute of each product/service they like or dislike. In this case study, we use both SAS® Text Miner and SAS® Sentiment Analysis Studio to perform sentiment mining. For this purpose, we use artificial data created based on actual online customer reviews of apps for mobile devices.

## Data

We modified and anonymized six hundred actual customer reviews posted online and created an artificial data set. Five hundred of these reviews are used for building models, and the remaining one hundred reviews are used for testing models. Raw textual data has been categorized into positive and negative groups based on 5-star numerical ratings given by the same consumers who posted those review comments on the site. Comments with ratings greater than or equal to 4 stars are considered as positive and less than or equal to 2 stars are considered as negative for the purpose of this case study. The comments with ratings in between 2 and 4 were excluded in this analysis. For modeling, we have two data sets for text mining, and two directories (folders) for sentiment mining as described below.

Text Mining data sets:

- Negative reviews: APP_TM\model\APP_neg.sas7bdat
- Positive reviews: APP_TM\model\APP_pos.sas7bdat

Sentiment Mining Model Data Folders:

- Negative reviews in folder: APP\APP_SM\model\neg
- Positive reviews in folder: APP\APP_SM\model\pos

For testing sentiment mining models, you have two directories that contain positive and negative documents.

Sentiment Mining Test Data Folders:

- Negative reviews in folder: APP\APP_SM\test\neg
- Positive reviews in folder: APP\APP_SM\test\pos

The above mentioned data sets and documents are available in the folder *Case Studies* ▶ *Case Study 3* in the data provided with this book. See the website for this book.

Display C3.1 illustrates a detailed flow chart of the steps involved in our analysis. In the following section, we present the details of the techniques applied in this case study.

**Display C3.1 Feature-based NLP Sentiment Mining flow chart**

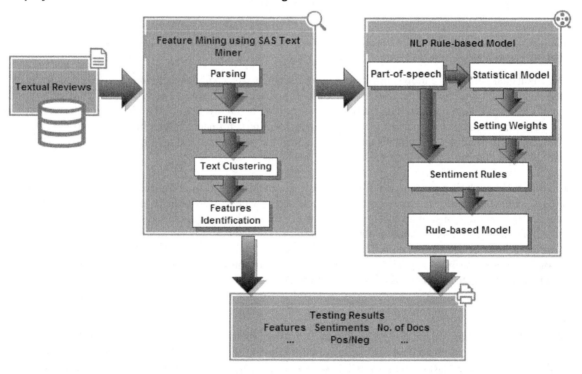

## Text Mining for Negative App Reviews

1.  Open SAS Enterprise Miner 12.1 and create a new project. Create a new library pointing to the folder (*Case Studies* ▶ *Case Study 3* ▶ *APP_TM* ▶ *model*) that contains input data sets for text mining models (App_neg.sas7bdat and App_pos.sas7bdat). Create two data sources with the two data sets in the library that you created. Set up data roles and levels as shown in Display C3.2. Each data set has two columns, "id" and "text". Column "id" is a unique nominal variable to uniquely identify each textual comment. Column "text" contains the textual comment.

**Display C3.2 Column roles and levels information**

| Name | Role | Level |
|------|------|-------|
| id | ID | Nominal |
| text | Text | Nominal |

2. Create a diagram and name it "product_TM". Drag the negative review data source (APP_NEG) into the diagram workspace. Add "Text Parsing", "Text Filter", "Text Topic", and "Text Cluster" nodes to the diagram and connect them as shown in Display C3.3. All of these nodes are run with default options unless specifically mentioned below.

**Display C3.3 Process Flow Diagram in SAS® Text Miner 12.1**

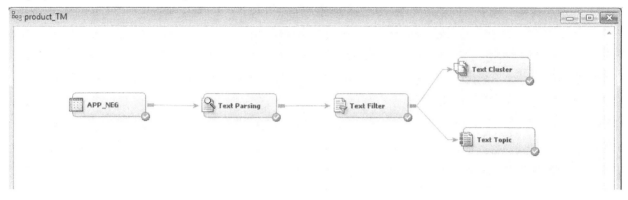

3. In the "Text Parsing" node, find the "Ignore Parts of Speech" option on the properties panel and click the ellipsis button. Hold "Crtl" key on your keyboard and select abbr, aux, conj, det, interj, num, part, prep, pron, and prop (Display C3.4). Click "OK" to save and exit.

**Display C3.4 Parts of Speech Ignore List**

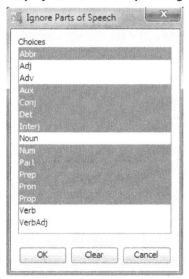

4. Text clustering is used for unsupervised grouping of text documents. In the properties panel of the "Text Cluster" node (Display C3.5), change the value of "Max SVD Dimensions" to 40. Singular Value Decomposition (SVD) is used to reduce dimensionality by transforming the term-by-document frequency matrix into a lower dimensional form. Smaller values of k (2 to 50) are thought to generate better results for text clustering using short textual comments such as the ones used in this case study.

Also, change the value of the property "Descriptive Terms" to 8. This setting ensures that eight terms are displayed to describe each cluster in the results.

**Display C3.5 Text cluster node property panel**

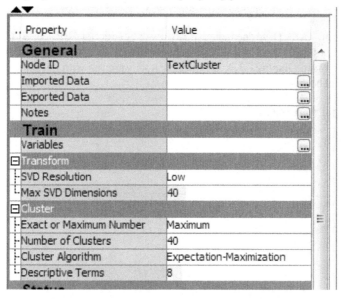

5. Right click on "Text Cluster" node and click "Run". After the run completion, click "Results" on the pop-up window. Display C3.6 shows the results from text clustering. Thirteen clusters are generated. Each cluster has a fair number of documents. The descriptive terms as shown in Display C3.6 for each cluster give us a sense of what terms are defining each of those clusters.

**Display C3.6 Cluster results**

| Cluster ID | Descriptive Terms | Frequency |
|---|---|---|
| 1 | skins +download installed +skin list downloaded +reboot always ... | 24 |
| 2 | waste money developers +install +thing +want abc 'a lot of' ... | 10 |
| 3 | 'great product' icons +icon great refresh smaller +happen +look ... | 12 |
| 4 | competitor free 'competitor products' better +find toggle products +nice ... | 32 |
| 5 | standard skins +load super +'weather product' +update home last ... | 21 |
| 6 | +lock +unlock broken +force animations +device +appear annoying ... | 16 |
| 7 | fixed +day great temperature +fix time working installed ... | 22 |
| 8 | +time +love properly geolocation updated updating broken updates ... | 22 |
| 9 | accounts permissions abc +remove reason +account +buy +developer ... | 10 |
| 10 | 'anonymous statistics' statistics ugly +option anonymous cents bought battery ... | 19 |
| 11 | +problem +version problems system anymore paid +year refresh ... | 16 |
| 12 | +screen space 'home screen' refund home 'a lot of' +animation bugs ... | 18 |
| 13 | +'battery product' latest 'latest update' stopped working +fix +update last ... | 28 |

6. From step 5 you got a basic idea of how consumers may be using this product and some of the features of this product. In this step, you will run the text topic node to get deeper insights about topics of conversation and features mentioned in those topics.

7. Right-click on the "Text Topic" node and select "Run". It will run the node with default settings. After the run is complete, click "Results". In the results window, maximize the "Topics" window and you can see the list of topics generated (Display C3.7). From Display C3.7, you can see that features such as "Weather", "Update", "Time/ Clock", or "Money/ Price" appear several times in the twenty-five

topics generated. Using the results from text clustering modify and customize the topics to narrow down these to a handful of relevant topics.

**Display C3.7 Text topic results obtained using default settings**

| Topic ID | Topic | # Docs |
|---|---|---|
| 1 | +update,last,+fix,+break,+skin | 37 |
| 2 | competitor,+competitor product,better,half,free | 14 |
| 3 | +screen,home,home screen,weather,space | 23 |
| 4 | +stop,working,battery,+time,+battery product | 26 |
| 5 | +skin,+install,+download,list,+show | 32 |
| 6 | +late,+late update,+fix,+update,+clock | 24 |
| 7 | money,waste,worth,+developer,+download | 15 |
| 8 | +time,wrong,wrong time,+update,weather | 33 |
| 9 | +permission,+account,+remove,abc,reason | 11 |
| 10 | +load,time,system,+clock,+download | 30 |
| 11 | +love,+update,+device,+product,+issue | 27 |
| 12 | +problem,+version,+pay,last,+year | 21 |
| 13 | +statistic,anonymous,anonymous statistics,+add,ugly | 18 |
| 14 | +device,+unlock,+appear,+animation,+delete | 23 |
| 15 | +fix,temperature,working,+bug,+pay | 26 |
| 16 | +freeze,+day,+clock,+find,toggle | 24 |
| 17 | +setting,+look,manually,+device,+star | 23 |
| 18 | standard,+skin,+find,+standard skin,toggle | 25 |
| 19 | +work,fine,system,+install,+fix | 30 |
| 20 | location,weather,back,geolocation,+show | 28 |
| 21 | great,great product,+look,weather,+developer | 14 |
| 22 | +look,+disappoint,+device,a bit,+icon | 30 |
| 23 | +fix,+force,+add,system,+close | 29 |
| 24 | +icon,+option,+screen,+work,+large | 22 |
| 25 | refund,+hour,+purchase,useless,+find | 24 |

8. In this step you can create customized "User Topics". Right-click and rename the current text topic node as a Text Topic (Default). Then add another text topic node and attach it to the text filter node. Right-click and rename this as a Text Topic (Custom). In the properties panel of the "Text Topic (Custom)" node, change the value of "Number of Multi-term topics" to "0", because you want to generate customized user topics and you do not want SAS Enterprise Miner 12.1 to do this for you automatically. Click the ellipsis button (highlighted) next to "User Topics" as shown in Display C3.8.

**Display C3.8 Text topic node property panel**

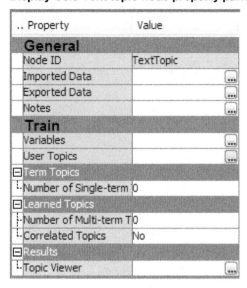

9. In the pop-up window (See Display C3.9), you can create custom topics of your interest. To add a new topic, click on the orange sun icon (pointed arrow); to delete an existing and unwanted item, highlight the row you want to remove, and then click delete icon.

**Display C3.9 User topics workspace**

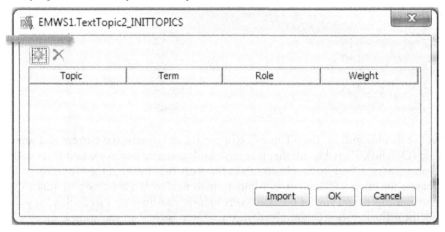

10. For negative reviews, you will create eight customized topics. Add topics as shown in Display C3.10. In this window, all rows for the same topic value represent a unique topic; For example, "battery" and "battery usage" are values for the topic "battery life". "Weight" indicates the importance of each term within its topic. In this case study, we give the same importance to each term. Hence, all weights will be equal to "1". In practice, you should experiment with different weights based on domain expertise and then test to see how those work. Click "OK" to save and exit. Instead of typing all of the topic terms, click on the Import button and select the SAS data set **texttopic_pos** from your library. Then, right click on the "Text Topic" node and select "Run". After the run completes, click "Results" to see user-defined topics along with the number of documents for each topic.

**Display C3.10 User topics with customized topics**

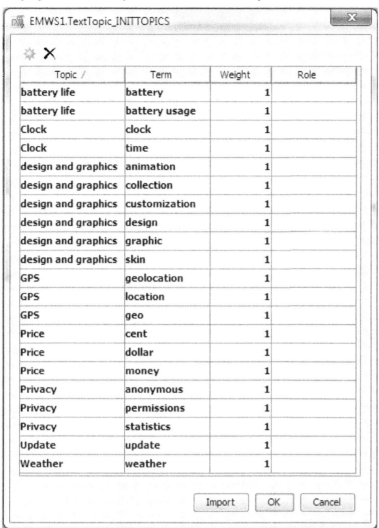

| Topic / | Term | Weight | Role |
|---|---|---|---|
| battery life | battery | 1 | |
| battery life | battery usage | 1 | |
| Clock | clock | 1 | |
| Clock | time | 1 | |
| design and graphics | animation | 1 | |
| design and graphics | collection | 1 | |
| design and graphics | customization | 1 | |
| design and graphics | design | 1 | |
| design and graphics | graphic | 1 | |
| design and graphics | skin | 1 | |
| GPS | geolocation | 1 | |
| GPS | location | 1 | |
| GPS | geo | 1 | |
| Price | cent | 1 | |
| Price | dollar | 1 | |
| Price | money | 1 | |
| Privacy | anonymous | 1 | |
| Privacy | permissions | 1 | |
| Privacy | statistics | 1 | |
| Update | update | 1 | |
| Weather | weather | 1 | |

11. From Display C3.11, you can see that the newly created topics are fairly distributed with the exception of the topic "privacy" which has a frequency of 5.

**Display C3.11 Text topic results**

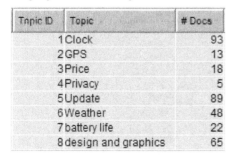

| Topic ID | Topic | # Docs |
|---|---|---|
| 1 | Clock | 93 |
| 2 | GPS | 13 |
| 3 | Price | 18 |
| 4 | Privacy | 5 |
| 5 | Update | 89 |
| 6 | Weather | 48 |
| 7 | battery life | 22 |
| 8 | design and graphics | 65 |

12. To get a better understanding of these topics and how they are related to the terms, open the interactive topic viewer by selecting the ellipsis button in the properties panel next to Topic viewer in the Results section, as shown below in Display C3.12.

**Display C3.12 Launch Interactive topic viewer**

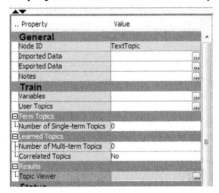

13. The interactive topic viewer allows users to select each topic and find the terms and documents that relate to that topic, as shown below in Display C3.13. As you can see, the bottom section (documents) of the window displays all text documents with the document topic weight for the topic "Clock" which you have selected from the top section (topics). The Interactive topic viewer also lets users analyze and manage term cut-off and document cut-off scores. See Chapter 6, *"Clustering and Topic Extraction,"* for more information on the Interactive topic viewer.

**Display C3.13 Interactive Topic Viewer**

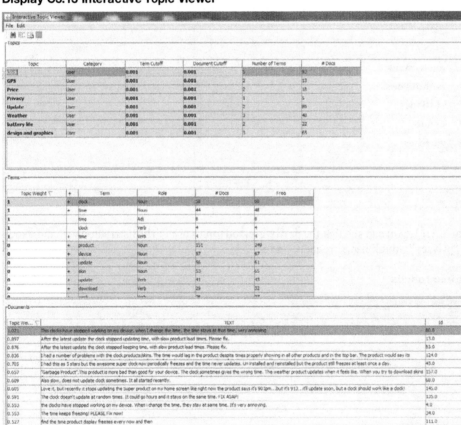

You should explore these results to get a better sense of how the customized topics enhance the analysis.

## Text Mining for Positive App Reviews

You will now create another process flow within the same diagram using the positive app reviews data set and repeat Steps 3 to 6 above in order to get a sense of what customers are talking about in the positive reviews.

1. Select all of the nodes except the APP_NEG (data source node) and then "copy" and "paste" in the same diagram (Display C3.14) workspace. By doing this, the property settings of all nodes are automatically retained so that you do not have to change each node's settings. Now add the positive review data (APP_POS) to the data sources and then click and drag the data source into the diagram and connect it to the "Text Parsing" node as shown in Display C3.14.

**Display C3.14 Diagram with APP_POS process flow nodes**

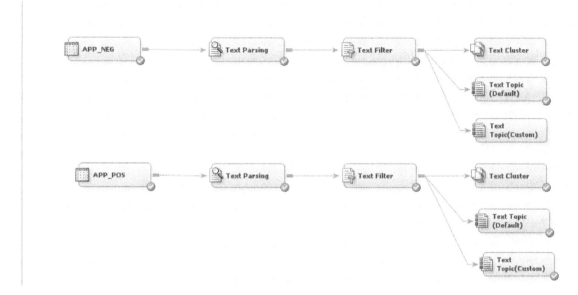

2. From the APP_POS process flow, right click on the "Text Cluster" node and click "Run". Display C3.15 shows the cluster results from the text cluster node. Eleven clusters are generated. Documents are fairly distributed among these clusters. The descriptive terms from each cluster give us a sense of the characteristics of grouped comments as shown in Display C3.15.

**Display C3.15 Text cluster results**

| Cluster ID | Descriptive Terms | | Frequency |
|---|---|---|---|
| 1 | latest different fixed +update +style +system +fix few | ... | 38 |
| 2 | +love +'weather product' wallpaper current favorite version weather paid | ... | 32 |
| 3 | choose awesome highly nice battery animated animations excellent | ... | 34 |
| 4 | +download skins options +theme love updates +match +easy | ... | 18 |
| 5 | worth well money full +old +keep +customization accurate | ... | 8 |
| 6 | +'great product' great +product battery +add years +time +old | ... | 26 |
| 7 | 'home screen' screen home +thing animations years 'best product' +day | ... | 12 |
| 8 | sense +developer +work +want +keep flawlessly works developers | ... | 20 |
| 9 | 'best products' 'good products' beautiful +good better best 'best product' +'good product' | ... | 23 |
| 10 | fixes working works great +run clocks accurate always | ... | 27 |
| 11 | fixing found +thing +match +skin +time +issue +find | ... | 12 |

3. From step 8, you have a basic idea of how consumers are using this product and some of the features of this product that they are talking about positively in their reviews. Next you will run the text topic node to get deeper insights about topics of conversation and features mentioned in those topics. Run the "Text Topic" node with default settings. The generated topics in Display C3.16 show that features

such as "Weather", "Update", "Time/ Clock", or "Money/ Price" appear several times in the twenty-five topics that are generated. The results show pretty much similar features as discovered for negative review data source. Therefore, you will continue to use the same "User Defined" topics and run the custom node.

**Display C3.16 Text Topic node results with default settings**

| Topic ID | Topic | # Docs |
|---|---|---|
| 1 | great,+great product,+product,+year,+issue | 29 |
| 2 | +fix,+update,latest,+issue,+device | 28 |
| 3 | +want,+developer,+work,+work,+pay | 23 |
| 4 | +weather product,weather,+option,accurate,fun | 22 |
| 5 | great,+work,working,nice,+star | 27 |
| 6 | +good,+well product,market,+good product,+problem | 38 |
| 7 | screen,home,home screen,weather,+thing | 13 |
| 8 | +system,different,+clock,+clock,+style | 29 |
| 9 | +skin,+download,+option,+device,weather | 28 |
| 10 | +find,+time,+clock,+thing,weather | 20 |
| 11 | +device,+developer,+day,always,+problem | 28 |
| 12 | choose,+skin,highly,battery,+problem | 21 |
| 13 | nice,+animation,+customize,love,+weather product | 27 |
| 14 | +love,+product,clock product,a lot,+clock | 29 |
| 15 | +update,last,+work,happy,weather | 27 |
| 16 | version,weather,live,few,+look | 28 |
| 17 | +update,competitor,+system,+competitor product,+work | 22 |
| 18 | +add,weather,nice,animated,+fix | 33 |
| 19 | beautiful,+good,weather,a lot,good | 12 |
| 20 | +clock,stock,+add,+easy,+option | 32 |
| 21 | +look,great,+customize,good,+work | 22 |
| 22 | worth,well,money,+keep,+work | 19 |
| 23 | excellent,excellent product,few,+device,+issue | 16 |
| 24 | awesome,battery,+well product,+device,+star | 16 |
| 25 | +work,fine,latest,love,+customize | 22 |

4.  Now right click on the "Text Topic (Custom)" node and select "Run". Open the results and see user-defined topics along with the number of documents. As shown in Display C3.17, the "Text Topic" node only generated six topics. Comparing the custom topics results between positive and negative reviews reveal that there was not a single document that talked about GPS or privacy topics in the positive reviews. The custom created topics are fairly distributed with the exception of the topic "price" which has a frequency of only 6.

**Display C3.17 Custom topic results for positive reviews**

| Topic ID | Topic | # Docs |
|---|---|---|
| 1 | Clock | 54 |
| 2 | Price | 6 |
| 3 | Update | 30 |
| 4 | Weather | 58 |
| 5 | battery life | 10 |
| 6 | design and graphics | 45 |

As explained in Step 6.g from the negative comments process flow, you can use interactive topic viewer from the text topic node for easier and better understanding of topics and cut-off scores. This analysis has greatly helped in getting a high-level idea of the sentiment expressed in the user reviews. These insights are very valuable for digging deeper into the text in order to extract sentiment at feature and document level. The next section will take you a step further into sophisticated sentiment mining using SAS® Sentiment Analysis Studio.

## NLP Based Sentiment Analysis

Parts-of speech (POS) tagging is often the most time consuming and challenging task before doing sentiment analysis of any text data. Online textual reviews are often short, non-grammatical sentences and contain slang terms, abbreviations, and symbols which make the POS tagging even more difficult. However, the text mining in SAS® Enterprise Miner 12.1 provided us with good insights into what consumers are talking about in these reviews. Those insights will come in very handy as you conduct sentiment analysis via SAS® Sentiment Analysis Studio 12.1.

1. Open SAS Sentiment Analysis Studio 12.1. Start a new project. Use default settings for both the rule-based model and the statistical model.
2. In the "Corpus" white workspace, right click and select "New Corpus." Type in a name such as App_Reviews.
3. Right click on "Positive" and choose "Add a Directory". Go to the folder (*Case Studies ▶ Case Study 3 ▶ APP_SM ▶ model ▶ pos*) where files for modeling the positive sentiments are stored and click "select folder." You will see the 250 positive reviews (text files) imported and listed in the "Reference Files" workspace. To see the contents of any particular file, click and select it in the Reference Files space (See Display C3.18).

**Display C3.18 Sentiment Analysis Studio corpus view panel**

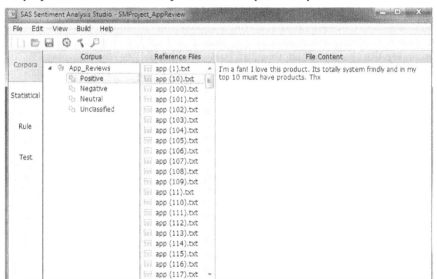

4. Repeat the previous step to add the Negative directory (*Case Studies ▶ Case Study 3 ▶ APP_SM ▶ model ▶ neg*) where files for modeling the negative sentiments are stored. You do not have "Neutral" and "Unclassified" documents in this case study.
5. Build the Statistical Model
   a. Click on the "Statistical" tab (left panel). Right click in "Statistical Models" white workspace and choose "New Model". Name the new model (such as Adv_App_Review), select "Advanced", and click "OK". On the statistical model configuration panel (See Display C3.19), you can make changes to improve your statistical model performance. Here, you will use default settings.

**Display C3.19 Statistical model configuration panel**

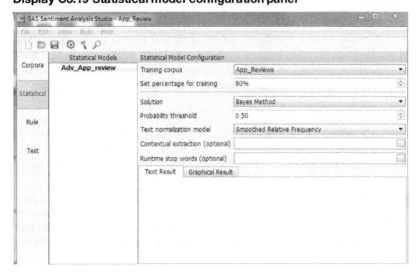

b.  On the top menu, click "Build" and select "Build Statistical Model". Select the model name that you just created and click OK to run. After the run completes, you will see the results as shown in Display C3.20. One of the major reasons for building a "Statistical Model" is to extract a list of most commonly used words and terms and reuse them as a start list for a rule-based model. From the result, you can see that the overall accuracy of "Statistical Model" is 76% (Display C3.20). Later you will compare the accuracy of both statistical model and rule-based model in the conclusion section.

**Display C3.20 Statistical model training results**

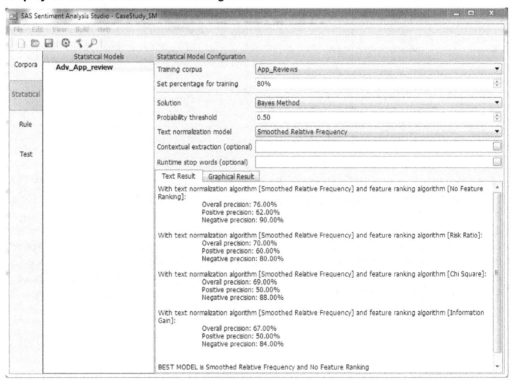

6.  Building a Rule-Based Model
    a.  Click on the "Rule" tab from the left panel. On the top menu, click "Build" and select "Import

Learned Features" to import all keywords to start building a rule-based model. You can load the features from a model or a data file. Let us start with importing from the "Statistical Model". Change the "Number of keywords to import" to "200" and click "OK". The reason to keep 200 keywords is simply because it is easier to review, edit, and change in later steps. Next click on "Tonal Keywords" and you will see 200 positive words and 200 negative words. You can scroll down and see different keywords for positive or negative reviews. While many of these keywords look reasonable, some do not. As an analyst, you often start with these keywords and then modify them to create your own list. Display C3.21 is an example of first four positive keywords that were imported into "Tonal Keywords". The weight of each word is auto-calculated based on the word's frequency.

**Display C3.21 Examples of positive tonal keywords**

| Positive | Negative | Neutral | |
|---|---|---|---|
| | Type | Body | Weight |
| 1 | CLASSIFIER | Thank | 666.185 |
| 2 | CLASSIFIER | Nice | 601.464 |
| 3 | CLASSIFIER | Awesome | 496.226 |
| 4 | CLASSIFIER | Beautiful | 307.695 |

    b.   It is recommended that you create multiple rules to achieve better results. At this stage of analysis, all terms and words are serving globally. Hence a rule-based model when tested on the test data will yield the same number of documents for each feature. It often works better if you divide words into different categories. You will create different categories under "Intermediate Entities". The reason for doing this is to correctly and thoroughly detect each sentiment word along with its feature.

    c.   You can create custom rules either globally or create rules for each feature. To create rules globally, click "Tonal Keyword". Under "Positive", "Negative", or "Neutral" tabs, first edit the "Body" section, and then change "Type" and "Weight". Note that rules will not work if you do not change the "Type" of rules. Display C3.22 is an example of creating rules in "Tonal Keywords". In this example, a global CONCEPT rule "love@" is created. Words such as "love", "loves", "loved", and "loving" are detected and counted for all features.

**Display C3.22 Example of creating Concept Rule under Tonal Keyword**

| Rules | Search Rules | | | All types | ▼ |
|---|---|---|---|---|---|
| Tonal Keyword | Positive | Negative | Neutral | | |
| ◢ Intermediate Entities | | Type | | Body | Weight |
| | 1 | CONCEPT | love@ | Global Rule | 1 |
| | 2 | CLASSIFIER | | | 1 |

Display C3.23 is an example of creating rules for a specific feature. This rule will be triggered only if the feature name "weather", an adverb and a positive sentiment adjective (we have defined a group of positive sentiment adjectives), happen to occur within a distance of seven words.

**Display C3.23 Example of creating rules for a specific feature**

| Rules | Search Rules | | | | All types | ▼ |
|---|---|---|---|---|---|---|
| Tonal Keyword | Definitions | Positive | Negative | Neutral | | |
| ▷ Intermediate Entities | | Type | | Body | | Weight |
| ◢ Products | 1 | PREDICATE_RULE | (DIST_7,"_def{Productweather}","_a{_def{ADV}}","_b{_def{POSADJ2}}") | | 2.5 |
| ◢ Product | 2 | CLASSIFIER | | Rule for "Weather" feature | | 1 |
| weather | 3 | CLASSIFIER | | | | 1 |

d.  For your convenience, we have created a set of rules and entities for use with this data set. Go to "File" and select "Import Rules". Choose the XML document rule file "App_Rule.XML" and click OK. All custom created rules will be imported into this project.

e.  Features discovered from text mining analysis are implemented in this section. Currently you can see all these features under "Product." If you want to create a new product, right click on "Products" and select "New Product." If you want to add features to a "Product", right click on "Product" and select "New Feature".

f.  Under "Intermediate Entities", you will find the entities "ADV", "NEGADJ1", "NEGADJ2", "NEGWORD", "OP", "POSADJ1", "POSADJ2", and "VERB".

g.  By clicking on "ADV", you can see a list of words that are used as adverbs. Words in "NEGADJ1" are negative adjectives with sentiment weight of 1. The same rules apply to "NEGADJ2". "POSADJ1" are positive adjectives with sentiment weight of 1. Same rules apply to "POSADJ2". NEGADJ1/POSADJ1 are the lists with all adjective words that we considered having less negative/positive sentiments than the words in NEGADJ2/POSADJ2. When "ADV" word and "NEGADJn"/"POSADJn" word occur together, then the sentiment weight becomes "n+0.5". Sentiment weights are often subjective and difficult to interpret as explained in Chapter 8, *"Sentiment Analysis."* Interested readers are advised to follow the references mentioned in the text.

"OP" contains other products that can be considered competitors. In "VERB", it contains verbs that are often used in these reviews. In this custom list you will find the type of "VERB" is "CONCEPT" with the symbol "@" added to the verbs. Symbol "@" will detect all verb forms. CLASSIFIER, CONCEPT, CONCEPT_RULE, and PREDICATE_RULE types were used for this case study. CLASSIFIER rules are used to match a term or a phrase. We used CLASSIFIER rules to match the words which can be used only for a feature. For example, "expensive" can only be used for feature "price". Display C3.24 shows an example of PREDICATE_RULE and CONCEPT_RULE. In "DIST_n", n is the number of words between matches on rules. The first match is tagged as position 1, the second match as position2 and it goes on until the last match (n). "_def" matches the definition for products or features. "_def{Productweather}" is a definition for the feature "weather" of "Product". "_a" and "_b" are arguments that match when these two arguments match in a document. "SENT" will match the words and definition only within the same sentence.

**Display C3.24 Examples of PREDICATE_RULE and CONCEPT_RULE**

| # | Type | Body | Weight |
|---|------|------|--------|
| 1 | PREDICATE_RULE | (DIST_7, "_def{Productweather}","_a{_def{VERB}}", "_b{_def{NEGWORD}}") | 1 |
| 2 | CONCEPT_RULE | (SENT,"_c{_def{NEGADJ1}}","_def{Productweather}") | 1 |
| 3 | PREDICATE_RULE | (DIST_7,"_def{Productweather}","_a{_def{ADV}}","_b{_def{NEGADJ2}}") | 2.5 |
| 4 | PREDICATE_RULE | (DIST_7,"_def{Productweather}","_a{_def{ADV}}","_b{_def{NEGADJ1}}") | 1.5 |
| 5 | CONCEPT_RULE | (SENT,"_c{_def{NEGADJ2}}","_def{Productweather}") | 2 |
| 6 | CLASSIFIER | | 1 |
| 7 | CLASSIFIER | | 1 |
| 8 | CLASSIFIER | | 1 |
| 9 | CLASSIFIER | | 1 |
| 10 | CLASSIFIER | | 1 |
| 11 | CLASSIFIER | | 1 |
| 12 | CLASSIFIER | | 1 |
| 13 | CLASSIFIER | | 1 |
| 14 | CLASSIFIER | | 1 |
| 15 | CLASSIFIER | | 1 |

h.  Next go to "Build" and select "Build Rule Based Model". After the model is built you will be notified with a successful message but no visible results.

i. Go to the "Test" tab. Right click in the "Test Data" white workspace and select the "New Test Directory". Go to test folders *Case Studies* ▶ *Case Study 3* ▶ *APP_SM* ▶ *model* ▶ *test* for sentiment mining and import both the "neg" and "pos" folders separately.

j. Right-click on each imported directory and select "Test in Rule Based Model". Display C3.25 and Display C3.26 show the testing results from a rule-based model for positive and negative test data, respectively. Both these results show 92% accuracy. Additionally, positive and negative sentiments toward to each feature is illustrated.

**Display C3.25 Testing results from Positive Directory**

■ Positive ■ Negative ■ Neutral

Results for selected folder:
This directory is Positive
Positive precision is 92.00%.
Number of articles:50
Number of positive articles:46
Number of negative articles:3
Number of neutral articles:1
Positive percent:92.00%.

(46 / 3 / 1)

| | |
|---|---|
| Product/performance | (117 / 17 / 0) |
| Product/clock | (14 / 6 / 0) |
| Product/update | (5 / 2 / 0) |
| Product/weather | (14 / 2 / 0) |
| Product/design | (12 / 0 / 0) |
| Product/price | (3 / 0 / 0) |
| Product/theme | (10 / 1 / 0) |
| Product/service | (3 / 0 / 0) |
| Product/battery | (1 / 1 / 0) |
| Product/privacy | (0 / 1 / 0) |
| Product/GPS | (1 / 0 / 0) |

**Display C3.26 Testing results from Negative Directory**

■ Positive ■ Negative ■ Neutral

Results for selected folder:
This directory is Negative
Negative precision is 92.00%.
Number of articles:50
Number of positive articles:3
Number of negative articles:46
Number of neutral articles:0
Positive percent:6.00%.

(3 / 46 / 1)

| | |
|---|---|
| Product/update | (3 / 6 / 0) |
| Product/clock | (3 / 14 / 0) |
| Product/performance | (14 / 86 / 0) |
| Product/privacy | (0 / 7 / 0) |
| Product/service | (0 / 2 / 0) |
| Product/battery | (0 / 3 / 0) |
| Product/theme | (2 / 8 / 0) |
| Product/price | (1 / 6 / 0) |
| Product/weather | (1 / 8 / 0) |
| Product/design | (1 / 0 / 0) |
| Product/GPS | (1 / 7 / 0) |

k. Expand the negative testing directory, right click on the text file 1.txt and select "Test in Rule-based Model". The testing results are shown in Display C3.27. This file has been classified as a negative comment. Four features are identified in this coment. Words in blue are identified as "features"; words in green are identified as positive sentiment; and words in red are detected as negative sentiment.

l. Expand the positive testing directory, right click on the text file 2.txt and select "Test in Rule-based Model". The testing results are shown in Display C3.28. This file has been classified as a positive comment. Four features have been detected.

**Display C3.27 Single text file test result (Negative Directory)**

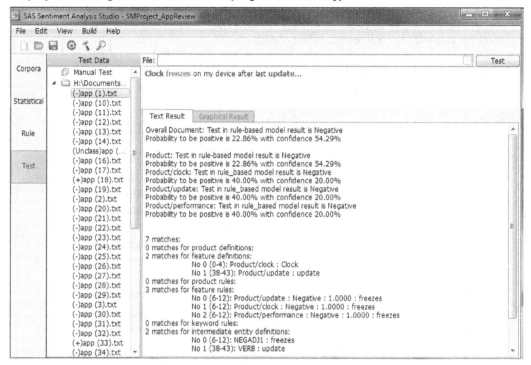

**Display C3.28 Single text file test result (Positive Directory)**

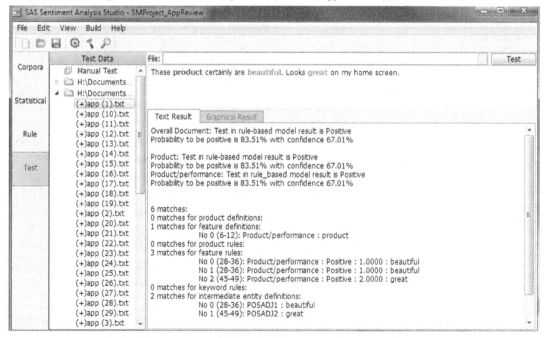

## Summary

The NLP rule based model provides deeper insights in understanding consumers' sentiments. Primarily, you understand that the product users are satisfied with product graphics designs, but are concerned and unhappy about privacy features. Also, looking at the model results, the rule-based model performs extremely better than a statistical model (Table C3.1).

**Table C3.1 Statistical and rule-based model comparison**

| Products | Statistical Model | | | Rule-based Model | | |
|---|---|---|---|---|---|---|
| | Positive Precision | Negative Precision | Overall Precision | Positive Precision | Negative Precision | Overall Precision |
| Product | 62% | 90% | 76% | 92% | 92% | 92% |

The statistical model was built without identifying the product features. On the other hand, a rule-based model was built using a list of product features. These product features were derived from text mining analysis using SAS® Text Miner and also from the terms extracted from the statistical model. Using both SAS Text Miner and rule-based Sentiment Analysis using SAS® Sentiment Analysis Studio helps to achieve better results.

# Case Study 4 Exploring Injury Data for Root Causal and Association Analysis

Mary Osborne
Murali Pagolu

## Introduction

The United States Consumer Product Safety Commission (CPSC) is an independent agency in the United States that regulates the sale and manufacturing of more than 15,000 products in the United States. CPSC conducts research on products in the market and issues bans or recalls on unsafe products from use to prevent potential hazards. Consumers use the CPSC hotline or their website to report injuries, incidents or issues concerning the safety of products. In addition to these, CPSC operates the National Electronic Injury Surveillance System (NEISS), which collects Emergency Room data from 100 hospitals and use that data to calculate probabilistic national estimates.

## Objective

In this case study, we attempt to explore, explain and analyze consumer complaints from the National Electronic Injury Surveillance System using **SAS® Text Miner and SAS® Enterprise Content Categorization Studio**. Through this analysis you can solve the following business problems and help to improve the search in a public records database about injuries:

1. Determine the types of injuries sustained by people.
2. Examine which actions lead to which types of injuries.
3. Identify which products are specifically associated with which type of injuries.

## Data Description

For this case study, data is extracted from NEISS online query system containing consumer reported injury data. The following SAS data sets are provided to you for this case study.

Neiss_school.sas7bdat – Data related to injuries and issues associated with Falls/Falling

Neiss_sports.sas7bdat – Data related to injuries and issues associated with Sports

Neiss_swallow.sas7bdat – Data related to injuries and issues associated with swallowing products

These data sets are available in the folder *Case Studies* ▶ *Case Study 4*. See the website for this book. In addition to these data sets, we have also provided you with the raw files for each of these injury categories.

SAS® Content Categorization project 'CPSC_CC' and SAS® Enterprise Miner project 'CPSC' are also made available in this folder for your reference.

## Step-by-Step Instructions

### Part 1: SAS Text Miner

1. Copy the folder *Case Study 4* to a location on your PC (e.g., D:\).
2. Launch SAS® Enterprise Miner and create a new project and name it 'CPSC'. Enter the SAS code shown below in the project start up code and click 'Run now' to assign 'Data' library.
   *libname CPSC "D:\Case Study 4";*
3. Create three diagrams 'school', 'sports' and 'swallow' in the new project, one for each data set to explore and analyze them individually. Register all the data sets specified in Step 1 with default roles assigned by the SAS® Enterprise Miner data source wizard.
4. Open the diagram 'school' and drag and drop the data source 'neiss_school' into the diagram. Also, drag and drop 'text parsing', 'text filter' and 'text topic' nodes into the diagram and connect them in a sequence as shown in Display C4.1 below. Right click on the 'Text Topic' node and choose 'Run'.

**Display C4.1: Text mining process flow to analyze NEISS complaints related to 'schools'**

5. Click on the ellipsis button next to **'Filter Viewer'** under 'Results' in the properties panel for 'Text Filter' node to open 'interactive filter viewer'. Within the Terms pane, you will find all the terms parsed by the text parsing node and subsequently chosen to be dropped/kept for further analysis by the 'Text Filter' node. Sort the list by the 'TERM' column and choose the row with "+fall" and click on the "+" next to the term *fall*. The list will expand and you can see all the verb forms of "fall" (See Display C4.2). Instead of counting these as separate terms, they are treated as the same term for the purposes of analysis. You can also see two words "fall" in separate entries, but they are of the roles noun and adjective, respectively. In addition to these, you can also see a term 'fal' which is most probably misspelt and should have been the same term fall. Hence, both 'fal' and 'fall' should be considered as synonyms though one of these terms is misspelled.

**Display C4.2: Partial screenshot of expanded "fall" term in interactive filter viewer**

| | TERM ▼ | FREQ | # DOCS | KEEP | WEIGHT | ROLE | ATTRIBUTE |
|---|---|---|---|---|---|---|---|
| | fall | 35 | 31 | ☑ | 0.575 | Noun | Alpha |
| | fall | 5 | 5 | ☑ | 0.799 | Adj | Alpha |
| ⊟ | fall | 568 | 533 | ☑ | 0.226 | Verb | Alpha |
| | fall | 5 | 5 | | | Verb | Alpha |
| | fell | 526 | 496 | | | Verb | Alpha |
| | falling | 37 | 36 | | | Verb | Alpha |
| | fal | 1 | 1 | ☐ | 0.0 | Prop | Alpha |

6. Repeat Step 5 for the word "+head" and you can see the number of documents in which the term 'heads' is found to be lot lesser compared to the term 'head' (See Display C4.3). This may indicate that accidents involving two or more heads such as bumping heads is less likely to occur than accidents involving s a single head.

**Display C4.3: Partial screenshot of expanded term "head" in interactive filter viewer**

| | TERM ▼ | FREQ | # DOCS | KEEP | WEIGHT | ROLE | ATTRIBUTE |
|---|---|---|---|---|---|---|---|
| ⊟ | head | 448 | 394 | ☑ | 0.257 | Noun | Alpha |
| | heads | 8 | 8 | | | Noun | Alpha |
| | head | 440 | 387 | | | Noun | Alpha |

7.  As you found in Step 5, there are many terms which are misspelled in the documents. Hence, the individual term frequency (FREQ) is not accurate since the misspelled terms attribute to noise in the data. The 'Text Filter' node offers a spell check feature to suggest the correct spellings of misspelled words. To enable spell check in the 'Text Filter' node, choose 'Yes' for the property "Check Spelling" and select an english dictionary in the 'Text Filter' properties (Display C4.4). For this case study, we have provided you a SAS data set (engdict. sas7bdat) pre-built with english words to serve as a dictionary. Copy this data set to the same location as you have used for creating a library in Sstep 2. Click on the ellipsis button next to 'Dictionary' and browse through the library CPSC to select the dictionary data set.

**Display C4.4: Properties in Text Filter node to enable spell check**

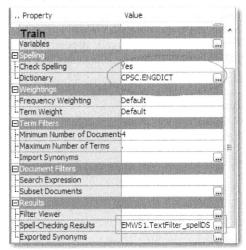

Once you choose the spell check options and rerun the 'Text Filter' node, the results of the spell check are stored in the data set specified against 'Spell-Checking Results' property under the Results section (Display C4.4). In this case, EMWS1 is the name of the workspace folder/library specific to this diagram and 'Text Filter_spellDS' is the name of the data set which contains these results.

8.  Navigate to the following path based on the Enterprise Miner project location that you chose to create this project.
    <....EMProject Location...>\Workspaces\EMWS1

    Make a copy of 'TextFilter_spellDS.sas7bdat' data set in the same location as the CPSC library.

9.  Go to the folder representing the CPSC library, right click on this data set and choose 'Open with SAS X' (where X is the version of SAS installed on your machine) to open it in Base SAS. By default the data set is opened in browse mode (Display C4.5). From the menu options, select Edit ▶ Edit mode. This will enable you to edit the entries in the data set. In the data set you will find that the incorrectly spelled words are caught by the spell checker and correct spellings are suggested (See Display C4.5). However, there may be some words which are appropriate in their original form and spell check is not required.

**Display C4.5: Output data set created as a result of using spell check in the Text Filter node**

| | Parent # Docs | Term | # Docs | Parent | Role | Parent Role |
|---|---|---|---|---|---|---|
| 1 | 9 | small | 4 | small | Noun | Adj |
| 2 | 2751 | schhol | 1 | school | Prop | Noun |
| 3 | 128 | maile | 1 | male | Prop | Noun |
| 4 | 4 | wire | 2 | wire | Adj | Noun |
| 5 | 21 | 86yof | 1 | 8yof | Noun | Prop |
| 6 | 21 | 81yof | 1 | 8yof | Noun | Prop |
| 7 | 21 | 84yof | 2 | 8yof | Noun | Prop |
| 8 | 10 | peice | 1 | piece | Prop | Noun |
| 9 | 27 | pencil | 1 | pencil | Verb | Noun |
| 10 | 54 | caugh | 1 | caught | Prop | Verb |

For example, terms such as 86yof means '86 year old female' when describing the age and gender of the person injured/affected. However, in this case the spell checker suggested that the correct form should be 8yof meaning '8 year old female' and this is not correct. Hence, you need to delete such mistakenly corrected entries from the data set manually. For the purpose of this case study, we provided you the cleaned data set (textfilter_synonyms_school.sas7bdat).

10. The results of spell checking can be used as a preliminary synonyms list for the purposes of parsing the text. By doing this, all those misspelled terms will be treated as synonyms with the correctly spelled word forms giving accurate calculation of unique term frequency. This is the most critical input for deriving text topics in further steps. By default, you will see 'SASHELP.ENGSYNMS' as the synonyms data set in this node. Click on the ellipsis button next to the 'Synonyms' property to open the ENGSYNMS data set in a pop-up window (Display C4.6). As you can see, it has only one entry for 'sas' representing 'sas institute' in it.

**Display C4.6: Window showing the current synonyms list (data set)**

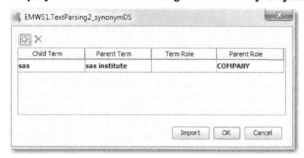

Now, click on the 'Import' button to browse through the library 'CPSC' and select the cleaned synonyms list specific to the schools data 'textfilter_synonyms_school.sas7bdat'. You can see the synonyms data set changed in the properties panel for the 'Text parsing' node (Display C4.7). Now that you have new synonyms listed for this diagram, you need to rerun the text parsing and text filtering processes again to pickup the changes. Right click on 'Text Filter' node and select 'Run'.

**Display C4.7: Properties in Text Parsing node to provide Synonyms data set**

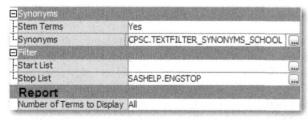

11. Open the 'Filter Viewer' in the Text Filter node again after you have rerun the process as described in the previous step. Expand the term '+fall' again and you should find that the misspelled words such as 'fal' and 'fekll' are treated as synonyms with its root word 'fall' (Display C4.8). This is one of the best

ways to quickly collect the synonyms list to use in stemming a lot of terms to their root forms when misspelled.

**Display C4.8: Properties in Text Parsing node to provide Synonyms data set**

| | TERM | FREQ | # DOCS | KEEP ▼ | WEIGHT | ROLE | ATTRIBUTE |
|---|---|---|---|---|---|---|---|
| ⊟ | fall | 546 | 513 | ☑ | 0.231 | Verb | Alpha |
| | fekll | 1 | 1 | | | Miscellaneous Pro... | Entity |
| | fall | 5 | 5 | | | Verb | Alpha |
| | falling | 37 | 36 | | | Verb | Alpha |
| | fel | 1 | 1 | | | Miscellaneous Pro... | Entity |
| | fell | 502 | 474 | | | Verb | Alpha |

12. You can also use the Terms pane to understand what are the keywords that are associated with specific types of accidents at school (See Display C4.9). For example, verb terms such as "hits" or "play" are useful to describe the cause of injury. Medical terms such as "laceration" and "contusion" can also help to understand types of injuries. Abbreviated terms such as "yom" and "yof" represent year-old-male and female, respectively. Even though these terms are unique to this set of documents, SAS® Text Miner understands the terms that are important and also identify their parts of speech (proper noun).

**Display C4.9: Key terms retained by the text filter node based on the term weights**

Terms

| | TERM | FREQ | # DOCS | KEEP ▼ | WEIGHT | ROLE | ATTRIBUTE |
|---|---|---|---|---|---|---|---|
| ⊞ | at school | 2179 | 2172 | ☑ | 0.039 | Organization | Entity |
| ⊞ | play | 811 | 792 | ☑ | 0.17 | Verb | Alpha |
| | dx | 801 | 771 | ☑ | 0.177 | Miscellaneous Pro... | Entity |
| | dx | 580 | 557 | ☑ | 0.219 | Prop | Alpha |
| ⊞ | fall | 546 | 513 | ☑ | 0.231 | Verb | Alpha |
| ⊞ | school | 370 | 369 | ☑ | 0.261 | Noun | Alpha |
| ⊞ | hit | 376 | 366 | ☑ | 0.264 | Verb | Alpha |
| ⊞ | contusion | 365 | 364 | ☑ | 0.262 | Noun | Alpha |
| | fell | 333 | 307 | ☑ | 0.302 | Noun | Alpha |
| | hit | 301 | 297 | ☑ | 0.288 | Noun | Alpha |
| ⊞ | pain | 335 | 293 | ☑ | 0.304 | Noun | Alpha |
| ⊞ | finger | 384 | 268 | ☑ | 0.309 | Noun | Alpha |
| ⊞ | injure | 276 | 262 | ☑ | 0.314 | Verb | Alpha |
| | pt | 283 | 260 | ☑ | 0.309 | Miscellaneous Pro... | Entity |
| ⊞ | football | 252 | 248 | ☑ | 0.311 | Noun | Alpha |
| | yom | 262 | 246 | ☑ | 0.324 | Miscellaneous Pro... | Entity |
| ⊞ | basketball | 239 | 230 | ☑ | 0.325 | Noun | Alpha |
| ⊞ | ankle | 288 | 216 | ☑ | 0.352 | Noun | Alpha |
| ⊞ | head | 231 | 210 | ☑ | 0.335 | Noun | Alpha |
| ⊞ | laceration | 204 | 197 | ☑ | 0.341 | Noun | Alpha |

13. Next explore the links between these words. Select "+fall" from the terms list. Right-click and select "View Concept Links" to open up the concept linking diagram (See Display C4.10). Concept links helps in explorative analysis of the data and shows how strongly certain terms are associated with other terms. This step can help to understand how various subjects/themes are discussed in the corpus of documents. These results can provide as input to formulate initial business and linguistic rules to further strengthen the classification schemes.

**Display C4.10: Partial screenshot of terms list showing how to open concept linking diagram**

| TERM | FREQ | # DOCS | KEEP ▼ | WEIGHT | ROLE | ATTRIBUTE |
|---|---|---|---|---|---|---|
| at school | 2179 | 2172 | ✓ | 0.039 | Organization | Entity |
| play | 811 | 792 | ✓ | 0.17 | Verb | Alpha |
| dx | 801 | 771 | ✓ | 0.177 | Miscellaneous Pro... | Entity |
| dx | 580 | 557 | ✓ | 0.219 | Prop | Alpha |
| fall | | | | 0.231 | Verb | Alpha |
| school | | | | 0.261 | Noun | Alpha |
| hit | | | | 0.264 | Verb | Alpha |
| contusion | | | | 0.262 | Noun | Alpha |
| fell | | | | 0.302 | Noun | Alpha |
| hit | | | | 0.288 | Noun | Alpha |
| pain | | | | 0.304 | Noun | Alpha |
| finger | | | | 0.309 | Noun | Alpha |
| injure | | | | 0.314 | Verb | Alpha |
| pt | | | | 0.309 | Miscellaneous Pro... | Entity |
| football | | | | 0.311 | Noun | Alpha |
| yom | | | | 0.324 | Miscellaneous Pro... | Entity |

Context menu (overlaid on terms list):
- Add Term to Search Expression
- Treat as Synonyms
- Remove Synonyms
- Keep Terms
- Drop Terms
- View Concept Links
- Find
- Repeat Find
- Clear Selection
- Print...

14. In the concept link diagram, the selected term appears at the center of the tree, surrounded by the terms that are most highly associated with that term (See Display C4.11). The width of the lines in the display represents the strength of association between terms; a thicker line indicates a stronger association. If you position the mouse pointer over a term, then a tooltip shows two numbers, separated by a forward slash (/). The first number is the number of documents in which the term appears with its immediate predecessor in the display. The second number represents the total number of documents in which the term occurs in the document collection.

In this example, you can see two things that SAS® Text Miner can do to help you initially explore the data. You can see that the software created some noun groups, or multi-word phrases, such as "+monkey bar". You can also begin to see some evidence that some falls involve things like monkeybars and floors. You can move the diagram around to help make it easier to see the words on the screen. By clicking anywhere on the screen and moving the mouse, the diagram also moves.

**Display C4.11: Concept linking diagram showing terms associated with "+fall"**

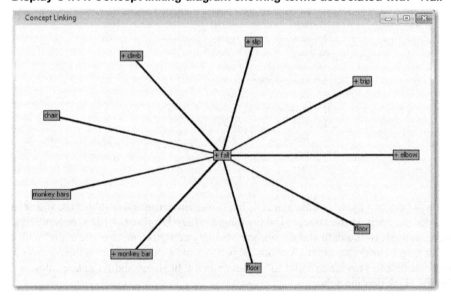

15. You can also learn more about falls by double-clicking on the words linked to "fall" to see their associated links. Double-click on "floor" to find the words most associated with falls on floor. As you can see there are two terms showing up here for 'floor'. One term represents 'floor' as a noun and another represents a verb. Move your pointer over to both these terms in the diagram and verify which

one shows the highest frequency. Double-click on the one which has more frequency and it will show more terms linked to 'floor'. These terms explain that the hits on heads are due to slippery floor when wet (See Display C4.12). Based on the thickness of the bars connecting the terms, you can find that more incidents are occuring on concrete, wet, and gym floors. Close the interactive filter viewer by pressing the X at the top and do not save changes.

**Display C4.12: Concept linking diagram showing terms associated with "floor"**

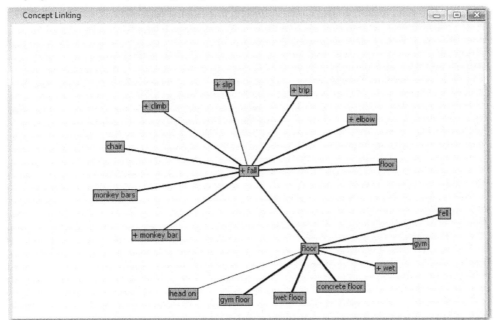

16. The Text Topic node is useful for assigning documents to multiple topics. It also has the nice feature of enabling you to explore how the different topics relate to the individual documents. This can be useful if you are trying to build a taxonomy, or classification scheme, to ensure documents are classified correctly or people can search and retrieve relevant information. Right-click on the Text Topic node and click 'Run.' Once the node has completed its run successfully, click "Results." You will see four panes in the window (See Display C4.13).

The Topics pane shows the keywords associated with each topic. The Number of Documents by Topics pane shows how many documents are classified with each topic. The Number of Terms by Topics pane shows how many terms from the Terms List are associated with each topic. There is also an Output pane with SAS output. Click on the first row in the Topics pane. This will cause the bars associated with that cluster to be marked in the other two non-output panes. The first Text Topic involves problems surrounding **finger cuts** in doors. Highlight the second topic and you will find that it seems to be about **ankle sprains** while playing **basketball**.

**Display C4.13: Results from Text Topic node for NEISS data related to 'School'**

This process is about trying to determine just which keywords are used and how they indicate which topics. This is one way to begin building a taxonomy, which will be discussed in-detail in the SAS® Enterprise Content Categorization section. Close the Results window for the Text Topics, minimize Text Miner and bring up Content Categorization.

**Note:** You may repeat this entire process for the other two categories 'swallow' and 'sports' in separate EM diagrams following Steps 4 through 16. This way, you should be able to identify key topics/themes for all the three categories to help build a taxonomy in the next step.

## Part 2: SAS Enterprise Content Categorization

SAS® Enterprise Content Categorization is used to organize and classify documents into different topic areas based upon certain features in the documents. In order to classify documents, the user has to define certain rules for each topic. This example focuses on a user-written taxonomy (or classification scheme) based upon the initial text mining results.

1. Launch SAS® Enterprise Content Categorization Studio and create a new project, name it **'CPSC_CC'**. Right click on 'CPSC_CC' within the taxonomy pane and click 'Add Language'. Select 'English' as the language and click Ok.
2. Right click on 'English' and select the option 'Create Categorizer from Directories'. Browse to the location on your PC where you have copied the folder 'docs' that we provided you for this case study. Navigate to that folder and click OK to create those categories based on the folder structure (See Display 14). The three categories, sports, swallowing, and school are created based on those three folders. For each of these categories, you can start creating subcategories that fit well with these categories. As a next step you can write Boolean rules to help define what goes into each subcategory based on the topics discovered using SAS® Text Miner in the first part of this case study.

**Display C4.14: Partial screen capture of categories and topical sub-categories**

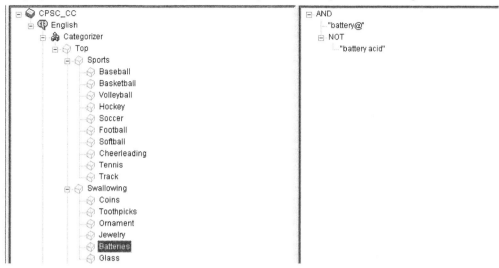

For example, you can include various forms of the word "battery" to tag documents as being about batteries. The "@" sign at the end of the word is a type of wildcard that accounts for the various expansions of the word "battery". By having the "NOT" clause with the phrase "battery acid" you can exclude documents that involve battery acid.

3. Click on "Coins" underneath "Swallowing" in the taxonomy pane. Enter the key terms that represent various types of US metallic currency and the word "coin". Because of the @ sign at the end of each term, the plural forms of the terms are captured (See Display C4.15).

**Display C4.15: Boolean rules for the subcategory 'Coins'**

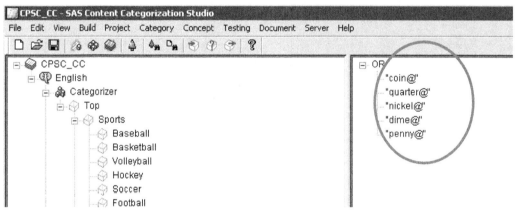

4. Click on "Swallowing" in the taxonomy pane. Include various synonyms for "swallow" in the taxonomy, such as ingest, eat, and drink (See Display C4.16).

**Display C4.16: Boolean rules for the category 'swallowing'**

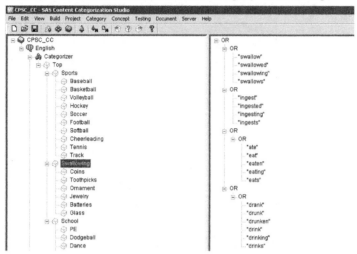

Underneath the topmost "OR" in the rules pane, right-click and select "Add Operator | OR" (Display C4.17).

**Display C4.17: List of Boolean Operators available at the root node in a category**

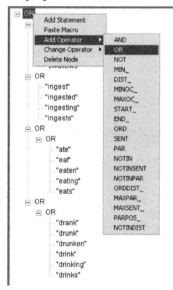

Now click on the new "OR" at the bottom and select "Add Statement" (See Display C4.18).

**Display C4.18: 'Add statement' option used to add a new rule**

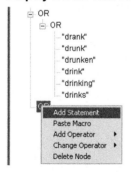

Type in "imbibe@v" here and it will add in all the verb conjugations of the term imbibe, such as 'imbibing' and 'imbibed' (See Display C4.19). Similarly, you can wildcard on specific parts of speech, such as nouns or adjectives if required.

**Display C4.19: Added Boolean rule to the existing rules set for the 'Swallowing' category**

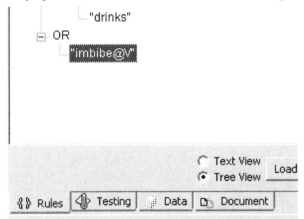

5.  Similarly, write the boolean rules based on the topic terms extracted for all categories and sub-categories. You may open the content categorization project 'CPSC_CC.tk' that we have provided you to understand how to write these rules.
6.  Set up the testing paths for all the categories and their sub-categories as shown in Table C4.1 below. Clear the training path if populated since it is not required. Switch to the Testing Tab while having the "Swallowing" node highlighted. This tab is next to the Rules tab. Click TEST at the top. This will prompt you that categorizer is not up to date and needs to be rebuilt. Click Yes. After the Categorizer is rebuilt, click TEST again.

**Table C4.1: Testing Paths by section category for categorizer**

| Category | Testing Path |
| --- | --- |
| Sports | D:\Case Study 5\CPSC\docs\Sports |
| Swallowing | D:\ Case Study 5\CPSC\docs\Swallowing |
| School | D:\ Case Study 5\CPSC\docs\School |

7.  The Testing tab gives the analyst an initial idea of how rules are working by checking for either the specified rule or all rules for a set of documents. In the testing pane there are three columns: the Test File, the Result, and the Relevancy. The test file lists the individual document tested. The Result here indicates whether the categorization rules made a PASS or FAIL on the document (See Display C4.20). Passing here is defined as having at least one example of a rule trigger within a document. The Relevancy shows the number of rule hits and whether or not the number of hits falls above a certain threshold. In this case, the project is set up to only look at the frequency of the hits, and the threshold is set to a single hit for a document to be 'relevant' or PASS. You may modify the category to account for factors such as document length, establishing minimum thresholds, weighing parts of the document more than others, and even weighing certain nodes of the taxonomy more than others.

**Display C4.20: Results from Testing Tab for 'Swallowing' category**

Scroll down the testing results and you will find that some reviews do not mention the "swallowing" issue. These are the documents that have a result of FAIL. These documents are all about incidents where someone swallowed something. Some of the documents fail because there are other ways to say swallow that have not yet been included in the rules (such as 'ingestion'), but another reason is that some documents are simply misclassified as being about "swallowing". A common finding in text analytics is that the unstructured data can disagree with the structured data at times. An analyst can use text analytics to validate structured data, find emerging issues, and enhance modeling for events.

8.  Double click on a document (say swallow889.txt) and this will automatically open up the Document tab (fourth on bottom). There you can see how the individual rules (highlighted in red) are triggering a category match for this particular document (See Display C4.21). Ingested, Eating, and Swallowed are the rules which triggered a match on this document.

**Display C4.21: Rules triggering the 'Swallowing' category match for a document**

53 YOF INGESTED FOREIGN OBJECT, PT WAS EATING A PIECE OF CHICKEN WHEN HER FORK BROKE AND SHE SWALLOWED A PEICE, + CHOKING SENSATIOON

Once you have reviewed the results in the "Testing" tab, you may go back to the "Rules" tab and modify the rules to include more terms such as 'ingestion'. Thus you can understand that the taxonomy creation is an iterative process. As the language evolves over time you will probably need to add new words or phrases to the taxonomy to ensure that you properly capture information of interest.

**Note:** For your reference we have provided you the SAS Content Categorization Studio project CPSC_CC.tk which you can open and verify the Boolean rules that are pre-built for all the three categories and their sub-categories. You may need to change the testing paths to ensure they are pointing to the correct folders/files on your machine.

## Summary

- SAS® Text Miner and SAS® Enterprise Content Categorization Studio can be used together to complement each other's unique linguistic capabilities for the purpose of exploring, analyzing, and extracting valuable information from textual data.

- The spell check feature in the text filter node can be utilized to provide the preliminary synonyms list for parsing the content and stemming terms. By following this method, you can significantly reduce the time required to verify misspelled words and accurately generate the term-by-document frequency occurrence matrix which is critical for deriving text topics.

- The Interactive filter viewer in the text filter node helps you to identify the most important terms in a document corpus without spending significant effort. Important terms are retained and unimportant terms are dropped automatically by the software based on their frequency of occurrence both within a document and the entire corpus.

- The Text topic node automatically generates the key terms/themes that are significant in defining the document collection at a very high level. These results can help you understand different areas of causes/concerns in the data without digging through each document. They are also pivotal in helping you build the taxonomy to classify the document collection based on type of injury/cause of injury.

- Content Categorizer in SAS® Enterprise Content Categorization Studio provides an easy-to-use graphical interface to quickly build a Boolean rule based model to identify and classify documents in various categories and sub-categories. The point and click feature helps you to avoid writing those rules in the syntax that the tool understands.

# Case Study 5 Enhancing Predictive Models Using Textual Data

Maheshwar Nareddy
Goutam Chakraborty

While text mining customer responses can reveal valuable insights about a customer, plugging the results from text mining into a typical data mining model can often significantly improve the predictive power of the model. Organizations often want to use customer responses captured in the form of text via emails, customer survey questionnaires, and feedback on websites for building better predictive models. One way of doing this will be to apply text mining to reveal groups (or, clusters) of customers with similar responses or feedback. This cluster membership information about each customer may then be used as an input variable to augment the data mining model. With this additional information, the accuracy of a predictive model can improve significantly.

The data used in this case study is hypothetical. But, it is created based on a real data set of a client company (Fuel Stop Company with 300+ gas stations in the US). Some of the text comments, variable names, and descriptions have been disguised to protect the identity of the client company.

The hypothetical case involves customers calling the fuel company's call center for many different reasons. Customers' comments via phone were captured by call-center reps and typed into a form. These comments were later merged with numeric variables from the fuel company's database about these customers (by matching them via the company's loyalty card number).

## Data Description

The merged data set (survey_text_and_numeric) is used in this case study. The data set is available in the Case Studies ▶ Case Study 5 folder of the data provided with the book. The purpose of this case study is to demonstrate how the use of textual data in conjunction with numeric data in a predictive model improves the performance of the predictive model.

## Step-by-Step Instructions

1. Create a SAS® Enterprise Miner project and a diagram within the project.
2. Create a library to point to a folder where the data set is located. Add the data source, **survey_text_and_numeric**, to the project (via your library).
3. In Step 4, ensure that you use Advanced Metadata Advisor Options as shown in Display C5.1.

**Display C5.1 Data Source creation**

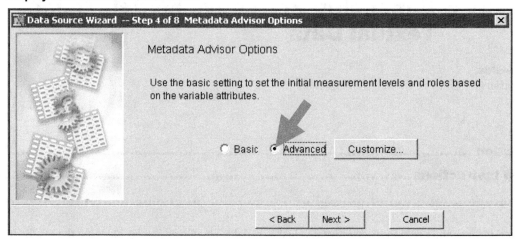

4. The variable roles and levels should be as shown in Display C5.2.

**Display C5.2 Data source creation: Variable properties**

| Name | Role | Level | Report | Order | Drop | Lower Limit | Upper Limit |
|------|------|-------|--------|-------|------|-------------|-------------|
| AcctType_flag | Input | Binary | No | | No | . | . |
| Choice_flag | Input | Binary | No | | No | . | . |
| Comment_1 | Text | Nominal | No | | No | . | . |
| Comment_2 | Text | Nominal | No | | No | . | . |
| Comment_all | Text | Nominal | No | | No | . | . |
| Comp_card_flag | Input | Binary | No | | No | . | . |
| Contact_Flag2 | Input | Binary | No | | No | . | . |
| Contact_flag | Input | Binary | No | | No | . | . |
| CustType_flag | Input | Binary | No | | No | . | . |
| Cust_ID | ID | Nominal | No | | No | . | . |
| HQ_flag | Input | Binary | No | | No | . | . |
| Loyal_Status | Input | Nominal | No | | No | . | . |
| Multi_flag | Input | Binary | No | | No | . | . |
| NewCust_Flag | Input | Binary | No | | No | . | . |
| Service_flag | Input | Binary | No | | No | . | . |
| Target | Target | Binary | No | | No | . | . |
| new_flag | Input | Binary | No | | No | . | . |

5. Click through and finish the next data creation steps by accepting the default options.
6. Drag the **survey_text_and_numeric** data to the diagram space.
7. Add a text parsing node, text filter node, and text cluster node, as shown in Display C5.3.

**Display C5.3 Process flow**

8. Right-click on the text parsing node and select 'Edit Variables'. Change the 'Use' field of Comment_1 and Comment_2 to "No," as shown in Display C5.4.

**Display C5.4 Text Parsing node input variables screen**

| Name | Use | Report | Role | Level |
|---|---|---|---|---|
| Comment_1 | No | No | Text | Nominal |
| Comment_2 | No | No | Text | Nominal |
| Comment_all | Default | No | Text | Nominal |

In this case study, you are using all of the comments together to create text clusters. It is, however, possible to create clusters separately for Comment_1 and Comment_2, and you should explore that on your own.

9. Run the flow from the Text Cluster node. Examine results.

You will find that there are many small clusters with few observations. This is not surprising, given the small data set. As a demonstration, you can ask text miner to create a maximum of 20 SVD dimensions and exactly 5 clusters, then describe those clusters using 10 terms.

10. Make changes in the properties panel of the Text Cluster node, as highlighted in Display C5.5.

**Display C5.5 Text Cluster node property panel**

| General | |
|---|---|
| Node ID | TextCluster |
| Imported Data | |
| Exported Data | |
| Notes | |
| **Train** | |
| Variables | |
| Transform | |
| SVD Resolution | Low |
| Max SVD Dimensions | 20 |
| Cluster | |
| Exact or Maximum Number | Exact |
| Number of Clusters | 5 |
| Cluster Algorithm | Expectation-Maximization |
| Descriptive Terms | 10 |
| **Status** | |

11. Run the cluster node and examine results.

It appears that five clusters with reasonable number of observations per clusters have been created by text cluster node as shown in Display C5.6. You should explore the cluster solution to get a feel for what these clusters might represent. Optionally, you may use a Segment Profile node to profile these clusters using the numeric variables available in the data.

**Display C5.6 Clusters frequencies and descriptive terms**

| Cluster ID | Descriptive Terms | | Frequency | Percentage |
|---|---|---|---|---|
| 1 | +restaurant cost restaurants facilities points +discount +shower +point +card cash | ... | 49 | 16% |
| 2 | +easy 'bad service' +service +discount bad +good productx +point better always | ... | 80 | 25% |
| 3 | parking food cleanliness cash +truck +stop +purchase compy convenience +hard | ... | 48 | 15% |
| 4 | 'free coffee' free drinks +drink rewards 'double points' double +expire showers clean | ... | 51 | 16% |
| 5 | 'unlimited showers' unlimited +buy people convenient +dirty +month +friendly clean showers | ... | 87 | 28% |

12. Add a Metadata Node to the diagram and connect it to Text Cluster node. Click on the ellipsis button next to Training in the Variables section of the Property panel of the Metadata node. Then, make changes as shown in Display C5.7.

**Display C5.7 Metadata node – Train variables**

| Name | Hidden | Hide | Role | New Role | Level | New Level | New Order | New Report |
|---|---|---|---|---|---|---|---|---|
| AcctType_flag | N | Default | Input | Default | Binary | Default | Default | Default |
| Choice_flag | N | Default | Input | Default | Binary | Default | Default | Default |
| Comment_1 | N | Default | Text | Default | Nominal | Default | Default | Default |
| Comment_2 | N | Default | Text | Default | Nominal | Default | Default | Default |
| Comment_all | N | Default | Text | Default | Nominal | Default | Default | Default |
| Comp_card_flag | N | Default | Input | Default | Binary | Default | Default | Default |
| Contact_Flag2 | N | Default | Input | Default | Binary | Default | Default | Default |
| Contact_flag | N | Default | Input | Default | Binary | Default | Default | Default |
| CustType_flag | N | Default | Input | Default | Binary | Default | Default | Default |
| Cust_ID | N | Default | ID | Default | Nominal | Default | Default | Default |
| HQ_flag | N | Default | Input | Default | Binary | Default | Default | Default |
| Loyal_Status | N | Default | Input | Default | Nominal | Default | Default | Default |
| Multi_flag | N | Default | Input | Default | Binary | Default | Default | Default |
| NewCust_Flag | N | Default | Input | Default | Binary | Default | Default | Default |
| Service_flag | N | Default | Input | Default | Binary | Default | Default | Default |
| Target | N | Default | Target | Default | Binary | Default | Default | Default |
| TextCluster_SVD1 | N | Default | Input | Default | Interval | Default | Default | Default |
| TextCluster_SVD2 | N | Default | Input | Default | Interval | Default | Default | Default |
| TextCluster_SVD3 | N | Default | Input | Default | Interval | Default | Default | Default |
| TextCluster_SVD4 | N | Default | Input | Default | Interval | Default | Default | Default |
| TextCluster_SVD5 | N | Default | Input | Default | Interval | Default | Default | Default |
| TextCluster_SVD6 | N | Default | Input | Default | Interval | Default | Default | Default |
| TextCluster_SVD7 | N | Default | Input | Default | Interval | Default | Default | Default |
| TextCluster_SVD8 | N | Default | Input | Default | Interval | Default | Default | Default |
| TextCluster_cluster_ | N | Default | Segment | Input | Nominal | Default | Default | Default |
| TextCluster_prob1 | N | Yes | Rejected | Default | Interval | Default | Default | Default |
| TextCluster_prob2 | N | Yes | Rejected | Default | Interval | Default | Default | Default |
| TextCluster_prob3 | N | Yes | Rejected | Default | Interval | Default | Default | Default |
| TextCluster_prob4 | N | Yes | Rejected | Default | Interval | Default | Default | Default |
| TextCluster_prob5 | N | Yes | Rejected | Default | Interval | Default | Default | Default |
| _document_ | N | Default | ID | Default | Nominal | Default | Default | Default |
| new_flag | N | Default | Input | Default | Binary | Default | Default | Default |

13. Drag a **Data Partition** tool from the Sample Tools palette into the diagram workspace and connect it with the Metadata node.

14. In the Property panel of the Data Partition, change Training to **80**, and Validation o **20**, and Testto **0** as shown in Display C5.8.

**Display C5.8 Data Partition node property panel**

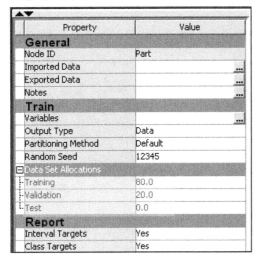

| Property | Value |
|---|---|
| **General** | |
| Node ID | Part |
| Imported Data | ... |
| Exported Data | ... |
| Notes | ... |
| **Train** | |
| Variables | ... |
| Output Type | Data |
| Partitioning Method | Default |
| Random Seed | 12345 |
| ☐ Data Set Allocations | |
| Training | 80.0 |
| Validation | 20.0 |
| Test | 0.0 |
| **Report** | |
| Interval Targets | Yes |
| Class Targets | Yes |

15. From the Model tab, drag a regression node and connect it to the data partition node. Rename the regression node as **Numeric Only (Reg).**
16. Right-click the Numeric Only (Reg) node and select Edit Variables.
17. Change the Use role of all cluster variables to No as shown in Display C5.9. Then, click OK.

**Display C5.9 Regression node (Numeric Only) – Input Variables**

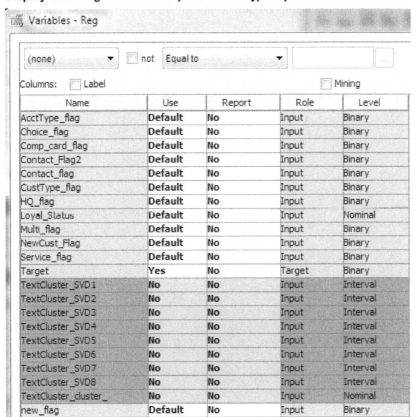

| Name | Use | Report | Role | Level |
|---|---|---|---|---|
| AcctType_flag | Default | No | Input | Binary |
| Choice_flag | Default | No | Input | Binary |
| Comp_card_flag | Default | No | Input | Binary |
| Contact_Flag2 | Default | No | Input | Binary |
| Contact_flag | Default | No | Input | Binary |
| CustType_flag | Default | No | Input | Binary |
| HQ_flag | Default | No | Input | Binary |
| Loyal_Status | Default | No | Input | Nominal |
| Multi_flag | Default | No | Input | Binary |
| NewCust_Flag | Default | No | Input | Binary |
| Service_flag | Default | No | Input | Binary |
| Target | Yes | No | Target | Binary |
| TextCluster_SVD1 | No | No | Input | Interval |
| TextCluster_SVD2 | No | No | Input | Interval |
| TextCluster_SVD3 | No | No | Input | Interval |
| TextCluster_SVD4 | No | No | Input | Interval |
| TextCluster_SVD5 | No | No | Input | Interval |
| TextCluster_SVD6 | No | No | Input | Interval |
| TextCluster_SVD7 | No | No | Input | Interval |
| TextCluster_SVD8 | No | No | Input | Interval |
| TextCluster_cluster_ | No | No | Input | Nominal |
| new_flag | Default | No | Input | Binary |

18. In the properties panel of the regression node, change Selection Model to **Stepwise** and Selection Criterion to **Validation Error,** as shown in Display C5.10.

**Display C5.10 Regression node property panel**

19. In the diagram space, right-click the Numeric Only (Reg) node and select Copy, and then right-click on the diagram workspace and select **Paste**. Change the name of the pasted node to **Numeric & Text (Reg)** and connect it with the data partition node.
20. Right-click the **Numeric & Text (Reg)** node and select **Edit Variables**.
21. Change the Use role of the cluster membership variable from "No" to **Default,** as shown in Display C5.11. Then click **OK**.

**Display C5.11 Regression node (Numeric and Text) – Input variables**

| Name | Use | Report | Role | Level |
|---|---|---|---|---|
| AcctType_flag | Default | No | Input | Binary |
| Choice_flag | Default | No | Input | Binary |
| Comp_card_flag | Default | No | Input | Binary |
| Contact_Flag2 | Default | No | Input | Binary |
| Contact_flag | Default | No | Input | Binary |
| CustType_flag | Default | No | Input | Binary |
| HQ_flag | Default | No | Input | Binary |
| Loyal_Status | Default | No | Input | Nominal |
| Multi_flag | Default | No | Input | Binary |
| NewCust_Flag | Default | No | Input | Binary |
| Service_flag | Default | No | Input | Binary |
| Target | Yes | No | Target | Binary |
| TextCluster2_SVD1 | No | No | Input | Interval |
| TextCluster2_SVD2 | No | No | Input | Interval |
| TextCluster2_SVD3 | No | No | Input | Interval |
| TextCluster2_SVD4 | No | No | Input | Interval |
| TextCluster2_SVD5 | No | No | Input | Interval |
| TextCluster2_SVD6 | No | No | Input | Interval |
| TextCluster2_SVD7 | No | No | Input | Interval |
| TextCluster2_SVD8 | No | No | Input | Interval |
| TextCluster2_cluster_ | Default | No | Input | Nominal |
| new_flag | Default | No | Input | Binary |

22. In the diagram space, copy and paste the Numeric Only (Reg) node. Change the name of the pasted node to **Numeric & SVD (Reg)** and connect it with the data partition node.
23. Right-click the Numeric & SVD (Reg) node and select Edit Variables.
24. Change the Use role of the SVD variables from "No" to **Default,** as shown in Display C5.12. Then, click **OK.**

**Display C5.12 Regression node (SVDs) – Input variables**

| Name | Use | Report | Role | Level |
|---|---|---|---|---|
| AcctType_flag | Default | No | Input | Binary |
| Choice_flag | Default | No | Input | Binary |
| Comp_card_flag | Default | No | Input | Binary |
| Contact_Flag2 | Default | No | Input | Binary |
| Contact_flag | Default | No | Input | Binary |
| CustType_flag | Default | No | Input | Binary |
| HQ_flag | Default | No | Input | Binary |
| Loyal_Status | Default | No | Input | Nominal |
| Multi_flag | Default | No | Input | Binary |
| NewCust_Flag | Default | No | Input | Binary |
| Service_flag | Default | No | Input | Binary |
| Target | Yes | No | Target | Binary |
| TextCluster_SVD1 | Default | No | Input | Interval |
| TextCluster_SVD2 | Default | No | Input | Interval |
| TextCluster_SVD3 | Default | No | Input | Interval |
| TextCluster_SVD4 | Default | No | Input | Interval |
| TextCluster_SVD5 | Default | No | Input | Interval |
| TextCluster_SVD6 | Default | No | Input | Interval |
| TextCluster_SVD7 | Default | No | Input | Interval |
| TextCluster_SVD8 | Default | No | Input | Interval |
| TextCluster_cluster_ | No | No | Input | Nominal |
| new_flag | Default | No | Input | Binary |

25. Drag a Model Comparison node from the Assess tab and connect it with all of the Regression nodes, as shown in Display C5.13.

**Display C5.13 Process flow**

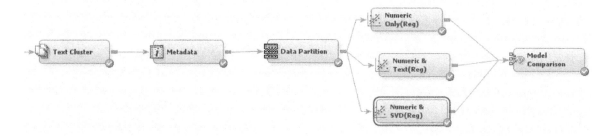

26. Right-click the **Model Comparison** node and run it. Examine the results.

**Display C5.14 Model comparision node results**

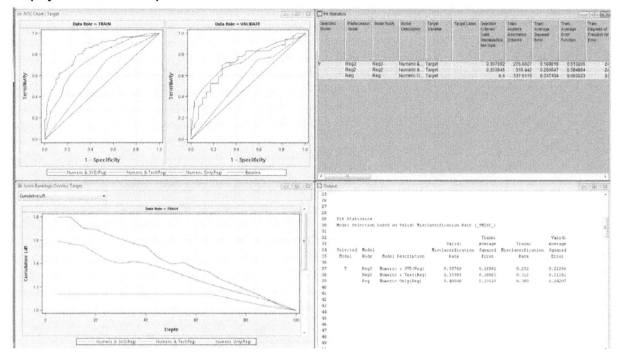

Notice that (Display C5.15) the Numeric & SVD (Reg) model has clearly outperformed the Numeric and Txt(Reg) model, which has outperformed the Numeric Only (Reg) model in model misclassification rate in validation data.

**Display C5.15 Model Fit Statistics from model comparison node**

| Selected Model | Predecessor Node | Model Node | Model Description | Target Variable | Target Label | Selection Criterion: Valid: Misclassifica tion Rate | Train: Akaike's Information Criterion | Train: Average Squared Error | Train: Average Error Function | |
|---|---|---|---|---|---|---|---|---|---|---|
| Y | Reg3 | Reg3 | Numeric & ... | Target | | 0.307692 | 276.6027 | 0.169618 | 0.513205 | |
| | Reg2 | Reg2 | Numeric & ... | Target | | 0.353846 | 310.442 | 0.200647 | 0.584884 | |
| | Reg | Reg | Numeric O... | Target | | 0.4 | 337.5115 | 0.235104 | 0.663023 | |

Thus, addition of SVDS or text clusters has improved the predictive ability of the model over a model that has only numeric variables. Interestingly, if you explore the results from each regression node, you will find that more numeric variables are playing significant roles in the predictive models when text clusters or text SVDS are included in the model.

# Summary

This is a quick demonstration using real-world data to show the improvement in predictive model performance by including textual comments. Unstructured data is powerful in providing a complete view of business processes and outcomes. Text mining results in various forms (SVDs, cluster memberships, text topics) that can be easily plugged into a predictive model and used as inputs. This capability will provide companies' unlimited opportunity to utilize the insights extracted from unstructured data in order to achieve better lift and accuracy in predictive models. SAS® Enterprise Miner makes this process relatively easy by allowing modeling nodes and text mining nodes to co-exist in one store.

# Case Study 6 Opinion Mining of Professional Drivers' Feedback

Mantosh Sarkar
Goutam Chakraborty

## Introduction

The widespread adoption of mobile applications has tremendously expanded the scope of obtaining timely customer feedback. While many companies collect feedback from their customers via mobile apps, they often restrict their analysis to numeric data and ignore analyzing customer feedback and sentiment from textual data because of perceived difficulties associated with analyzing text data. In this case study, you will analyze customer feedback by professional drivers sent via mobile devices. Currently, he company experts manually classify these textual feedbacks into positive and negative groups. Here, you will learn how to use SAS® Text Miner to automatically generate and summarize topics from positive and negative feedbacks, as well as classify comments into positive and negative groups. In addition, we also help you learn how to build rules in SAS Sentiment Analysis Studio to predict customer's sentiment automatically, so that experts' time can be used for more strategic purposes.

## Data

A leading retail & energy company with hundreds of truck stops located in most of the U.S. has kindly agreed to provide the data for this case study anonymously. It offers fuel, fast food, and convenience store services to its customers. In addition, it offers amenities such as food from national restaurant chains as well as trucking supplies, showers, and RV dump stations. The data used in this case study is feedback obtained via a mobile phone app that is used by professional drivers to locate a store, find nearby fuel stations, and check loyalty program reward point balances. This feedback is free-form text comments from professional truck drivers who talk about various topics related to their experience at the store, questions on required services, etc. Currently, company experts classify driver's feedback into positive and negative by reading and analyzing it.

For reasons of anonymity, the name of the company is masked in the text comments. After data cleaning and validation, 2,335 text comments were used for this case study. We decided to keep the data and analysis separate for positive and negative comments. You will start with SAS® Text Miner to extract insights from topics the professional drivers are talking about in their feedback.

### Analysis Using SAS® Text Miner

The data sets used for this analysis are:

- neg_model_new.sas7bdat
- pos_model_new.sas7bdat
- engdict.sas7bdat

The data sets are available in the Case Studies ▶ Case Study 6 folder of the data provided with the book. The following sections will walk you through a detailed step-by-step text mining process.

1. Create a new project in SAS® Enterprise Miner and create a library pointing to the location where the data resides using the project start code or File ▶ New ▶ Library menu.
2. Create two data sources using the provided input data sets: "pos_model_new", "neg_model_new".
3. Create a new diagram and drag the two input data sources onto the diagram space. Connect the data sources with text miner nodes as shown in Display C6.1. The nodes with "Default" in their names are run with default settings in their properties panel.

**Display C6.1 Text mining process flow**

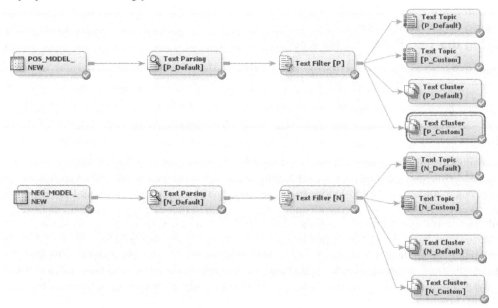

4. Change the default dictionary setting in both text filter nodes by selecting the "engdict.sas7bdat" from your library. Click the ellipsis button next to the 'Dictionary' property setting. Right-click the text filter nodes and run both the process flows. An interesting output of the text filter node are concept links. They help to identify relationships between terms. You can view concept links by first clicking the ellipsis button next to Filter viewer in the properties panel of the text filter node, then right clicking any term in the Filter Viewer results and selecting "View Concept Links". Display C6.2 shows the concept link diagram for the word "shower" obtained from the text filter node in the negative feedback process flow. The thickness of the lines connecting the words signifies the strength of the association between terms. The term "shower" is strongly associated with terms such as key, water, toilet, shower head, clean, sink, towel, and hair. Reading through the comments that contain these terms, you will find that customers complain about showers not being clean, cold water, dirty sinks, the shower room keys0 not working properly, and not having enough towels available.

**Display C6.2. Concept link diagram for the term "shower"**

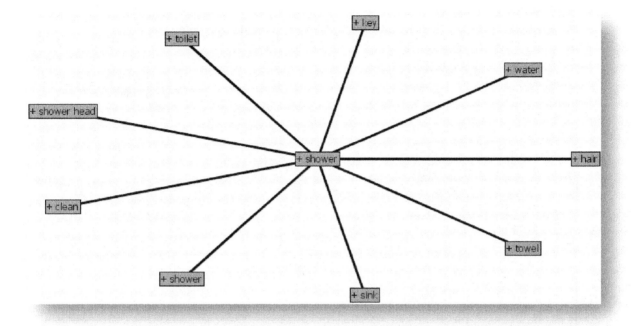

5. Modify the properties of Text Topic (Custom) node as below.
6. Change "Number of multi-term topics" to 13 for negative comments and 3 for positive comments (as shown in Display C6.3).
7. Run the process flow from the Text Topic nodes.

**Display C6.3 Properties panel of Text-topic (custom) node for positive feedbacks and negative feedbacks**

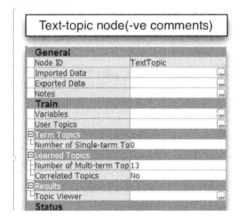 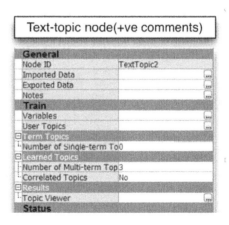

The data is first parsed and then fed into the text filter node. The output from text filter node is then fed to the text topic node, which generates 25 multi-term topics by default (as shown in Display C6.4 and Display C6.6). Given the small data size, 25 topics are deemed too many. Also, these 25 multi-terms have redundant themes. By changing "Number of Multi-term topics" in the property panel in a trial-and-error method, only 13 multi-term topics were retained for custom text topic node which uses the negative comments as input (as shown in Display C6.7) and only 3 for the custom text topic node (as shown in Display C6.5) which uses the positive comments as input. In this data, the numbers of topics for complaints outnumber the topics for appreciation, which is consistent with prior consumer research findings (conducted with the help of an external research agency) that people are more likely to complain than compliment. Note that in this case, you are configuring the text topic node to pick the number of topics and use appropriate terms to describe those topics. However, it is also possible to specify custom topics using user defined terms via the 'user topics' functionality.

**Display C6.4. Text topic node results from positive comments with default settings**

| Topic ID | Category | Topic | Document Cutoff | Term Cutoff | Number of Terms | # Docs |
|---|---|---|---|---|---|---|
| 1 | Multiple | +water,+hot,+shower,+pressure,hot water | 0.443 | 0.192 | 22 | 52 |
| 2 | Multiple | +service,+customer,+great service,+excellent,+good | 0.455 | 0.192 | 18 | 115 |
| 3 | Multiple | +clean,+shower,+clean shower,+restroom,+friendly | 0.501 | 0.191 | 22 | 124 |
| 4 | Multiple | +help,+employee,help,+find,helpful | 0.420 | 0.202 | 49 | 103 |
| 5 | Multiple | +work,good work,+good,keep up,+keep | 0.417 | 0.190 | 25 | 63 |
| 6 | Multiple | +employee,+smile,+customer,+know,+greet | 0.426 | 0.198 | 39 | 114 |
| 7 | Multiple | fuel,+desk,fuel desk,+lady,+young | 0.353 | 0.191 | 19 | 73 |
| 8 | Multiple | +great,great store,+great service,+shower,+store | 0.433 | 0.185 | 19 | 99 |
| 9 | Multiple | +shower,+clean,+towel,+shower,+hook | 0.396 | 0.189 | 22 | 110 |
| 10 | Multiple | +stop,+truck,+truck stop,always,polite | 0.387 | 0.184 | 22 | 84 |
| 11 | Multiple | +staff,+friendly,+friendly staff,+location,+great | 0.434 | 0.181 | 17 | 102 |
| 12 | Multiple | +nice,helpful,+store,+manager,+night | 0.436 | 0.185 | 26 | 126 |
| 13 | Multiple | always,+stop,+place,+clean,+restroom | 0.380 | 0.181 | 26 | 113 |
| 14 | Multiple | +tire,+shop,+fast,+man,+guy | 0.352 | 0.176 | 36 | 60 |
| 15 | Multiple | +manager,+want,+location,+lady,+know | 0.333 | 0.183 | 44 | 100 |
| 16 | Multiple | people,+excellent,+friendly,friendly people,+good | 0.383 | 0.174 | 30 | 89 |
| 17 | Multiple | +store,+good,great store,+staff,+clean store | 0.354 | 0.174 | 25 | 133 |
| 18 | Multiple | +love,+awesome,+app.,+place,+store | 0.355 | 0.171 | 26 | 80 |
| 19 | Multiple | +job,great job,good job,+great,+good | 0.333 | 0.172 | 33 | 72 |
| 20 | Multiple | +clean,+customer,+gas,+good,coffee | 0.348 | 0.176 | 47 | 110 |
| 21 | Multiple | fstore,+nice,+great service,+nice fstore,+clean | 0.324 | 0.169 | 22 | 125 |
| 22 | Multiple | +park,+lot,+place,+park lot,coffee | 0.281 | 0.166 | 30 | 65 |
| 23 | Multiple | great,+place,+stop,+look,+employee | 0.306 | 0.165 | 26 | 79 |
| 24 | Multiple | +location,+stop,+customer,+staff,+store | 0.275 | 0.167 | 33 | 98 |
| 25 | Multiple | +time,+drive,+professional,+stop,+week | 0.245 | 0.164 | 52 | 87 |

**Display C6.5. Text topic node results from positive comments with custom settings**

| Topic ID | Topic ▲ | Category | Document Cutoff | Term Cutoff | Number of Terms | # Docs |
|---|---|---|---|---|---|---|
| 2 | +shower,+clean,+hot,+nice,+water | Multiple | 0.784 | 0.270 | 30 | 120 |
| 1 | +store,+great,+service,+friendly,+clean | Multiple | 0.927 | 0.304 | 36 | 129 |
| 3 | fstore,+employee,+want,fuel,+help | Multiple | 0.607 | 0.274 | 45 | 124 |

**Display C6.6. Text topic node results from negative comments with default settings**

| Topic ID | Topic | Category | Document Cutoff | Term Cutoff | Number of Terms | # Docs |
|---|---|---|---|---|---|---|
| 1 | +shower,+clean,+floor,+towel,+hair | Multiple | 0.486 | 0.141 | 65 | 201 |
| 2 | +park,+park lot,+lot,+truck,+driver | Multiple | 0.395 | 0.126 | 42 | 108 |
| 3 | +water,+hot,+hot water,+shower,+cold | Multiple | 0.346 | 0.115 | 39 | 85 |
| 4 | +wait,+shower,+hour,+minute,+clean | Multiple | 0.393 | 0.122 | 67 | 160 |
| 5 | +card,+reader,+pump,+fuel,+card reader | Multiple | 0.361 | 0.120 | 66 | 122 |
| 6 | +coffee,+creamer,+empty,+area,+pot | Multiple | 0.334 | 0.118 | 70 | 91 |
| 7 | +truck,+stop,+truck stop,+driver,+stop | Multiple | 0.364 | 0.120 | 60 | 168 |
| 8 | +sandwich,+subway,food,+order,+eat | Multiple | 0.327 | 0.121 | 91 | 115 |
| 9 | +island,+fuel,+fuel island,+fuel,+driver | Multiple | 0.341 | 0.115 | 49 | 134 |
| 10 | +dirty,+bathroom,+store,+restroom,+clean | Multiple | 0.357 | 0.116 | 60 | 177 |
| 11 | desk,fuel desk,+fuel,+lady,+service | Multiple | 0.366 | 0.117 | 76 | 124 |
| 12 | +service,+customer,+poor,+bad,customer service | Multiple | 0.336 | 0.114 | 59 | 159 |
| 13 | +know,+stop,+want,+person,+day | Multiple | 0.353 | 0.116 | 99 | 168 |
| 14 | +tire,+hour,+shop,tire shop,+want | Multiple | 0.292 | 0.110 | 68 | 56 |
| 15 | +card,fstore,+spend,+charge,+pay | Multiple | 0.338 | 0.116 | 105 | 172 |
| 16 | +line,+customer,+manager,+store,+walk | Multiple | 0.341 | 0.113 | 94 | 171 |
| 17 | +toilet,+paper,+stall,+toilet paper,+man | Multiple | 0.271 | 0.111 | 67 | 111 |
| 18 | +customer,+know,+number,always,+long | Multiple | 0.314 | 0.114 | 121 | 153 |
| 19 | +rude,+employee,+manager,back,fstore | Multiple | 0.293 | 0.108 | 43 | 179 |
| 20 | water,+pressure,water pressure,+shower,+head | Multiple | 0.262 | 0.104 | 50 | 68 |
| 21 | +price,+pump,+app.,+high,gas | Multiple | 0.275 | 0.108 | 81 | 121 |
| 22 | +app.,+show,+phone,+manager,+coupon | Multiple | 0.285 | 0.106 | 65 | 104 |
| 23 | +time,+last,+employee,+store,+week | Multiple | 0.290 | 0.108 | 81 | 170 |
| 24 | +staff,fstore,+stop,+fuel,+location | Multiple | 0.270 | 0.105 | 80 | 199 |
| 25 | +towel,fstore,+smell,+wash,+stop | Multiple | 0.215 | 0.101 | 90 | 136 |

**Display C6.7. Text topic node results from negative comments with custom settings**

| Topic ID | Topic | Category | Document Cutoff | Term Cutoff | Number of Terms | # Docs |
|---|---|---|---|---|---|---|
| 1 | +shower,+floor,+clean,+dirty,+bathroom | Multiple | 0.556 | 0.155 | 75 | 194 |
| 2 | +park,+park lot,+lot,+truck,+driver | Multiple | 0.449 | 0.137 | 44 | 121 |
| 3 | +shower,+wait,+clean,+shower,+minute | Multiple | 0.468 | 0.135 | 67 | 162 |
| 4 | +customer,+service,+rude,+employee,+manager | Multiple | 0.461 | 0.136 | 63 | 180 |
| 5 | +sandwich,+subway,food,+order,+order | Multiple | 0.424 | 0.138 | 98 | 141 |
| 6 | +card,+pump,+reader,+fuel,+know | Multiple | 0.448 | 0.135 | 72 | 145 |
| 7 | fstore,+stop,+stop,+truck,+location | Multiple | 0.457 | 0.135 | 89 | 193 |
| 8 | +coffee,+creamer,+empty,+store,fstore | Multiple | 0.397 | 0.130 | 79 | 118 |
| 9 | +fuel,+island,desk,+fuel island,fuel desk | Multiple | 0.393 | 0.127 | 56 | 164 |
| 10 | +water,+hot,+hot water,+shower,+cold | Multiple | 0.363 | 0.120 | 47 | 88 |
| 11 | +app.,+price,fstore,+card,+show | Multiple | 0.359 | 0.126 | 90 | 149 |
| 12 | +tire,+hour,+shop,+want,+employee | Multiple | 0.349 | 0.121 | 82 | 100 |
| 13 | +pressure,water,+shower,water pressure,+head | Multiple | 0.280 | 0.112 | 67 | 93 |

8. Attach a cluster node to both process flows (Positive and Negative), run them with default settings, and examine the results.
   With default settings, 8 clusters are generated for positive comments and 9 clusters are generated for negative comments. While these cluster sizes look reasonable, after trying different cluster numbers and using domain expertise, the number of clusters is changed as follows.

9. Modify the properties of 'the Text Cluster (Custom)' node by changing the property "Exact or Maximum Number" to "Exact" and "Number of clusters" to 13 for negative comments and 3 for positive comments (as shown in Display C6.3). Run the text cluster nodes and examine the results.

Cluster analysis with custom settings results in 12 clusters of reasonable sizes for negative comments (as shown in Display C6.8). Results show that the top factors that customer complained about were dirty shower rooms, trash all over the parking lot, rude employees, etc. These results are similar to those obtained via topic mining (text topic node). For positive comments, cluster analysis results in 3 groups, and their interpretations are similar to the topics found for the positive comments. This is not surprising, because you have deliberately forced the number of clusters and number of topics to be similar in this case.

**Display C6.8. Summary of results from SAS® Text Miner**

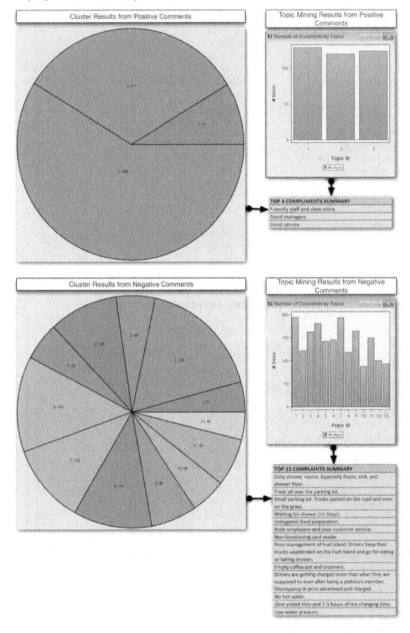

While Expectation-Maximization (EM) cluster algorithm is generally the preferred method for clustering textual data, it is possible to use hierarchical clustering or even SOM/Kohonen (once the SVDs have been calculated). Let us see how hierarchical clustering performs on the negative comments and how the hierarchical clusters relate to the EM clusters.

10. Attach another cluster node to the process flow for negative comments. Change the cluster algorithm to Hierarchical in the properties panel. Run this node and examine the results. The results reveal nine clusters with a reasonable number of documents per cluster, as shown below:

**Display C6.9. Hierarchical clustering results for negative comments**

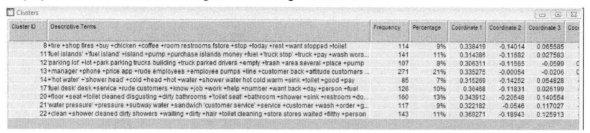

| Cluster ID | Descriptive Terms | Frequency | Percentage | Coordinate 1 | Coordinate 2 | Coordinate 3 | Coo |
|---|---|---|---|---|---|---|---|
| 8 | +tire +shop tires +buy +chicken +coffee +room restrooms fstore +stop +today +rest +want stopped +toilet | 114 | 9% | 0.338419 | -0.14014 | 0.065585 | |
| 11 | 'fuel islands' +'fuel island' +island +pump +purchase islands money +fuel +'truck stop' +truck +pay +wash wors... | 141 | 11% | 0.314386 | -0.11682 | 0.027583 | |
| 12 | 'parking lot +lot +park parking trucks building +truck parked drivers +empty +trash +area several +place +pump | 107 | 8% | 0.306311 | -0.11565 | -0.0599 | C |
| 13 | +manager +phone +price app +rude employees +employee pumps +line +customer back +attitude customers ... | 271 | 21% | 0.335275 | -0.00054 | -0.0206 | C |
| 14 | +hot water' +'shower head' +cold +head +hot +water +shower water hot cold warm +sink +toilet +good +pay | 85 | 7% | 0.315269 | -0.14252 | 0.054828 | |
| 17 | 'fuel desk' desk +service +rude customers +know +job +work +help +number +want back +day +person +fuel | 126 | 10% | 0.36468 | -0.11831 | 0.026199 | |
| 20 | +floor +seat +toilet cleaned disgusting +dirty bathrooms +toilet seat +bathroom +shower +sink +restroom +do... | 160 | 13% | 0.343912 | -0.20548 | 0.140554 | |
| 21 | 'water pressure' +pressure +subway water +sandwich 'customer service' +service +customer +wash +order +g... | 117 | 9% | 0.322182 | -0.0546 | 0.117027 | |
| 22 | +clean +shower cleaned dirty showers +waiting +dirty +hair +toilet cleaning +store stores waited +filthy +person | 143 | 11% | 0.369271 | -0.18943 | 0.125913 | |

To compare the clustering results between EM and hierarchical algorithm, you can read the descriptive terms of each cluster identified by the nodes and if needed, skim through the actual documents for each cluster. It is also useful to compare the cluster memberships of each document using the two methods. This is described next.

Attach a merge node to the process flow for negative comments and connect the two clustering nodes (N_Default and N_Hierarchical) to the merge node.

1. In the property panel of the merge node, select the ellipsis button for variables. Change the Merge Role of ID to "By". Run the merge node.
2. Attach a metadata node to the merge node. In the property panel of the metadata node, select the ellipsis button for Train under Variables.
3. In the pop-up window, sort (by clicking) the table by Role. You will find two variables with the role 'Segment' (each name ending with _cluster_). Those two variables represent cluster memberships of each document from the two cluster nodes. Select those two variables together and click the Explore button.

**Display C6.10 Cluster membership variables**

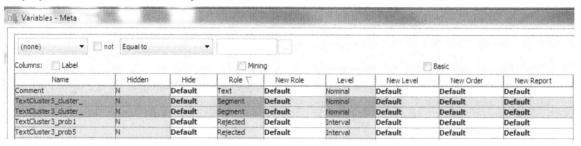

| Name | Hidden | Hide | Role | New Role | Level | New Level | New Order | New Report |
|---|---|---|---|---|---|---|---|---|
| Comment | N | Default | Text | Default | Nominal | Default | Default | Default |
| TextCluster5_cluster_ | N | Default | Segment | Default | Nominal | Default | Default | Default |
| TextCluster3_cluster_ | N | Default | Segment | Default | Nominal | Default | Default | Default |
| TextCluster3_prob1 | N | Default | Rejected | Default | Interval | Default | Default | Default |
| TextCluster3_prob5 | N | Default | Rejected | Default | Interval | Default | Default | Default |

4. In the Explore results, click each bar to understand the relationships between clusters from the two algorithms.
   Clicking the highest bar in the bottom panel shows that very little commonality exists between cluster memberships from the two algorithms. The purpose of this demonstration is not to prove that one algorithm is better than the other. At the end of the day, clustering is a partitioning tool, and any partitioning may be useful. It is up to the user to carefully examine each cluster solution, using domain expertise to determine which solution is more usable.

**Display C6.11 Relationships between clusters from the two algorithms**

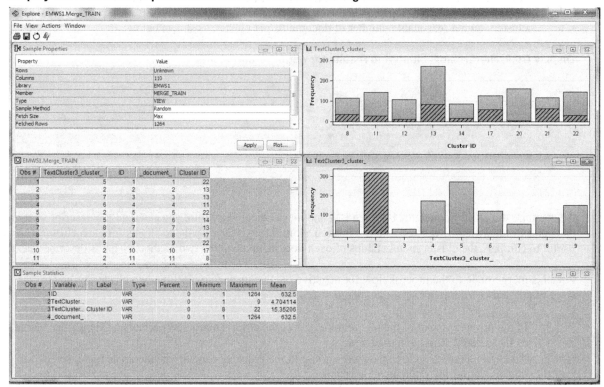

## Analysis Using the Text Rule-builder Node

The Text Rule-builder node is available in the SAS® Enterprise Miner version 12.1 and above. The Rule-builder node creates an ad hoc set of rules with user-definable outcomes. These rules are basically Boolean if-then-else type rules that can be exported to SAS® Content Categorization Studio

For this part of demonstration, you will use the following data sets.

- All_model.sas7bdat (a data set that combines positive and negative comments for building models. This data set has 90% of all comments)
- All_test.sas7bdat (a data set that combines positive and negative comments for testing models built. This data set has 90% of all comments)
- Engdict.sas7bdat

1. Create a new project in SAS® Enterprise Miner and create a library pointing to the location where the data resides using the project start code or File ▶ New ▶ Library menu. Create a new diagram.
2. Create a new Data source using All_model.sas7bdat. Make sure to change the 'Role' of variable 'Sentiment' in Step 5 of 'Data Source Wizard' to target as shown in Display C6.12. The sentiment variable reflects whether company experts judge the comment is as positive or negative.

**Display C6.12. Data source creation process for data set All_model**

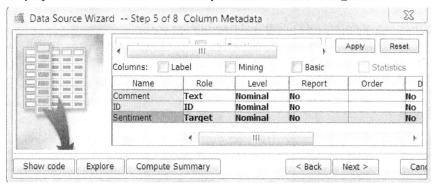

3. Create another data source using the **All_Test** data set. Select the role of the variable *sentiment_original* to text as shown in Display C6.13. The *sentiment_original* variable reflects whether the comment is judged as positive or negative by experts in the company.

**Display C6.13. Data source creation process for All_Test data set**

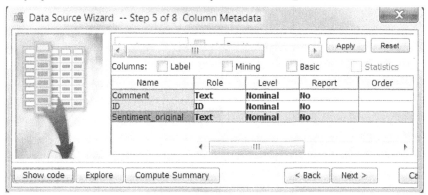

4. In step 8 of the data source wizard, select the role type for this data set as "Score" instead of "Raw" as shown in Display C6.14. For help importing the data set as a scoring data set, please see the help section of SAS Enterprise Miner 12.1 (Help->Contents->Create a score data set).

**Display C6.14. Data source creation process for data set All_Test**

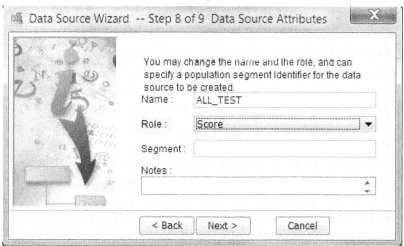

5. Now drag the data set **All_Model** onto the diagram space. Connect the data set with nodes as shown in

Display C6.15.

**Display C6.15. Text mining process of comments using text rule-builder node**

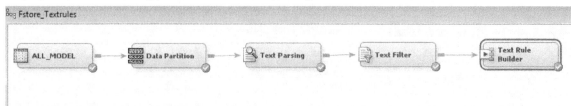

6. In the property panel of data partition node, change the value of "Data Set Allocations" as shown in Display C6.16.

**Display C6.16. Data partition node property panel**

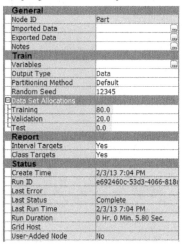

7. Use default settings for the text parsing node.
8. In the property panel of 'Text Filter' node (as shown in Display C6.17), import the dictionary engdict from the SAS library where you saved this data set.

**Display C6.17. Text Filter node property panel**

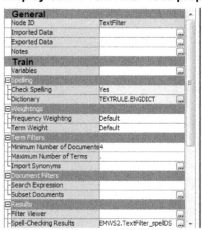

9. Use default properties for 'Text Rule-Builder' node. Run the node and examine the results. Results show a pretty good validation misclassification rate of 0.2156, as shown below.

**Display C6.18. Overall model classification results from the rule builder node**

| Target | Target Label | Fit Statistics | Statistics Label | Train | Validation | Test |
|---|---|---|---|---|---|---|
| Sentiment | | _ASE_ | Average Squared Error | 0.066913 | 0.081408 | |
| Sentiment | | _DIV_ | Divisor for ASE | 3358 | 844 | |
| Sentiment | | _MAX_ | Maximum Absolute Error | 0.999981 | 0.999687 | |
| Sentiment | | _NOBS_ | Sum of Frequencies | 1679 | 422 | |
| Sentiment | | _RASE_ | Root Average Squared Error | 0.258675 | 0.285321 | |
| Sentiment | | _SSE_ | Sum of Squared Errors | 224.6934 | 68.70848 | |
| Sentiment | | _DISF_ | Frequency of Classified Cases | 1679 | 422 | |
| Sentiment | | _MISC_ | Misclassification Rate | 0.1757 | 0.21564 | |
| Sentiment | | _WRONG_ | Number of Wrong Classifications | 295 | 91 | |

While the model seems to be performing reasonably from looking at the overall misclassification rate, it is also important to evaluate how well the model classifies each outcome (positive or negative). The numbers reported below show that the model does about equally well in predicting positive versus negative cases.

**Display C6.19. Model Classification Results from the Rule-Builder Node for Positive and Negative comments**

```
 89
 90    Classification Table
 91
 92    Data Role=TRAIN Target Variable=Sentiment Target Label=' '
 93
 94                         Target      Outcome    Frequency     Total
 95    Target    Outcome   Percentage  Percentage    Count     Percentage
 96
 97    NEGATIVE  NEGATIVE   83.2558     88.6139       895       53.3055
 98    POSITIVE  NEGATIVE   16.7442     26.9058       180       10.7207
 99    NEGATIVE  POSITIVE   19.0397     11.3861       115        6.8493
100    POSITIVE  POSITIVE   80.9603     73.0942       489       29.1245
101
102
103    Data Role=VALIDATE Target Variable=Sentiment Target Label=' '
104
105                         Target      Outcome    Frequency     Total
106    Target    Outcome   Percentage  Percentage    Count     Percentage
107
108    NEGATIVE  NEGATIVE   79.8535     85.8268       218       51.6588
109    POSITIVE  NEGATIVE   20.1465     32.7381        55       13.0332
110    NEGATIVE  POSITIVE   24.1611     14.1732        36        8.5308
111    POSITIVE  POSITIVE   75.8389     67.2619       113       26.7773
112
113
114
```

10. Click on the ellipsis button next to content categorization code in the properties panel of the text rule-builder node. You will see the rules that the text builder node has created from this data (as shown in Display C6.20). Being Boolean in nature, these rules are easy to understand.

**Display C6.20 Content categorization code obtained from the text rule-builder node**

```
Content Categorization Code                                                                    [x]

F_Sentiment =POSITIVE ::
(OR
, (AND, (OR, "helpfull" , "helpful" ))
, (AND, (NOT, "fstore" ), (OR, "greatest" , "great" , "grest" ))
, (AND, (NOT, (OR, "fule" , "fuel" , "feul" )), (OR, "friendliest" , "friendly" , "freindly" ))
, (AND, (OR, "nice" , "nicest" , "nice" , "nicer" , "nices" ))
, (AND, (OR, "awesome" , "awsome" , "awsome" ))
, (AND, (OR, "excellent" , "exellent" ))
, (AND, (NOT, (OR, "shower" , "showeres" , "showers" , "shower" , "shower6" )), (NOT, (OR, "last" , "last" )), (NOT, (OR, "feling" , "fuel" , "fueled" , "fuelings" , "fe
))
, (AND, (OR, "love" , "loving" , "loved" ))
, "love"
, (AND, (OR, "oustanding" , "outstanding" ))
, "great"
, "kindness"
, "grest"
, (AND, (OR, "greatest" , "great" , "grest" ))
, (AND, (OR, "remembered" , "remember" )))
F_Sentiment =NEGATIVE ::
(OR
, (AND, (OR, "dirty" , "dirtiest" ))
, (AND, (OR, "rude" , "rudest" ))
, "dirty"
, (AND, (OR, "prices" , "price" , "price" ))
, (AND, (OR, "filty" , "filthy" , "filthiest" , "filth" ))
, (AND, (OR, "waiting" , "waitting" , "wainting" ))
, (AND, (OR, "heads" , "head" , "head" ))
, (AND, (OR, "wait" , "waiting" , "waits" , "waisted" , "waited" ))
, (AND, (OR, "toilet" , "toilet" ))
, (AND, (OR, "floor" , "floor" , "floors" ))
, (AND, (OR, "fuel islands" , "fuel island" ))
, (AND, (OR, "worst" , "bad" , "bad" ))
, (AND, (OR, "showeres" , "shower" , "shower6" , "shower" , "showers" ), (NOT, (OR, "clean" , "clean" , "cleanest" )))
, (AND, (OR, "know" , "knowing" , "known" , "knew" , "knows" ))
, (AND, (OR, "fuel" , "feul" , "fule" )))

                                                        [ Save As ]  [ OK ]  [ Cancel ]
```

11. Attach a score node to the text rule-builder node.
12. Drag the All_Test data source to the diagram workspace and connect it to the score node.
13. Run the score node and examine results.

The scoring results shown below look reasonable, since the % of positive and negative in the scored data is similar to those from the training and validation data. However, in this data set (unlike in real scoring cases), you have the actual sentiment values, and those can be compared against the predicted sentiment from the text rule-builder model via a cross-tab. The cross-tab between the two results can be generated easily by using a SAS code node in this diagram space.

**Display C6.21 Scoring results**

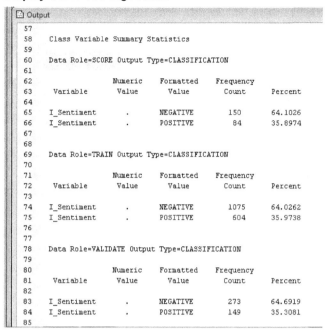

```
57
58    Class Variable Summary Statistics
59
60    Data Role=SCORE Output Type=CLASSIFICATION
61
62                 Numeric    Formatted    Frequency
63    Variable     Value      Value        Count      Percent
64
65    I_Sentiment    .         NEGATIVE      150       64.1026
66    I_Sentiment    .         POSITIVE       84       35.8974
67
68
69    Data Role=TRAIN Output Type=CLASSIFICATION
70
71                 Numeric    Formatted    Frequency
72    Variable     Value      Value        Count      Percent
73
74    I_Sentiment    .         NEGATIVE     1075       64.0262
75    I_Sentiment    .         POSITIVE      604       35.9738
76
77
78    Data Role=VALIDATE Output Type=CLASSIFICATION
79
80                 Numeric    Formatted    Frequency
81    Variable     Value      Value        Count      Percent
82
83    I_Sentiment    .         NEGATIVE      273       64.6919
84    I_Sentiment    .         POSITIVE      149       35.3081
85
```

14. Attach a SAS Code node to the rule-builder node. In the property panel of 'SAS Code' node, click the ellipsis button next to 'Code Editor.'

**Display C6.22. SAS Code node property panel**

| .. Property | Value |
|---|---|
| **General** | |
| Node ID | EMCODE |
| Imported Data | |
| Exported Data | |
| Notes | |
| **Train** | |
| Variables | |
| Code Editor | |
| Tool Type | Utility |
| Data Needed | No |
| Rerun | No |
| Use Priors | Yes |
| **Score** | |
| Advisor Type | Basic |
| Publish Code | Publish |
| Code Format | DATA step |
| **Status** | |
| Create Time | 2/3/13 8:03 PM |
| Run ID | 53e4ed21-7465-430c-913! |
| Last Error | |
| Last Status | Complete |
| Last Run Time | 2/3/13 10:29 PM |
| Run Duration | 0 Hr. 0 Min. 6.65 Sec. |
| Grid Host | |
| User-Added Node | No |

15. In the pop-up code window, type the code as shown in the Training Code window in Display C6.23.

**Display C6.23. Code editor window**

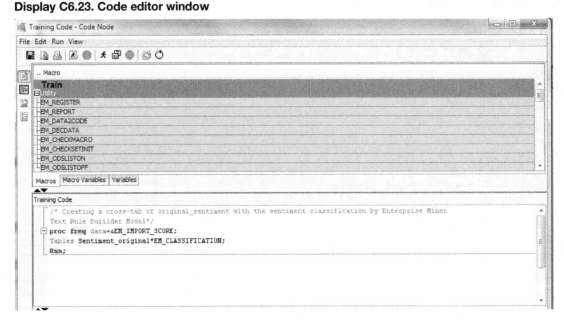

16. Close the SAS code node (select save), then run it and examine results.

It seems that 126 out of 141 negative comments (89.36%) were correctly classified, and 69 out of 93 positive comments (74.19%) were also correctly classified. Overall, 195 out of 234 (83.33%) comments were correctly classified by the text rule builder model. These are pretty good results.

**Display C6.24. Comparing scoring results with known values.**

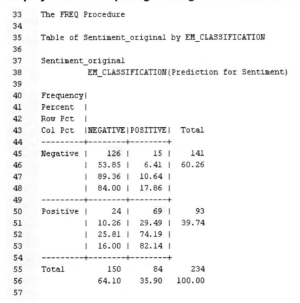

```
33    The FREQ Procedure
34
35    Table of Sentiment_original by EM_CLASSIFICATION
36
37    Sentiment_original
38            EM_CLASSIFICATION(Prediction for Sentiment)
39
40    Frequency|
41    Percent  |
42    Row Pct  |
43    Col Pct  |NEGATIVE|POSITIVE|  Total
44    ---------+--------+--------+
45    Negative |   126  |    15  |   141
46             |  53.85 |   6.41 | 60.26
47             |  89.36 |  10.64 |
48             |  84.00 |  17.86 |
49    ---------+--------+--------+
50    Positive |    24  |    69  |    93
51             |  10.26 |  29.49 | 39.74
52             |  25.81 |  74.19 |
53             |  16.00 |  82.14 |
54    ---------+--------+--------+
55    Total        150       84      234
56               64.10    35.90   100.00
57
```

## Analysis Using SAS® Sentiment Analysis Studio

SAS® Sentiment Analysis Studio classifies the feedback into positive and negative sentiments and categorizes them into features. SAS® Sentiment Analysis Studio has three different types of models: statistical model, rule based model, and hybrid model. Hybrid model is a combination of a statistical model and a rule based model. In practice, a statistical model is often used as a starting point in sentiment mining, because it provides a baseline model that can be set up quickly. Using it as a starting point, a rule based model can then be built, after which each lexicon rule can be modeled and analyzed over and over again. A successful rule based model often can explain results very intuitively for managers who may not like to delve deeply into detailed statistical analysis.

## Building a Statistical Model

1. Create a new project in Sentiment Analysis Studio using the default settings for rule-based and statistical models.
2. Select Corpora tab. Right-click in the white space in Corpus and select new corpus (name it as 'Fstore'). Right-click Positive (as shown in Display C6.25) under Fstore and select Add a Directory. Point to the folder Case Studies ▶ Case Study 6 ▶ Fstore Sentiment Mining Data ▶ Model data ▶ Positive and click Select Folder.

**Display C6.25. Sentiment Analysis Corpus view panel (Positive)Q**

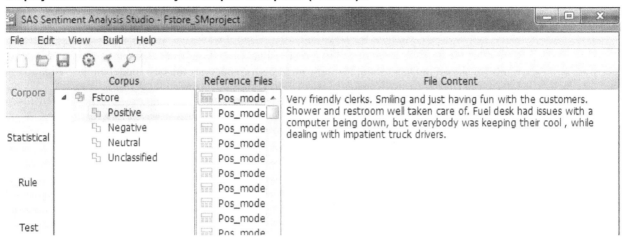

3. Right-click Negative (as shown in Display C6.26) under Fstore and select Add a Directory. Point to the folder Case Studies ▶ Case Study 6 ▶ Fstore Sentiment Mining Data ▶ Model Data ▶ Negative and click Select Folder.

**Display C6.26. Sentiment Analysis Corpus view panel (Negative)**

4. Select Statistical tab. Right-click in the white space and select New Model (as shown in Display C6.27). Create a new statistical model with advanced settings by selecting the 'advanced' option in the 'Add New Model' pop-up box. Click OK.

**Display C6.27. Creating advanced statistical model**

5.  Right-click the newly created model in the Statistical Models and select the option "Train Model". The results (as shown in Display C6.28) show good performance for the best model on the validation data.

**Display C6.28. Statistical model training results**

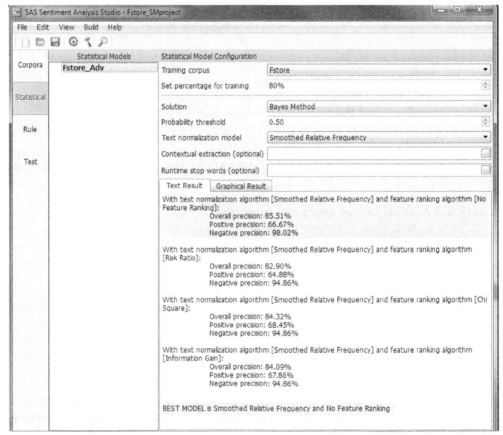

6.  Right-click the model name one more time and select Validate Model. Note that these (as shown in Display C6.29) match the results reported for the best model in the earlier step.

**Display C6.29. Statistical model validation results**

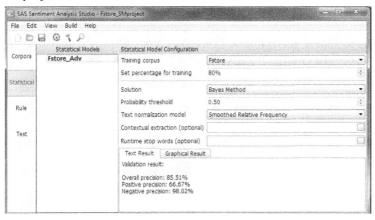

7.  Select the Test tab, right-click in the white space of Test Data, and select new test directory. Point to the negative folder in the Testing Data directory.
8.  Right-click the folder under Manual Test and select Test, in Statistical Model. The results look good, as shown below.

**Display C6.30. Statistical model testing results for negative directory**

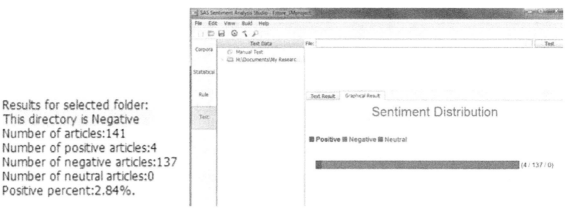

Results for selected folder:
This directory is Negative
Number of articles:141
Number of positive articles:4
Number of negative articles:137
Number of neutral articles:0
Positive percent:2.84%.

9.  Repeat the testing steps with the positive directory. Results are as shown below.

**Display C6.31. Statistical model testing results for negative directory**

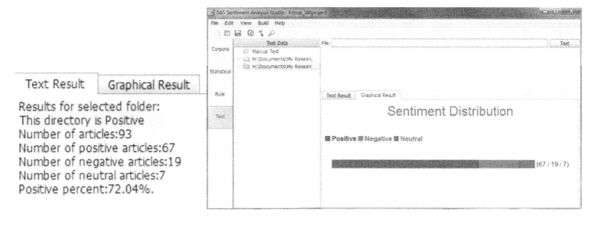

Text Result    Graphical Result

Results for selected folder:
This directory is Positive
Number of articles:93
Number of positive articles:67
Number of negative articles:19
Number of neutral articles:7
Positive percent:72.04%.

On the surface the statistical models seem to be performing well! But, when you extract the rules from the statistical model via import learned features, some of the rules are difficult to understand.

10. Click the Rule tab, and from the top menu, select Import Learned Features. Select the statistical model created in prior steps. Click on Tonal Keyword, then select Positive or Negative tab (as shown in Display C6.32) to get a sense of the rules.

**Display C6.32. Imported learned features from a statistical model**

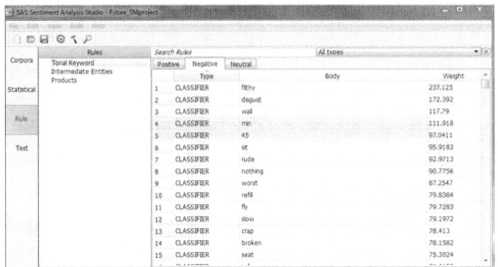

For example, numbers were classified as negative sentiment rules. The imported list was also very long and unwieldy. Generally speaking, these rules can be improved by editing/modifying imported rules from the statistical models. For the purpose of this case study, through trial and error, we have built a set of rules (filename: Fstore_CustomRule.xml) that you will use next.

## Building a Rule-based Model

11. Import the rule file Fstore_CustomRule.xml from File Menu ▶ Import rules (as shown in Display C6.33). This XML file is available in the Case Studies ▶ Case Study 6 folder.

**Display C6.33. Imported rules from file Fstore_CustomRule.xml**

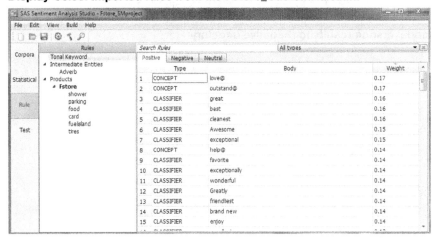

12. If you want to create your own rules, you can either create rules globally or create rules for each feature. To create rules globally, click "Tonal Keyword". Under "Positive", "Negative", or "Neutral"

tabs, first edit "Body", and then change the "Type" and "Weight". Note that the rules will not work if you do not select the "Type" of rules. An example of creating rules in "Tonal Keywords" follows. In this example, the term Thank@ will detect "thank", "thanks", "thanked", and "thanking" and count for all features.

**Display C6.34. Examples of positive tonal keyword**

| Rules | Search Rules | | All types | | |
|---|---|---|---|---|---|
| Tonal Keyword | Positive | Negative | Neutral | | |
| ◢ Intermediate Entities | | **Type** | **Body** | **Weight** | |
| Adverb | | | | | |
| ◢ Products | 1 | CONCEPT | Thank@ | 0.12 | |

Another example of creating rules for a specific feature follows. In this example, the rule will be triggered only if the term "not clean" is mentioned with possible definitions of feature "shower" as defined within 'Definitions' in a single sentence.

**Display C6.35. Example of creating concept rule for a specific feature**

| Rules | Search Rules | | | All types | | |
|---|---|---|---|---|---|---|
| Tonal Keyword | Definitions | Positive | Negative | Neutral | | |
| ◢ Intermediate Entities | | **Type** | | **Body** | | **Weight** |
| Adverb | | | | | | |
| ◢ Products | 1 | CONCEPT_RULE | | (SENT,"_c{not clean}","_def{Fstoreshower}") | | 0.08 |
| ◢ **Fstore** | 2 | CLASSIFIER | | | | 1 |
| shower | ? | CLASSIFIER | | | | ? |

13. Build the rule-based model (Build-> Build Rule-based Model).
14. Now test the folders in Rule-based Model and examine the results.

   After scoring these rules on the negative test directory, which contains 141 negative comments, 112 comments were classified as negative, 19 comments as positive, and 10 as neutral, which is a very good result. As you carefully read all the comments in negative directory, you will find that company experts actually misclassified some of the mixed comments which the rule based model identified as neutral! You will also discover that most number of the complaints were regarding *shower* followed by *food, parking, Fuel Island, tires,* and *card*. Display C6.36 shows the exact number of comments that were classified for each feature.

**Display C6.36. Testing results of negative feedbacks from a rule-based sentiment model**

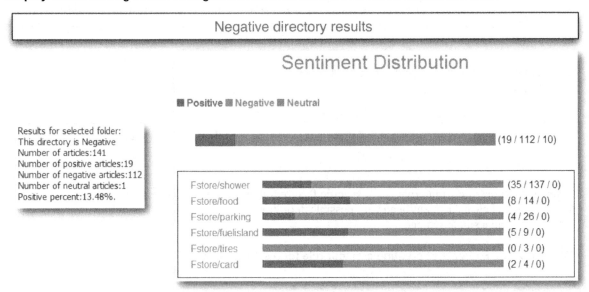

The rule-based model also did a good job in classifying positive comments. In the test directory containing 93 positive comments, 81 comments were classified correctly as positive which resulted in accuracy of 87.10 %( as shown in Display C6.37). "Food" received highest numbers of positive feedback followed by *showers, parking, tires,* and *card.*

**Display C6.37. Testing results of positive feedbacks from a rule-based sentiment model**

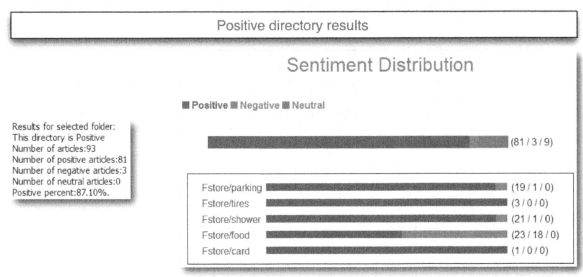

To know more granular details on the model performance, each comment can be individually tested and examined. The example below shows why a negative comment was classified as negative by the rule-based model. The model chooses an interesting color code. For instance, the words 'Don't', 'not', 'never' are shown in red, which indicates they are negative, whereas word 'good' is shown in green, which indicates a positive word. Words in blue are features or attributes.

## Display C6.38. Single Text File Testing Result (negative directory)

Don't understand why the gas is so high here when it is not at your store in Amarillo. We had good gas prices here until your store came to town, now they are the highest in the Panhandle! Wish your firm had never come to town. In 30 mi;es the gas is 21 ce nts cheaper. You are greedy!

| Text Result | Graphical Result |
| --- | --- |

Overall Document: Test in rule-based model result is Negative
Probability to be positive is 46.66% with confidence 6.68%

Fstore: Test in rule-based model result is Negative
Probability to be positive is 48.07% with confidence 3.85%


9 matches:
2 matches for product definitions:
       No 0 (68-72): Fstore : store
       No 1 (126-130): Fstore : store
0 matches for feature definitions:
3 matches for product rules:
       No 0 (0-4): Fstore : Negative : 0.1900 : Don't
       No 1 (56-58): Fstore : Negative : 0.1900 : not
       No 2 (94-97): Fstore : Positive : 0.1900 : good
0 matches for feature rules:
4 matches for keyword rules:
       No 0 (0-4): Negative : 0.1000 : Don't
       No 1 (56-58): Negative : 0.0700 : not
       No 2 (94-97): Positive : 0.1000 : good
       No 3 (208-212): Negative : 0.0700 : never
0 matches for intermediate entity definitions:

Product prominence information:
       Fstore: Bottom 80%
Product dominance information:
       Fstore: Exclusive

## Display C6.39. Single Text File Testing Result (Positive directory)

| File: | | Test |
| --- | --- | --- |

I would just like to say that corey and melody gave me great customer service!

| Text Result | Graphical Result |
| --- | --- |

Overall Document: Test in rule-based model result is Positive
Probability to be positive is 52.53% with confidence 5.06%


2 matches:
0 matches for product definitions:
0 matches for feature definitions:
0 matches for product rules:
0 matches for feature rules:
2 matches for keyword rules:
       No 0 (13-16): Positive : 0.0900 : like
       No 1 (55-59): Positive : 0.1600 : great
0 matches for intermediate entity definitions:

Product prominence information:
Product dominance information:

## Summary

Analyzing reviews from customers can provide insightful information, which in turn helps companies to improve their quality of service and help to differentiate their services from their competitors.

Prompt response is the key to win customer loyalty. This can only be done when there is an automatic system in place which can immediately classify customer's feedback into positive and negative and provide appropriate response. Once the rules have been developed, it will be easy to set up an automatic scoring system for future comments that can be classified in real time when appropriate management interventions can happen.

# Case Study 7 Information Organization and Access of Enron Emails to Help Investigation

Dan Zaratsian
Murali Pagolu
Goutam Chakraborty

## Introduction

Across industries, text analytics is opening the door for new, innovative ways to understand behavior, motives, interests, and trends buried within rich textual data. In most instances, this type of data is most advantageous when it is paired with structured data, such as demographic or transactional data, to enhance your view of the customer or party of interest. Text analytics is used across industries and can be applied to a variety of data. This data may consist of survey responses, call center notes, emails or chat messages, social media, news posts, forums, blogs, a document collection, warranty claims, and many more. This is why it is critical for an organization to take advantage of text analytics tools to extract relevant, insightful information from this type of data.

In the world of fraud and criminal activity, investigators benefit from having as much as information as they can in order to solve the puzzle. But there are really two fundamental problems with this. First, they only have so much time in the day to review the information, so having the right tools to analyze and sort through the volumes of data in the least amount of time is essential. Second, most of this information is in the form of documents, emails, and other textual formats.

The Enron Corporation was an American energy, commodities, and services company based in Houston, Texas. In 2001, it filed for bankruptcy. Before Enron's, bankruptcy filing, Enron employed approximately 20,000 staff. It was one of the world's leading electricity, natural gas, communications, and pulp and paper companies, with claimed revenues of nearly $101 billion in 2000. Later, it was discovered that many of Enron's recorded assets and profits were inflated, or even wholly fraudulent and nonexistent. For example, in 1999, Enron promised to pay back a Merrill Lynch investment with interest in order to show a profit on its books. Debts and losses were placed in offshore accounts that were not included in the firm's financial statements. More sophisticated and mysterious financial transactions between Enron and related companies were used to take unprofitable entities off the company's financial records. These "offshore" entities were limited partnerships between Enron and LJM Cayman LP and LJM2 Co-Investment LP, created to buy Enron's poorly performing stocks to improve its financial statements. These two partnerships received funding of approximately $390 million from a group of investors. Enron created entities it called "Raptors" and transferred more than $1.2 billion in assets into Raptor accounts, including millions of shares of Enron common stock, long-term rights to purchase millions more shares, plus $150 million of Enron notes payable. It capitalized the Raptors and booked the notes payable issued as assets on its balance sheet while increasing the shareholders' equity for the same amount.

## Objective

Understanding the actions and motives of fraudsters is a vital part of predicting (and preventing) fraudulent activity. To gain an advantage, you must incorporate analytics within your investigative process. The goal of analytics is to turn data into actionable information. Unfortunately, most data is unstructured, making it more challenging to analyze than traditional structured data (such as demographic or transactional data). SAS Text

Analytics processes large volumes of text and provides useful analytics to support investigations. The objective of this case study is to demonstrate how SAS Text Analytics can be leveraged to extract meaningful information from textual data and to identify patterns and trends across thousands of Enron email correspondences.

## Data Description

The Enron email data, used in this case study, is publicly released as part of FERC's Western Energy Markets investigation and has been converted to industry standard formats by EDRM. The original data set consists of 1,227,255 emails with 493,384 attachments covering 151 custodians. The email is provided in Microsoft PST, IETF MIME, and EDRM XML formats. This data was downloaded from http://www.edrm.net/resources/data-sets/edrm-enron-email-data-set-v2 in an XML format.

This case study was performed using the following SAS software:

- SAS® Enterprise Miner
- SAS® Text Miner
- SAS® Enterprise Content Categorization
- SAS® Crawler (SAS Information Retrieval Studio)

## Step-by-Step Software Instruction with Settings/Properties

### Step 1: Information Retrieval and Parsing

As described above, we chose to download the files in XML format from the website http://edrm.net. The XML file looked something like this:

```
<?xml version="1.0" encoding="UTF-8"?>
- <article>
    <body>Attached is the Weekly deal report from 7/26/01-8/1/01. Lex Carroll Enron Power Marketing, Inc.
        Midwest Region 713-853-5426</body>
    <title>Weekly deal report</title>
    <id>60068</id>
    <msgdate> 02AUG2001 :06:39:26</msgdate>
    <from>Lex Carroll </from>
    <to>William Abler </to>
</article>
```

Our first objective is to convert the XML files into a structured format (a SAS data set) by parsing out the individual fields such as body, title, id, msgdate, from, and to. We chose to use SAS® Information Retrieval Studio to crawl the XML file directory and parse the XML using the built-in "parse_xml" post-processor. Below are step-by-step instructions and screenshots that help to clarify this process:

1. Launch the SAS® Information Retrieval Studio and create a new project. Name it 'Enron'.
2. Go to 'File Crawler' and change the 'General Settings' to reflect the properties as shown in Display C7.1. Under the "Paths" section, provide the complete path of the folder which contains the XML files you downloaded. Under the "Paths to Exclude' tab, add 'XML' as the filename extension you wish to exclude from file crawling. Click 'Apply Changes' to save settings.

**Display C7.1: Settings for File Crawler in SAS® Information Retrieval Studio**

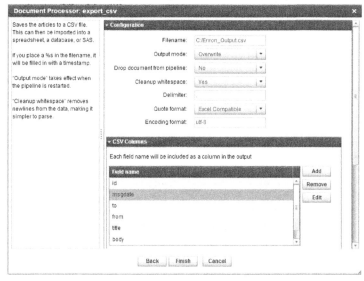

3. Go to 'Pipeline Server' and add the document processor – 'parse_xml' from the list of available document processors using the default settings.
4. Add the 'export_csv' document processor and make the changes as shown in Display C7.2. Ensure all the fields: id, msgdate, to, from, title and body are included in the columns for output. Click on 'Apply Changes' to save the settings.

**Display C7.2: Settings for 'export_csv' document processor**

5. Go to the 'File Crawler' and click on 'Start' to kick start the file crawling, extract the contents of XML files, and export the data into structured columns of a CSV file.
6. Rename the fields 'id' to '_document_', 'msgdate' to 'date', 'to' to 'recipients', 'from' to 'sender,' and 'title' to 'subject'.
7. You can then use the 'Import Data' task in SAS® Enterprise Guide to import the data in the output CSV file. The output SAS data set will look like the below (Display C7.3).

**Display C7.3: Snapshot of the Enron emails data imported from output CSV file**

| date | sender | recipients | subject | body | _document_ |
|---|---|---|---|---|---|
| Fri, 19 Oct 2001 | ken.shulklapper... | mike.grigsby@e... | Forbes Article-Gas Fired Power Plants | Don't know if you... | 1 |
| Wed, 24 Oct 2001 | members@realm... | members@realm... | Have you checked your credit rating lately | TheStreet.com a... | 2 |
| Wed, 4 Oct 2000 | john.arnold@enr... | jennifer.fraser@... | Re: | Premonition?... | 4 |
| Mon, 15 Oct 2001 | m..schmidt@enr... | m..schmidt@enr... | Enron Mentions - 10-15-01 | COMPANIES &... | 5 |
| Thu, 18 Oct 2001 | swl@winelibrary... | jarnold@enron.c... | 95 Pointer and more | To Place an orde... | 6 |
| Fri, 5 Oct 2001 | m..schmidt@enr... | m..schmidt@enr... | Enron Mentions | USA: Northwest... | 8 |
| Mon, 26 Nov 2001 | swl@winelibrary... | jarnold@enron.c... | 44% off wine, Beaucastel and more | Wine Library.co... | 9 |
| Wed, 4 Oct 2000 | john.arnold@enr... | jennifer.fraser@... | Re: | Premonition?... | 10 |
| Wed, 4 Oct 2000 | john.arnold@enr... | jennifer.fraser@... | Re: | Premonition?... | 11 |
| Wed, 14 Nov 200... | djtheroux@indep... | lighthouse@inde... | THE LIGHTHOUSE: November 14, 2001 | THE LIGHTHOU... | 12 |
| Fri, 16 Nov 2001 | holger.fahrinkrug... | harora@ect.enro... | UBSW: Preview of the confidence vote | Vote of confiden... | 13 |
| Fri, 23 Nov 2001 | holger.fahrinkrug... | harora@ect.enro... | UBSW: German inflation falls more strongly | *** PDF version i... | 14 |
| Wed, 14 Nov 200... | chairman.ken@e... | dl-ga-all_enron_... | Overview of Investor Conference Call | Today, Enron ho... | 15 |
| Wed, 14 Nov 200... | chairman.ken@e... | dl-ga-all_enron_... | Overview of Investor Conference Call | Today, Enron ho... | 16 |
| Thu, 7 Mar 2002 | john.monaco@ci... | robert.badeer@e... | RE: RE: Whats up | Good, you can g... | 17 |
| Fri, 8 Mar 2002 | john.monaco@ci... | robert.badeer@e... | RE: RE: Whats up | Texans, Titans w... | 18 |
| Fri, 8 Mar 2002 | robert.badeer@e... | Badeer, Robert | RE: RE: Whats up | Monaco, Trader... | 19 |
| Fri, 11 Jan 2002 | sbailey@crusesc... | susan.bailey@e... | FW: 22 things all dogs need to know | -----Original Mes... | 20 |

**Note:** For this case study, we are only providing you with a small sample since the original data is very large (~210 GB).

## Step 2: Data Exploration and Analysis

1. Launch SAS® Enterprise Miner, create a new project, and name it 'TM_Demo'. Alternatively, you can also open the project "Enron_TM" that we have provided. This is available in the folder *Case Studies* ▶ *Case Study 7* in the data provided with this book.
2. Register the SAS data set that we have provided to you (ENRON_SAMPLE.sas7bdat). This data set is also available in the folder *Case Studies* ▶ *Case Study 7* in the data provided with this book. Ensure that the variable 'body' is in 'Text' role, '_document_' in 'ID' role and the rest of variables are rejected.
3. Create a new diagram and name it 'Enron Analysis'. Drag the registered SAS data set into the diagram.
4. Connect the 'Text Parsing', 'Text Filter', and 'Text Topic' nodes in series as shown in the Display C7.4.

**Display C7.4: Text mining process flow for 'Enron Analysis'**

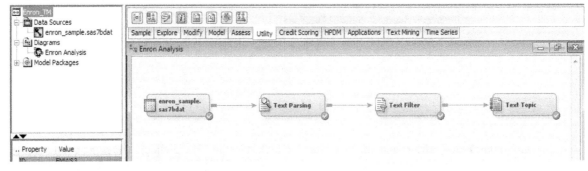

5. Click 'Text Parsing' node and change the following settings in the property pane.
   a. Find Entities – Standard
   b. Ignore Types of Entities – Company, Currency, Date, Person Address, Internet, Phone, and SSN
6. Click 'Text Filter' node and change the property 'Check Spelling' to 'Yes'.
7. Click 'Text Topic' node, right click, and select 'Run' to execute the text mining process flow.
8. Once the process flow is run successfully, right click and open the text parsing node results. You will find that all of the terms in the email content are parsed. They are either kept or dropped from the terms list based on the default stop list and property settings such as ignore types of entities, attributes,

and parts of speech. From this list, you can find lot of terms and phrases which you can use in building taxonomy for the classification of Enron emails.

9. Click on the Text Topic node and verify the results by looking at the multi-term text topics generated from the Enron sample data set (Display C7.5). You can make sense of these topics by looking at the terms describing each of the individual topics. For example, Topic 14 is about the Al Gore election, #16 is about trading telecommunications stock, #3 is about oil production, and #11 is tied to California electricity, which are all part of the Enron scandal. These terms can be used to build the categories/subcategories that will capture the e-mails related to those topics.

**Display C7.5: Text Topics emerged from mining the Enron sample data set**

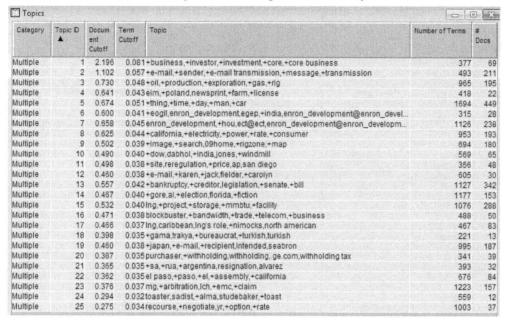

| Category | Topic ID ▲ | Document Cutoff | Term Cutoff | Topic | Number of Terms | # Docs |
|---|---|---|---|---|---|---|
| Multiple | 1 | 2.196 | 0.081 | +business,+investor,+investment,+core,+core business | 377 | 69 |
| Multiple | 2 | 1.102 | 0.057 | +e-mail,+sender,+e-mail transmission,+message,+transmission | 493 | 211 |
| Multiple | 3 | 0.730 | 0.048 | +oil,+production,+exploration,+gas,+rig | 965 | 195 |
| Multiple | 4 | 0.641 | 0.043 | eim,+poland,newsprint,+farm,+license | 418 | 22 |
| Multiple | 5 | 0.674 | 0.051 | +thing,+time,+day,+man,+car | 1694 | 449 |
| Multiple | 6 | 0.600 | 0.041 | +eogil,enron_development,egep,+india,enron_development@enron_devel... | 315 | 28 |
| Multiple | 7 | 0.658 | 0.045 | enron_development,+hou,ect@ect,enron_development@enron_developm... | 1126 | 238 |
| Multiple | 8 | 0.625 | 0.044 | +california,+electricity,+power,+rate,+consumer | 953 | 193 |
| Multiple | 9 | 0.502 | 0.039 | +image,+search,09home,+rigzone,+map | 694 | 180 |
| Multiple | 10 | 0.490 | 0.040 | +dow,dabhol,+india,jones,+windmill | 569 | 65 |
| Multiple | 11 | 0.498 | 0.038 | +site,reregulation,+price,ap,san diego | 356 | 48 |
| Multiple | 12 | 0.460 | 0.038 | +e-mail,+karen,+jack,fielder,+carolyn | 605 | 30 |
| Multiple | 13 | 0.557 | 0.042 | +bankruptcy,+creditor,legislation,+senate,+bill | 1127 | 342 |
| Multiple | 14 | 0.467 | 0.040 | +gore,al,+election,florida,+fiction | 1177 | 153 |
| Multiple | 15 | 0.532 | 0.040 | lng,+project,+storage,+mmbtu,+facility | 1076 | 288 |
| Multiple | 16 | 0.471 | 0.038 | blockbuster,+bandwidth,+trade,+telecom,+business | 488 | 50 |
| Multiple | 17 | 0.466 | 0.037 | lng,caribbean,lng's role,+nimocks,north american | 467 | 83 |
| Multiple | 18 | 0.398 | 0.035 | +gama,trakya,+bureaucrat,+turkish,turkish | 221 | 13 |
| Multiple | 19 | 0.460 | 0.038 | +japan,+e-mail,+recipient,intended,seabron | 995 | 187 |
| Multiple | 20 | 0.387 | 0.035 | purchaser,+withholding,withholding,.ge.com,withholding tax | 341 | 39 |
| Multiple | 21 | 0.365 | 0.035 | +sa,+rua,+argentina,resignation,alvarez | 393 | 32 |
| Multiple | 22 | 0.362 | 0.035 | el paso,+paso,+el,+assembly,+california | 676 | 84 |
| Multiple | 23 | 0.376 | 0.037 | mg,+arbitration,lch,+emc,+claim | 1223 | 157 |
| Multiple | 24 | 0.294 | 0.032 | toaster,sadist,+alma,studebaker,+toast | 559 | 12 |
| Multiple | 25 | 0.275 | 0.034 | recourse,+negotiate,yr,+option,+rate | 1003 | 37 |

## Step 3: Building Taxonomy (Categories and Concepts)

1. In this step, you can start building the taxonomy in SAS® Enterprise Content Categorization for the purpose of classifying the Enron email content into various categories and subcategories.

2. For this purpose, you may either choose to use an industry standard taxonomy such as IPTC and modify it to ensure it is customized for this particular case. This process is lot easier than building it from the scratch.

3. For learning purposes, you can open the "Enron_CC" content categorization project we have provided to you. This is available in the folder *Case Studies* ▶ *Case Study 7* in the data provided with this book. As you look at the taxonomy and the categories/sub-categories we have built for this case study, you will find that a lot of terms defining the categories/sub-categories are pre-existing from the IPTC taxonomy (See Display C7.6). However, you can leverage the features such as text parsing and text topic extraction in SAS® Text Miner to enhance the terms list which will efficiently categorize Enron e-mails.

**Display C7.6: Taxonomy created for classifying Enron e-mails into categories**

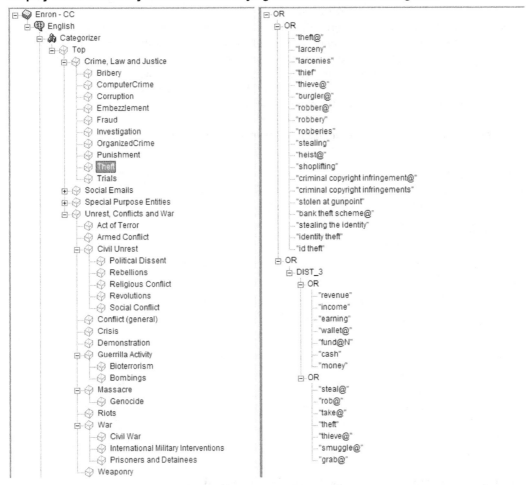

4. Similarly, you can write rules in ECC to capture simple classifier concepts such as the names of companies that can help your investigation. You can also write complex definitions to capture facts that are related to financial gains or losses involved in the scandal. You can rely on regular expression-based concept definitions for this purpose as shown in Display C7.7.

**Display C7.7: Concepts created to capture facts using regular expressions**

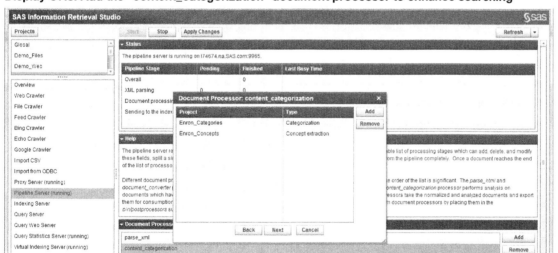

## Step 4: Indexing and Search

1. As a next step, you can include the content categorization project (categories & concepts) that you have built as a post processor in SAS® Information Retrieval Studio (Display C7.8). You need to use the "content_categorization" document processor from the list of available processors. This will facilitate facetted search and navigation through the matched concepts or categories found in the e-mail content.

**Display C7.8: Add the "content_categorization" document processor to enhance searching**

2. Upload the compiled categorization and concepts files to the SAS Content Categorization Server using the "Upload Concepts" & "Upload Categories" options available in the "Build" tab of SAS® Enterprise Content Categorization Studio.

3. Update the .desc file located in the installation path under the descriptors folder (Example: C:\Program Files\Teragram\Teragram Catcon Server\descriptors) and specify the path to the uploaded project files on the server under the models folder. Also, give meaningful names such as "Enron_Categories" and "Enron_concepts" to uniquely identify these projects.

4.  Restart the SAS® Content Categorization Server to make sure all the changes that you have made to the descriptor files are reflected and that the projects are available as document processors in the SAS® Information Retrieval Studio.

5.  You can also customize your search to either include/exclude specific concepts or categories from the available list in the taxonomy you have developed. Display C7.9 shows how various tabs in the document processor wizard for "content_categorization" allow you to do this.

**Display C7.9: Custom select concepts and categories to use in the search**

6.  After you finish adding the "categories" and "concepts" in the document processor, make sure it is placed between the "parse_xml" and "export_csv" processors that you have already created in Step 1. You can use the "Move up"/"Move Down" buttons available under the "Document Processors" pane. Click "Apply Changes" to make sure all the changes are saved.

7.  Now, go back to the File Crawler and kick start the crawling again. Once the emails are indexed by the SAS indexing server, you can query the indexed Enron emails by using the SAS query interface.

8.  Launch the SAS Query interface using the link provided under the Query Web Server. You will find that the categories and concepts are available to perform facetted search on the parsed Enron emails (Display C7.10).

**Display C7.10: SAS Query interface showing the search results of indexed Enron emails**

# Summary

Investigations and audits are typically accompanied by large volumes of text from multiple sources, and as the investigator, you face time-sensitive deadlines and an overwhelming amount of information. Perhaps one of the most time-consuming tasks is finding relevant information when you need it. SAS® Text Analytics uses natural language processing and pattern matching to facilitate timely fact extraction and data organization, giving you a holistic view of the information pertaining to your case.

SAS empowers intelligent search and advanced text analytics, supporting key functions such as:

- Multisource information retrieval and integration.
- Advanced natural language processing to parse information.
- Term stemming and misspelling identification.
- Content categorization based on statistical and linguistic rules.
- Fact and entity extraction based on part-of-speech tagging and pattern recognition.
- Document filtering.

What does this give you? Powerful tools that aid in the organization, exploration, and root-cause analysis of time-sensitive case evidence. This case study focuses on findings from analysis and information extraction of the Enron email archive. Playing the role of investigator, the goal is to extract key evidence of the Enron Corporation's suspicious accounting practices. The Enron email archive contains more than 500,000 emails from 159 personal email accounts. The task of reading all these emails is unimaginable and impractical, not to mention how difficult it would be to identify patterns among key email accounts. SAS Text Analytics uses both statistical and linguistic technology to parse, explore and categorize the email collection.

By using the Enron emails as a test case, we have identified new opportunities for investigators who need to understand large volumes of data as integrated evidence for their cases. Digging through large volumes of information is now an advantage for investigators because they are able to use SAS® Text Analytics to understand and categorize these documents.

# Case Study 8 Unleashing the Power of Unified Text Analytics to Categorize Call Center Data

Arila Barnes
Saratendu Sethi
Jared Peterson

## Introduction

HP reports more than 2.5 billion customer transactions per year, and health insurer Humana's provider call center handles more than 1 million calls per month. The sheer volume of data makes it cost-prohibitive to rely on humans alone to analyze the information in the call center agents' notes. Companies large and small are eager to find common customer issues early while keeping costs down. SAS® Text Analytics solutions such as SAS® Text Miner and SAS® Content Categorization are invaluable tools for tackling such problems. SAS continually strives to provide easier and speedier access to text analysis of unstructured data, addressing both the volume and the variety of big data business problems. Last year, SAS® Text Miner debuted the Text Rule Builder in support of active learning. This year, SAS is introducing these powerful features in a new Unified Text Analytics Interface (UTAI), which is aimed at the business analyst who wants to understand unstructured data in a more automated and meaningful way.

The SAS® UTAI combines the sophisticated term-clustering statistical methods of SAS® Text Miner with the rule-based natural language processing techniques of SAS® Content Categorization. Although those solutions are usually used by experts, the new web application provides a single, convenient interface that enables both business analysts and experts to interactively discover topics and build categorization models so that they can better respond to the problem of automatically detecting consumer issues in a timely and efficient manner.

This case study uses a call center scenario for a fictitious online printing company to guide you through the process of topic discovery, rule generation, model refinement, and deployment. This example also shows how to use Boolean rules to tune the model and how to use the DS2 procedure to deploy this model. The case study also demonstrates how to view the results of the analysis in SAS® Visual Analytics.

You can easily apply the process outlined in this case study to similar issues in other industries whenever data includes unstructured text that contains valuable information. Patient electronic records, physician notes, police records, and insurance claims are just a few examples in which potentially valuable information is available only in unstructured text.

## Data Description

The scenario in this case study analyzes about 15,000 free-form survey responses from the call center of an online printing company. The goal is to find customers' most common complaints about the company's products. This information enables you to classify future calls more accurately so that a call center agent can better resolve a customer issue on the first try. If the data is originally stored in a directory on the file system, the UTAI application automatically converts the files to text and loads them into a SAS data set library. The document conversion feature supports several common document formats: HTML, PDF, and Microsoft Office formats.

### Step-by-step Software Instruction with Settings/Properties

The UTAI provides a convenient wizard that enables you to quickly set up a project to explore unstructured data.

1.  To create a new project, click the "New Project" icon  to start the New Project wizard. In the wizard (shown in Display 1), type a name for your project in the **Name** field.
2.  To choose the data that you want to explore, select the document type from the **Document Source** list, where you can either browse for files or select a SAS data set from a library. If you select a data set, you can focus on a particular column by selecting it from the **Column** list. Display 1 shows that this scenario focuses on the 'oe_comment' column in the OPREPORTING data set in the TM library. The system is automatically preconfigured with smart defaults for processing natural language in your text. An algorithm randomly splits the specified data into a training set and a testing set for use in this model-building exercise.

**Display C8.1: New Project Wizard**

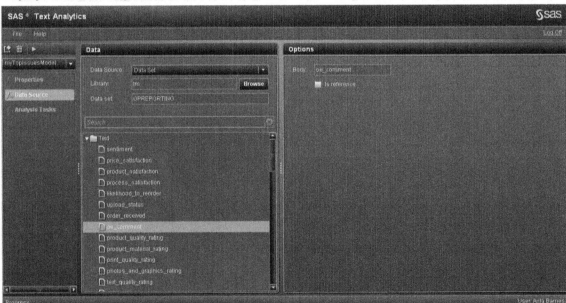

3.  After you create the project, you can run it in the background. Behind the scenes, the software automatically converts the files to text if needed, loads them into data sets, analyzes them for part-of-speech disambiguation and entity extraction, and runs SAS® Text Miner to discover topics from the raw document text. To explore the results, click **Tasks** under the **myTopIssuesModel** project in the left navigation pane. The UTAI suggests topics as shown in Display C8.2.

**Display C8.2: Topics Discovered Automatically**

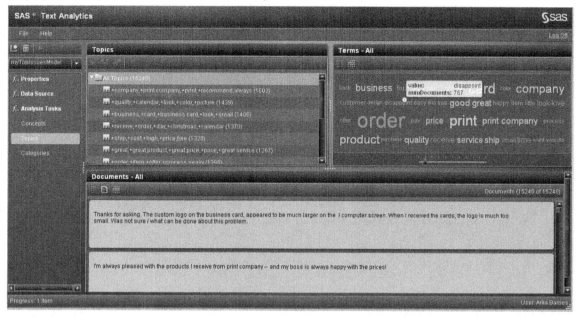

## Examining Topics

You can quickly browse the terms and their associated documents to get a feel for their importance. By default, each generated topic is named by using the five terms that occur most frequently in that set of documents. For each topic, the number of matching documents is displayed.

When you select a topic in the **Topics** pane, you can view the results in the following ways:

- A phrase cloud visualization, as shown in the upper right pane in Display C8.3. If the phrase cloud visualization is not displayed, you can display it by clicking the ⬚ icon.
- A table view, as shown at the bottom of Display C8.3. The table view is designed for expert users and is not described in this case study. If the table view is not displayed, you can display it by clicking the ( ⊞ ) icon.
- A concept map view, which is not shown in Display C8.3. For more information about the concept map, see the section "CONCEPT MAP VISUALIZATION." If the concept map view is not displayed, you can display it by clicking the ⋰ icon.

If you select the topic **+quality,+calendar,+look,+color,+picture** from the **All Topics** list, you see the phrase cloud that is shown in Display C8.3. This phrase cloud shows that "choice," "disappoint," and "business card" are potential issues to use for the classification model. You can control the number of terms displayed in the phrase cloud by moving the slider. Click **Apply** to see the terms highlighted in the **Documents** pane. Both the **Terms** pane and the **Documents** pane are updated based on selections in the **Topics** pane. When you interact with the options in the **Terms** pane, the **Documents** pane content is updated accordingly.

**Display C8.3: Phrase Cloud Visualization**

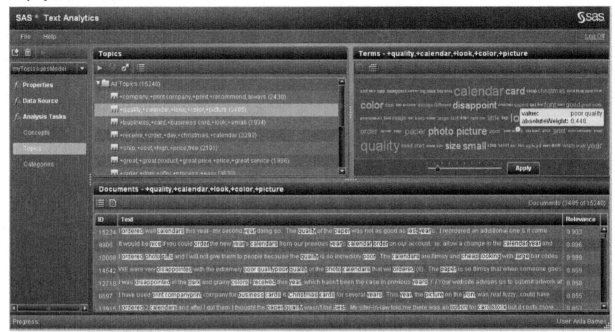

The **Documents** pane helps you understand how these topics are related to your data. You can review the results by investigating the **Relevance** score for each document in the **Documents** pane in various levels of detail, from viewing a quick concordance match, as shown in Display C8.5, to reading the entire document. Also, the **Terms** table and the phrase cloud visualization add more cues to help you decide which topics to use and which ones to ignore.

You can improve the accuracy of the topic discovery process by adjusting the individual term weights or by using "stop term" lists. A "stop term" list indicates which words or phrases to ignore. These are words that do not add value to the analysis; examples are the company name, agents' names, and common phrases. Display C8.4 illustrates how to drop the term "able" in the **Concepts** pane.

**Display C8.4: Keep or Drop Terms**

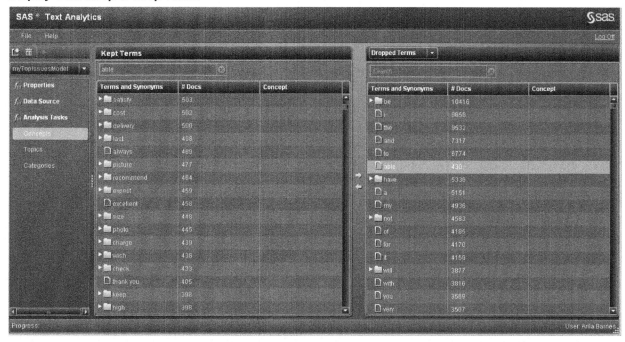

After exploring the matches in your documents, you can decide whether to combine topics or move to rule generation. A quick reading of the matching documents in Display 5 shows the following issues:

- choice of printing quantities
- choice of colors
- blurry photo
- disappointed with quality

**Display C8.5: Matching Documents**

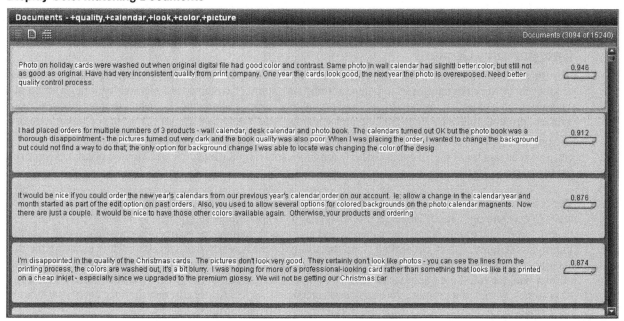

Notice that "a bit blurry" was not highlighted by the automatic topic discovery, so you can add a rule manually to catch that issue in the model. For more information, see the section "Editing or Adding Rules."

## Merging or Splitting Topics

You can merge topics to simplify results when terms are similar. In this example, it makes sense to merge the topics that contain negative terms such as "order." To merge topics, select two or more topics and either click the 🏿 icon or right-click and select **Merge Topics** from the context menu (as shown in Display C8.5).

Alternatively, you can split a topic. To split a topic, select it, and then either click the 🏿 icon or right-click and select **Split Topic** from the context menu.

**Display C8.6: Splitting and Merging Topics**

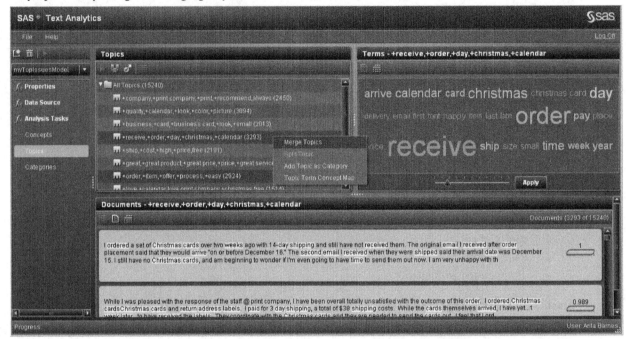

## Categorizing Content

Now that you have examined topics to understand what is going on in the data, you can decide to promote some or all of those topics as categories to begin building taxonomy. Taxonomy is a hierarchical organization of categories that are useful for classifying (and organizing) unstructured data according to your business needs. You can promote topics by right-clicking and selecting **Add Topic as Category** from the context menu or by clicking the 🗗 icon, as shown in Display C8.6. In the **Categories** pane, you can rename and add new categories as you build the taxonomy. The taxonomy captures your business domain and organizes your rules for future categorization.

The research community has approached taxonomic classification through a variety of techniques from the areas of text mining, natural language processing, and computational linguistics. With the explosion of unstructured data, organizations are eager to find automated methods of taxonomic classification in order to improve scalability, speed, consistency, and the ability to consistently reprocess data while ensuring the desired accuracy levels within the subjective context of their business. The UTAI combines the power of automatic text mining with the rule-based approach of SAS® Content Categorization to provide you with finer-grained control of accuracy. When you promote your topics as categories, the UTAI uses the rule-builder functionality of SAS® Text Analytics to automatically generate initial rules in Boolean syntax.

**Display C8.7: Generate Boolean Rules**

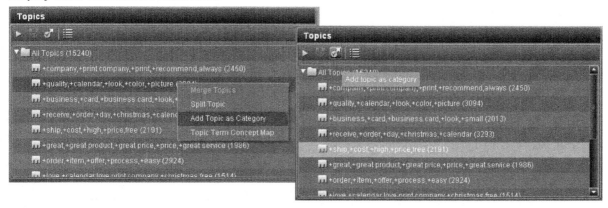

The generated rules can be further edited and reviewed in the Categories view, as shown in Display C8.8.

**Display C8.8: Build the Taxonomy**

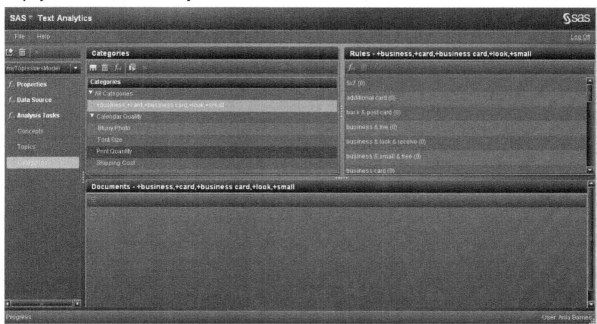

By convention, the generated rule names use "&" for AND Boolean rules, "- " for NOT rules, and spaces for OR rules. You can add new categories by clicking the ▥ icon, and you can add new rules by clicking the ƒₐ icon. You can rename both categories and rules, and you can arrange them hierarchically to build the taxonomy. A list of Boolean rules defines each category, and matches for the rules are displayed in the **Documents** pane.

## Concept Map Visualization

The concept map is a powerful visual tool that enables you to further tune the process of creating rules. The map helps you apply your domain knowledge by exploring the topics and their associated terms to generate rules that are based only on the relationships that apply to your domain. Display C8.9 shows a concept map for topic "+quality+calendar+look+color+picture." With one glance, you see strong connections with "color", "card," "font size," and "quality." If you explore "color," you see terms such as "size," "small," and "disappoint." There is definitely an issue with the color of the printed cards and the size of the font.

**Display C8.9: Concept Map Visualization**

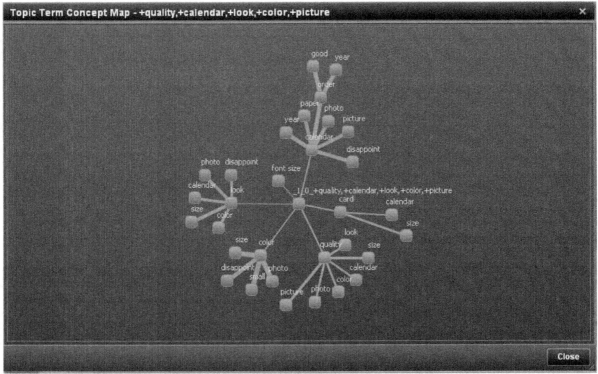

## Editing or Adding Rules

The UTAI uses the advanced linguistic technologies in SAS® Enterprise Content Categorization to provide the following sets of Boolean operators that offer the most flexibility:

- Boolean operators: **AND, OR, NOT**
- counting operators: number of occurrences and count of distinct terms
- proximity operators: operators that are based on word distance and the scope of the sentence or paragraph
- contextual operators: operators that are based on order, XML fields, position within the document, alignment or overlap, and so on

In the **Rules** pane, you can add new rules or edit the ones that are generated by the application. By interacting with the concept map, you can adjust the rules and categories to arrive at the final classification scheme. You can check syntax by clicking the 🖳 icon in the rule editor pop-up window, as shown in Display C8.10. After all the rules are validated, you can build the model again.

**Display C8.10: Syntax Validation**

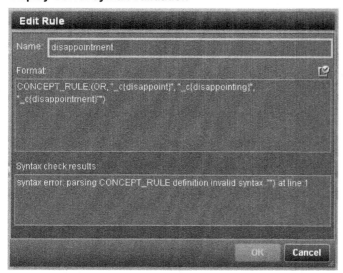

Earlier you noticed that issues such as "blurry photo" and "disappointment" were not highlighted by the automatic topic discovery process in Display 5. Certainly, these are issues that should not persist for the printing company customers. You can create simple Boolean rules for these issues, as shown in Display C8.10 and Display C8.11.

**Display C8.11: Create a Simple Boolean Rule**

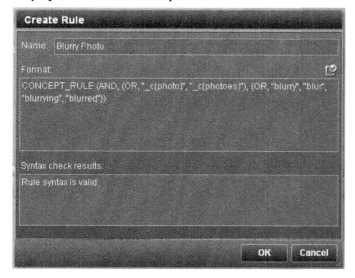

When you are satisfied with all the rules, you can build the complete taxonomy model by clicking the Compile categories icon 📑 as shown in Display C8.12.

**Display C8.12: Rebuild Categories**

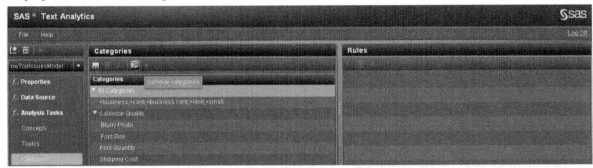

You can find the generated model (rules.li) in the SAS folder for the project. You can import this model to the SAS® Content Categorization Server in order to score new content within its existing SAS® Content Categorization deployment.

## Deploying a Model

When you created the project, you specified a SAS library. When you add new documents to this library, you can score them interactively by rerunning the project for the model that you created. You can deploy this model in more automated environments by using a DS2 program and the DS2 procedure, or you can deploy the model on the grid for big data scenarios. The following section shows how you can use PROC DS2.

## Using PROC DS2 for DEPLOYMENT

DS2 is a new SAS programming language that uses packages and methods to provide data abstraction. DS2 either executes by using PROC DS2 within a SAS session or executes directly within selected databases where SAS® Embedded Process is installed. The DS2 packages included in UTAI are TKCAT and TKTXTANIO. TKTXTANIO is a utility package that is required by TKCAT.

### TKCAT Source Code Sample

The following code illustrates how to use the TKCAT package to apply the model that you build in the SAS UTAI:

```
libname mydata 'C:\SAS Data Sets';
proc ds2;
 require package tkcat; run;
 require package tktxtanio; run;
 /* oe_comment is the column that contains text in your data set */
 table result(drop=(oe_comment status current_concept total_concepts transact
document settings model));

 dcl package tkcat cat(); /*TKCAT categorization engine package */
 dcl package tktxtanio txtanio(); /*Utility package required by TKCAT*/
 dcl binary(8) transact;
 dcl binary(8) document;
 dcl binary(8) settings;
 dcl binary(8) model;
 retain transact;
 retain settings;
 retain model;

 method init();
 /* Create a transaction to use for scoring documents */
 transact = cat.new_transaction();
 /* Create the default settings for the transaction */
 settings = cat.new_apply_settings();
 /* Set the model created in UTAI to apply to new documents */
```

```
model = txtanio.new_local_file('..\mytopissuesmodel\conf\rules.li');
 status = cat.set_apply_model(settings, model);
 if status NE 0 then put 'ERROR: set_binary fails';
 /*    Initialize the categorization engine with your model */
 status = cat.initialize_concepts(settings);
 if status NE 0 then put 'ERROR: initialize_concepts fails';
 end;

method run();
 set mydata.opreporting; /* Data set that contains text to analyze */
 /*Specify document to score */
 document = txtanio.new_document_from_string(oe_comment);
 status = cat.set_document(transact, document);
 if status NE 0 then put 'ERROR: set_document fails';
 /* Apply the model to the document transaction*/
 status = cat.apply_concepts(settings, transact);
 if status NE 0 then put 'ERROR: apply_concepts fails';
/* Iterate for each concept to get details */
 total_concepts = cat.get_number_of_concepts(transact);
 current_concept = 0;
 do while (current_concept LT total_concepts);
 myterm = cat.get_concept(transact, current_concept);
 tag = cat.get_concept_name(transact, current_concept);
 parent = cat.get_parent(transact, current_concept);
 sentence = cat.get_sentence(transact, current_concept);
 output;
 current_concept = current_concept+1;
 end;
 /* Reset document transaction for each document */
 cat.clean_concepts(settings, transact);
 txtanio.free_object(document);
 end;

method term();
 /* Clean up variables*/
 cat.free_transaction(transact);
 cat.free_apply_settings(settings);
 txtanio.free_object(model);
 end;
 run;
```

# Integrating with SAS® Visual Analytics

The UTAI web application has several built-in interactive reports, including the concept map and the phrase cloud. In SAS® Visual Analytics you can easily create additional reports from the data that are scored by the UTAI. From the SAS® Visual Analytics hub, you can select **Create Report** in the SAS® Visual Analytics Designer. Display C8.13 shows how to access SAS® Visual Analytics Designer to build a custom report. You can build a simple histogram, as shown in Display C8.14, to display the main categories into which new documents are placed when the model is applied to new data, and you can compare that histogram to the similar report on previous data. This type of report is useful for monitoring whether certain issues persist.

**Display C8.13: SAS® Visual Analytics Hub**

**Display C8.14: Custom Report Frequency of Root Cause**

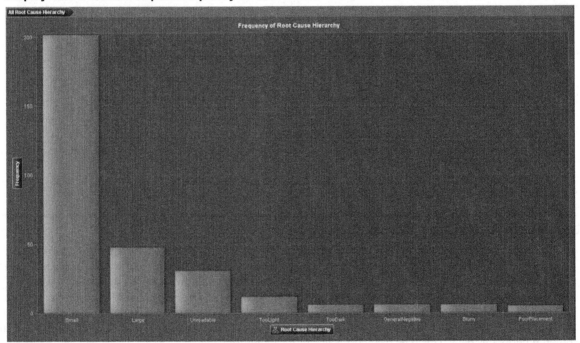

## Summary

Text is a largely unused asset in many organizations. Firms need to interpret, summarize, and report on information that is contained in documents. This case study demonstrates the convenience of a single user interface in a familiar call center scenario. It also helps you understand the benefits of a single web application that provides a framework for interactively discovering and building a content categorization model. In a single installation, the SAS® Unified Text Analytics Interface enables multiple users to access the combined power of the algorithms in SAS® Text Analytics and SAS® Content Categorization. It is a complementary solution in the

SAS® Text Analytics product suite that is also tightly integrated with SAS® High-Performance Analytics technologies. SAS® Unified Text Analytics Interface is formally planned to be released under the name SAS® Contextual Analysis in Q3 2013.

Although this case study focuses on call center data, the same principles apply to the collection of unstructured data in any organization. Furthermore, the rich analytical tools that SAS offers can augment the analysis of unstructured text to help organizations understand the virtues of moving beyond reporting to proactive, forward-looking business analytics that reduce uncertainty, predict with precision, optimize performance, and minimize risks in their business.

# Case Study 9 Evaluating Health Provider Service Performance Using Textual Responses

Gary Gaeth
Satish Garla
Goutam Chakraborty

## Introduction

The service industry has always relied on customer feedback for improving day-to-day business operations. Customer feedback is collected through surveys sent to the customer via mail, telephone and email. A survey can be used to collect different types of information. Typically a large portion of a survey contains closed-ended questions. When collecting service feedback, these questions expect customers to rate their experience on a scale, for example a Likert-scale. In addition, a small portion of the survey questionnaire is devoted for open-ended questions where a customer is allowed to write about things not included in the survey. Until the last decade these textual comments, commonly referred as unstructured data, was considered junk that was eating away storage space. With the growth of text analytics technology, companies have realized the value of textual responses. Companies across industries are successfully answering various business questions using text analytics applications like information retrieval, trend analysis, sentiment analysis, etc.

Sentiment analysis is used to classify a document as either positive, negative, or neutral based on the valence of the writer's opinion in the text. In many situations a single textual comment such as a product review cannot be classified as purely positive or purely negative. A reviewer might be talking about two different features of a product where he expresses satisfaction with one of the features and is disappointed with another. Extracting sentiment at the sentence level would require sophisticated tools like SAS® Sentiment Analysis Studio. However in the presence of data with pre-classified sentiment at the document level, a model can be developed that can be used to understand reasons for positive and negative sentiment and score new comments for sentiment without manual intervention. The Text Rule Builder node in SAS® Text Miner can be used to develop rules that identify the reasons for sentiment in the text. These rules can be used as a starting point for further sentiment analysis using SAS Sentiment Analysis Studio or for content categorization.

In this case study we explore patient response collected via a survey by one of the largest university hospitals in USA. For privacy concerns the name of the hospital in the text is changed to "UCARE". Names of places, persons and other entities are also anonymized in the data. The response to the question 'What was your overall experience with the service' is analyzed using SAS® Text Miner. The textual responses analyzed in this study are related to one of the hospital departments and were collected during a specific time period. Each response was manually classified as positive, negative, or neutral by experts. Whenever a comment had both a positive and negative sentiment, it was rated as neutral. This classification will be used as the target variable for training a model that can be used to classify new survey data. The Text rule builder node in SAS® Text Miner extracts rules from text that can further be used in SAS® Content Categorization Studio.

The data used in this case study cannot be provided with this book due to confidentiality reasons. However by following below steps you can analyze textual feedback where the comments are pre-classified into categories such as positive, negative and neutral.

1. Create a new project in SAS® Enterprise Miner and create a data source for the UCARE data.
2. Select the roles for the variables as shown in the Display C9.1.

**Display C9.1 Variables list window in data source creation**

| Name | Role | Level | Report |
|------|------|-------|--------|
| OVL_CMT | Text | Nominal | No |
| Rating | Target | Nominal | No |
| survey_id | ID | Nominal | No |

3. The variable OVL_CMT has the textual response and the variable rating has the values positive, negative, or neutral.
4. Create a diagram and drag the UCARE data set to the diagram workspace.
5. Connect a data partition node and the set the partitions to 70:30:0 as shown in Display C9.2.

**Display C9.2 Data partition node property panel**

| Train | |
|-------|--|
| Variables | ... |
| Output Type | Data |
| Partitioning Method | Default |
| Random Seed | 12345 |
| Data Set Allocations | |
| Training | 70.0 |
| Validation | 30.0 |
| Test | 0.0 |

6. Connect a text parsing node to the data partition node and change the settings as below:
   a. Parts of Speech: Select as shown in Display C9.3 and click OK

**Display C9.3 Ignore parts of speech window**

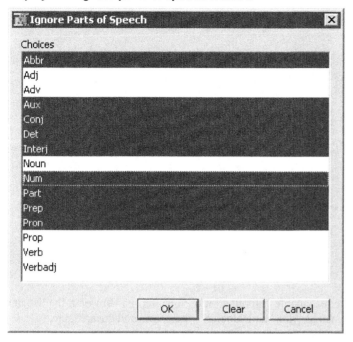

    b.   Ignore Types of Entities: Select all except Person, Product and Prop_Misc as shown in Display C9.4 and click OK.

**Display C9.4 Ignore types of entities window**

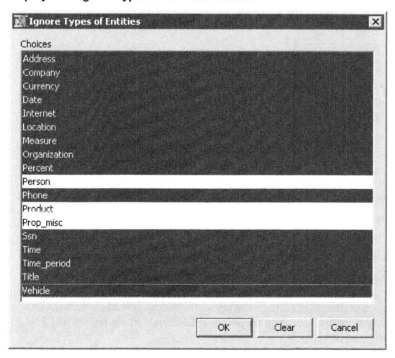

7.   Run the text parsing node and click on the results after the completion of the run.
8.   Sort the Terms table in the results window by clicking the keep column twice to view the terms with keep status 'Yes' as shown in Display C9.5.

**Display C9.5 Terms table from text parsing results window**

| Term | Role | Attribute | Freq | # Docs | Keep ▼ | Parent/Child Status | Parent ID | Rank for Variable numdocs |
|------|------|-----------|------|--------|--------|---------------------|-----------|---------------------------|
| + good | ... | Alpha | 238 | 222Y | | + | 28 | 4 |
| + experience | ... | Alpha | 123 | 119Y | | + | 63 | 7 |
| + care | ... | Alpha | 119 | 110Y | | + | 108 | 9 |
| + clinic | ... | Alpha | 122 | 110Y | | + | 10 | 9 |
| + doc | ... | Alpha | 118 | 102Y | | + | 112 | 11 |
| + wait | ... | Alpha | 137 | 101Y | | + | 21 | 12 |
| + time | ... | Alpha | 106 | 97Y | | + | 31 | 13 |
| + staff | ... | Alpha | 84 | 80Y | | + | 236 | 15 |
| ucare | ... | Alpha | 75 | 64Y | | | 357 | 17 |
| + hospital | ... | Alpha | 64 | 62Y | | + | 237 | 19 |
| + patient | ... | Alpha | 61 | 57Y | | + | 310 | 21 |
| + visit | ... | Alpha | 65 | 56Y | | + | 43 | 23 |
| always | ... | Alpha | 58 | 55Y | | | 70 | 24 |
| excellent | ... | Alpha | 55 | 55Y | | | 1 | 24 |
| + great | ... | Alpha | 56 | 52Y | | + | 427 | 26 |
| bad | ... | Alpha | 44 | 43Y | | | 62 | 28 |
| + good experience | ... Noun Group | Alpha | 40 | 40Y | | + | 216 | 31 |
| + appointment | ... | Alpha | 43 | 39Y | | + | 541 | 33 |
| + feel | ... | Alpha | 43 | 39Y | | + | 529 | 33 |
| helpful | ... | Alpha | 39 | 39Y | | | 238 | 33 |

The terms retained in the analysis are reasonable without a lot of noisy terms. However, the list can be improved by identifying synonyms such as "doc" and "doctor", "hospital" and "ucare", etc.

9.  Connect a text filter node to the text parsing node and change the settings as shown in Display C9.6.

**Display C9.6 Text filter node train properties panel**

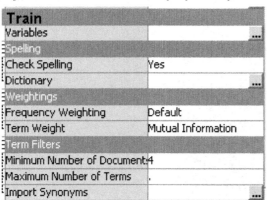

Since you have a target variable, the Mutual Information term weighting technique can be used to derive meaningful weights to the terms.

10. Run the text filter node.
11. After completion of the run, click on the ellipsis next to filter viewer in the property panel.

This is the step where an analyst spends significant amount of time processing the terms. Primary tasks performed at this stage are excluding irrelevant terms and creating custom synonyms. It is difficult to list all the operations performed at this stage. The following steps discuss few of those tasks and you are provided with a custom synonym list for use. You might see a difference in the results which is due to the difference in the terms dropped or kept.

12. In the Terms table, sort the KEEP column by kept terms. Select the terms "ucare" and "hospital", then right-click and select Treat as synonyms as shown in Display C9.7. In most of the comments ucare is referred to as hospital by patients. However, there could be instances when the hospital is used to refer to other hospitals. This is usually the choice left to the analyst to create such synonyms. Also, by

reading through the list you will see short forms of terms such as 'appt' for appointment and 'doc' for doctor. You can treat all these terms as synonyms.

**Display C9.7 Terms table in interactive filter viewer of the text filter node**

| | TERM | FREQ | # DOCS | KEEP ▼ | WEIGHT | ROLE | ATTRIBUTE |
|---|---|---|---|---|---|---|---|
| ⊞ | doctor | 81 | 73 | ☑ | 0.103 | Noun | Alpha |
| ⊞ | staff | 71 | 69 | ☑ | 0.135 | Noun | Alpha |
| | ucare | 75 | 64 | ☑ | 0.073 | Noun | Alph |
| ⊞ | hospital | 64 | 62 | ☑ | 0.077 | Noun | Alph |
| | excellent | 55 | 55 | ☑ | 0.21 | Adj | Alph |
| | always | 58 | 55 | ☑ | 0.148 | Adv | Alph |
| ⊞ | patient | 55 | 51 | ☑ | 0.242 | Noun | Alph |
| ⊞ | wait | 55 | 48 | ☑ | 0.41 | Verb | Alph |
| ⊞ | visit | 48 | 46 | ☑ | 0.36 | Noun | Alph |
| | good | 47 | 46 | ☑ | 0.197 | Noun | Alph |
| ⊞ | nurse | 43 | 42 | ☑ | 0.167 | Noun | Alph |
| ⊞ | good experience | 40 | 40 | ☑ | 0.192 | Noun Group | Alph |
| | helpful | 39 | 39 | ☑ | 0.179 | Adj | Alph |

Context menu overlay:
- Add Term to Search Expression
- Treat as Synonyms
- Remove Synonyms
- Keep Terms
- Drop Terms
- View Concept Links
- Find
- Repeat Find
- Clear Selection
- Print...

You can work through the whole list yourself or you can start with some of the synonyms you already created in other projects.

13. Copy the synonym data set to the library that you created for this project.
14. In the text parsing node property panel, click on the ellipsis next to Synonyms.
15. Click on Import, then click on the library that you create for the project
16. Select the data set "Syn" and click OK.
17. You will see all the synonyms imported into the custom synonyms window as shown in Display C9.8.

**Display C9.8 Custom synonyms pop-up window**

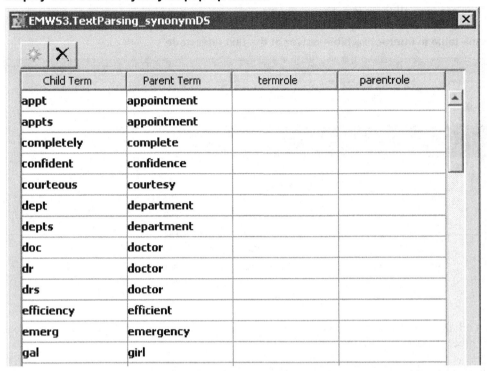

| Child Term | Parent Term | termrole | parentrole |
|---|---|---|---|
| appt | appointment | | |
| appts | appointment | | |
| completely | complete | | |
| confident | confidence | | |
| courteous | courtesy | | |
| dept | department | | |
| depts | department | | |
| doc | doctor | | |
| dr | doctor | | |
| drs | doctor | | |
| efficiency | efficient | | |
| emerg | emergency | | |
| gal | girl | | |

18. Connect a text rule builder to the text filter node as shown in Display C9.9 and rename the node to "TRB – Default".

**Display C9.9 Text mining diagram process flow**

19. Leave the properties of the rule builder node to default values. All the three properties, generalization error, purity of rules and, exhaustiveness have a default value of **Medium**.

In the results window (See Display C9.10), you will find typical data mining model fit statistics since this is similar to a structured model with a target variable.

**Display C9.10 Text rule builder node results window**

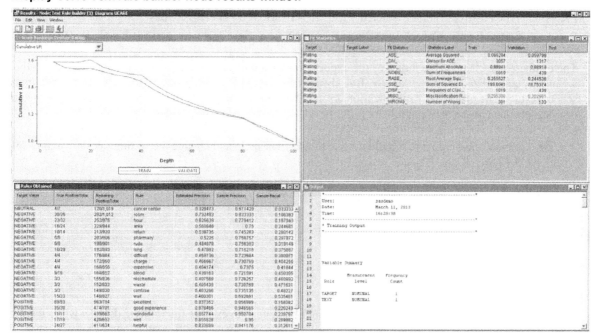

From the fit statistics table, you will find that the model misclassification rate in training and validation data is 29% and 30%, respectively. With a three-level target variable, the misclassification rate around 30% may sound reasonable. Other important results of the text rule builder node are the rules extracted from text. The rules are nothing but key terms that were identified to be significantly associated with a particular level of target variable. These rules are listed in the 'Rules Obtained' table. You will see terms hour, room, pharmacy, rude, etc. being identified as rules for the target level negative. In this data, all the rules are made of single terms. Rules could also be conjunctions of terms and their negations. The order of rules listed in the table is very important. The second rule in the table is extracted using the documents that were not satisfied with the first rule. Similarly the third rule is extracted using documents that were not covered by the first two rules. On a scoring data set, the rules are applied in the same order.

The rules table includes valuable statistics for each rule which indicate rule generalizability. Consider the rule, "rude" (See Display C9.11). The "Remaining Positive/Total" column has a value 198/901. This value indicates that the rule is extracted using 901 documents which are not covered by any of the previous rules in the table and among these documents there are 198 documents with level Negative. The "True positive/Total" column has a value 6/8 for this rule. This means, of the 901 documents, there are 8 documents that had the term "rude" and 6 of these documents have the target as "Negative." Using these values, you can derive precision and recall statistics. Precision measures the fraction of predicted documents that are true positives, and recall measures the fraction of actual documents that are true positives. Both these statistics use the results of the rules in the table up to the current rule.

**Display C9.11 Rules obtained from text rule builder with default settings**

| Target Value | True Positive/Total | Remaining Positive/Total | Rule | Estimated Precision | Sample Precision | Sample Recall |
|---|---|---|---|---|---|---|
| NEUTRAL | 4/7 | 120/1,019 | cancer center | 0.329473 | 0.571429 | 0.033333 |
| NEGATIVE | 30/36 | 282/1,012 | room | 0.732483 | 0.833333 | 0.106383 |
| NEGATIVE | 23/32 | 252/976 | hour | 0.626639 | 0.779412 | 0.187943 |
| NEGATIVE | 16/24 | 229/944 | area | 0.560646 | 0.75 | 0.244681 |
| NEGATIVE | 10/14 | 213/920 | return | 0.538735 | 0.745283 | 0.280142 |
| NEGATIVE | 5/5 | 203/906 | pharmacy | 0.5225 | 0.756757 | 0.297872 |
| NEGATIVE | 6/8 | 198/901 | rude | 0.484878 | 0.756303 | 0.319149 |
| NEGATIVE | 16/29 | 192/893 | long | 0.47892 | 0.716216 | 0.375887 |
| NEGATIVE | 4/4 | 176/864 | difficult | 0.469136 | 0.723684 | 0.390071 |
| NEGATIVE | 4/4 | 172/860 | charge | 0.466667 | 0.730769 | 0.404255 |

For the rule, "rude",

Precision: True Positives/ Total Predicted = (30+23+16+10+5+6)/119=0.756303

Recall: True Positive/ Total Actual: 90/282 = 0.319149

The estimated precision value indicates the expected precision of this rule in the hand-out data set using the setting for generalization error property. It is always advised to play with property settings and explore multiple sets of results. The properties generalization error and exhaustiveness control model over train, while the purity of rules lets you choose between few high-purity rules vs. many low-purity rules.

20. Connect another text rule builder node to text filter node and change the settings as shown in Display C9.12.

**Display C9.12 Text rule builder node train property panel**

| Train | |
|---|---|
| Variables | ... |
| Generalization Error | Very Low |
| Purity of Rules | Very Low |
| Exhaustiveness | Very High |

This selection of settings will work well when your objective is to explore the training data set without worrying about model performance on hold-out data. Selecting low on the generalization error and exhaustiveness properties will over train the model (fits very well in the training data). And selecting low on the purity of rules will extract rules that handle most terms and could generate very long rules.

21. After completion of the run, open the results window of the text rule builder node and view the results The fit statistics table shows improvement in the misclassification rate in the training data set with misclassification at 27.38%. However, the misclassification rate in the validation data set is 37.13%. This is clearly due to model over-training. The rules obtained table shows more rules compared to the rules extracted with default settings. For Target Value of negative, you will find one of the rules as "wait & ~excellent & ~always & ~care & hour" as shown in Display C9.13. Clearly, patients are unhappy with the wait times. Comments that do not have the terms 'excellent, always, care', but have the terms 'wait' and 'hour' contain negative sentiment. Similarly, you can change the settings to other values and explore your results for more insights.

**Display C9.13 Rules obtained from the text rule builder with modified settings**

| Target Value | True Positive/Total | Remaining Positive/Total | Rule | Estimated Precision | Sample Precision | Sample Recall |
|---|---|---|---|---|---|---|
| NEUTRAL | 5/17 | 102/981 | few | 0.274103 | 0.4 | 0.184874 |
| NEUTRAL | 10/33 | 97/964 | visit & ~care & ~ucare & ~experience & ~staff | 0.26665 | 0.363636 | 0.268908 |
| NEUTRAL | 11/56 | 87/931 | appointment & ~ucare | 0.190501 | 0.298611 | 0.361345 |
| NEGATIVE | 26/29 | 228/875 | room | 0.855521 | 0.896552 | 0.114035 |
| NEGATIVE | 5/5 | 202/846 | pharmacy | 0.782506 | 0.911765 | 0.135965 |
| NEGATIVE | 8/8 | 197/841 | wait & ~excellent & ~always & ~care & hour | 0.776784 | 0.928571 | 0.171053 |
| NEGATIVE | 7/8 | 189/833 | restroom | 0.745378 | 0.92 | 0.201754 |
| NEGATIVE | 4/4 | 182/825 | difficult | 0.740202 | 0.925926 | 0.219298 |
| NEGATIVE | 4/4 | 178/821 | wrong | 0.738936 | 0.931034 | 0.236842 |
| NEGATIVE | 4/4 | 174/817 | expensive | 0.737658 | 0.935484 | 0.254386 |
| NEGATIVE | 3/3 | 170/813 | lack | 0.683641 | 0.938462 | 0.267544 |
| NEGATIVE | 3/3 | 167/810 | cold | 0.682469 | 0.941176 | 0.280702 |
| NEGATIVE | 3/3 | 164/807 | disappoint | 0.681289 | 0.943662 | 0.29386 |
| NEGATIVE | 7/9 | 161/804 | colonoscopy | 0.672773 | 0.925 | 0.324561 |
| NEGATIVE | 3/3 | 154/795 | time & ~good & home | 0.661659 | 0.927711 | 0.337719 |
| NEGATIVE | 4/5 | 151/792 | later | 0.625902 | 0.920455 | 0.355263 |
| NEGATIVE | 3/3 | 147/787 | wait & ~excellent & ~always & patient | 0.619833 | 0.923077 | 0.368421 |
| NEGATIVE | 11/15 | 144/784 | wait & ~excellent & ~always & long | 0.614344 | 0.896226 | 0.416667 |
| NEGATIVE | 5/7 | 133/769 | scan | 0.593989 | 0.884956 | 0.438596 |
| NEGATIVE | 5/7 | 128/762 | rude | 0.592884 | 0.875 | 0.460526 |

The list of rules extracted provides excellent insights on the key service aspects of the hospital that are influencing patient experience. From the two rule builder node results, you can clearly see that the patients seem to have negative experiences with restroom, rescheduling, pharmacy, etc. And they are happy with the care, staff, overall experience, etc.

However, there are terms that need to be further investigated for understanding the key service aspects. For example, just the term "colonoscopy", shown in Display C9.13, that shows up as a rule in the results for negative rules does not provide sufficient information on what is going wrong with colonoscopy procedures. The text filter node can help in searching for documents that contain the term "colonoscopy". Looking at these documents can help in understanding the negative issues with this procedure.

22. Open the Interactive Filter viewer of the text filter node.
23. In the search box available in the top section of the report type '>#colonoscopy' and click apply as shown in Display C9.14.

**Display C9.14 Search box of interactive filter viewer**

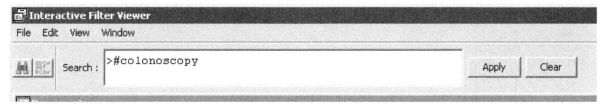

24. While searching for documents, including the symbols '>#' before the term will retrieve documents that that include the term or any of the synonyms that have been assigned to the term.
25. Right-click on the results and click 'toggle full-text' to view the full text

The results show that negative issues with colonoscopy are related to staff at registration, delay in sending results and wait time. Similarly, other rules from the rule-builder node can be further explored using the search feature in text filter node.

26. In the diagram, from the properties panel of the rule builder node with modified settings, click on the ellipsis next to Content Categorization code.

Here you will find the code that can be used with SAS® Content Categorization Studio. This code lists all the rules as Boolean rules with the synonyms and stemmed terms embedded in the rules as shown in Display C9.15. You can both save the code or copy and paste it in SAS® Content Categorization Studio. Refer to Case Study 2, "Automatic Detection of Section Membership for SAS Conference Paper Abstract Submissions" for a detailed demonstration on the utility of SAS® Content Categorization Studio in enhancing the performance of a classification model built using the text rule builder node.

**Display C9.15 Boolean rules generated by the text rule builder node**

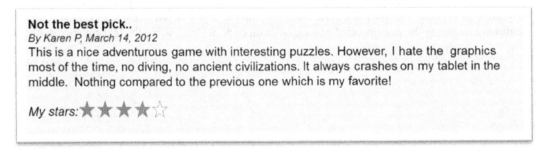

The other very useful feature of the rule-builder node is the active-learning functionality. In cases where humans are involved in rating textual comments, there is very high likelihood of wrongly classifying a comment either purely due to manual error or misinterpretation of the comment by the expert. A similar situation is commonly observed when using product reviews by users for sentiment analysis. In supervised sentiment analysis a product review's numeric rating is used to classify the sentiment of the comment. For example, consider the review for a newly launched video game as shown in Display C9.16. The reviewer seems to have more negatives than positives. However, he tends to give 4 stars for the product. The rating might have been influenced by his love for an earlier version of the game. Considering a rating above 4 as positive will be misleading. This will result in the extraction of wrong rules. Mistakes such as this can be corrected using the change target feature of the text rule builder node.

**Display C9.16 Sample video game review**

> **Not the best pick..**
> By Karen P, March 14, 2012
> This is a nice adventurous game with interesting puzzles. However, I hate the graphics most of the time, no diving, no ancient civilizations. It always crashes on my tablet in the middle. Nothing compared to the previous one which is my favorite!
>
> My stars: ★★★★☆

27. Click the ellipsis button next to Change Target Values property of the rule builder node with default settings.

The change target value table lists only wrongly classified comments. Often you will find that some comments should have been rated as negative, but were actually rated as positive by experts. Those types of mistakes can now be modified by changing the value in the last column to NEGATIVE from POSITIVE.

After correcting few target values and rerunning the node, the misclassification rates are changed as shown in Display C9.17. The change has improved the misclassification rate in training data slightly but made it poorer in the validation data set. Slight changes in the target variable can bring in significant changes in the results.

**Display C9.17 Fit statistics with modified target values**

| Target | Target Label | Fit Statistics | Statistics Label | Train | Validation | Test |
|--------|--------------|----------------|------------------|-------|-----------|------|
| Rating | | _ASE_ | Average Squared ... | 0.062506 | 0.059107 | |
| Rating | | _DIV_ | Divisor for ASE | 3057 | 1317 | |
| Rating | | _MAX_ | Maximum Absolute... | 0.998637 | 0.959627 | |
| Rating | | _NOBS_ | Sum of Frequencies | 1019 | 439 | |
| Rating | | _RASE_ | Root Average Squ... | 0.250012 | 0.243119 | |
| Rating | | _SSE_ | Sum of Squared Er... | 191.0805 | 77.84341 | |
| Rating | | _DISF_ | Frequency of Clas... | 1019 | 439 | |
| Rating | | _MISC_ | Misclassification R... | 0.2895 | 0.328018 | |
| Rating | | _WRONG_ | Number of Wrong ... | 295 | 144 | |

28. We will now score the model on a different data set "ucare_score". This data set only has an ID variable and a text variable.
29. Connect a score node to the rule builder node with default settings (TRB – Default)
30. Drag and drop the score data source on to the diagram and connect it to the score node as shown in Display C9.18.

**Display C9.18 Partial diagram process flow with the score node and score data set**

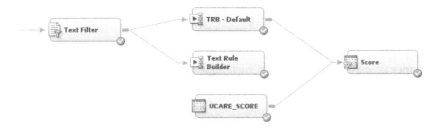

31. Run the score node.
32. After the completion of the run, click on the ellipsis button next to exported data property from the score node property panel.
33. Highlight score and click on "Browse" as shown in Display C9.19.

**Display C9.19 List of exported data sets from score node**

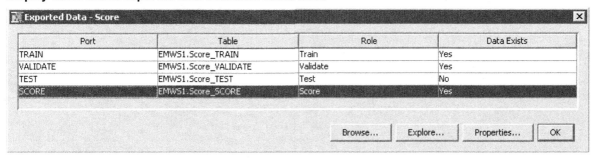

| Port | Table | Role | Data Exists |
|------|-------|------|-------------|
| TRAIN | EMWS1.Score_TRAIN | Train | Yes |
| VALIDATE | EMWS1.Score_VALIDATE | Validate | Yes |
| TEST | EMWS1.Score_TEST | Test | No |
| SCORE | EMWS1.Score_SCORE | Score | Yes |

You will find the documents in the score data set scored as positive, negative, or neutral in the Into: Rating column. This way the model can be used to identify the sentiment in any data set using key rules extracted from training data set.

In the presence of a target variable, another approach to build a predictive model with textual data is by using the text topic node/text cluster node along with any typical modeling technique. Text topics from a topic node or SVDs from the text cluster node can be used as inputs to a decision tree or neural network or any modeling node. However, a text rule builder provides an added advantage of having an active-learning facility and generates Boolean rules that can be used as starting points for content categorization. In this data set, text rule builder actually outperforms basic models built using the text cluster and text topic node as shown next.

34. Attach the following nodes: Text Cluster, Text Topic (from Text Mining Tab), Metadata (from Utility Tab), Decision Tree, Regression, Decision Tree, and Neural Network (from Model Tab) as shown below to the Text Filter node (See Display C9.20).

**Display C9.20 Text Mining process flow diagram**

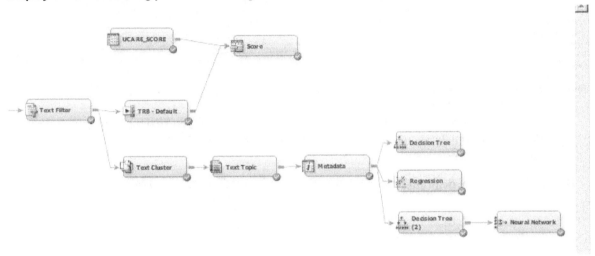

35. Make the following changes to the properties panel of each node (these changes are made based on trial-and-error with the data and using domain knowledge to get meaningful results).
    a.  Text Cluster: Max SVD Dimensions 25, Number of Clusters 8
    b.  Text Topic: Number of multi term topics 15
36. Run the flow from the Text topic node.

The Metadata node is useful in changing the roles of variables in the middle of a flow diagram. The Text Cluster and Text Topic nodes have created many different variables (such as cluster membership, topic flags etc.) that are now available as potential input variables in any predictive model.

37. In the Metadata node make following changes (click on the ellipsis button for variables under train) in the variable roles as shown in Display C9.21.

**Display C9.21 Changing Variable Roles via Metadata Node**

| Name / | Hidden | Hide | Role | New Role | Level | New Level | New Order | New Report |
|---|---|---|---|---|---|---|---|---|
| _dataobs_ | N | Default | ID | Default | Interval | Default | Default | Default |
| _DOCUMENT_ | N | Default | ID | Default | Nominal | Default | Default | Default |
| OVL_CMT | N | Default | Text | Rejected | Nominal | Default | Default | Default |
| Rating | N | Default | Target | Default | Nominal | Default | Default | Default |
| survey_id | N | Default | ID | Default | Nominal | Default | Default | Default |
| TextCluster_cluster_ | N | Default | Segment | Input | Nominal | Default | Default | Default |
| TextCluster_prob1 | N | Default | Rejected | Default | Interval | Default | Default | Default |
| TextCluster_prob2 | N | Default | Rejected | Default | Interval | Default | Default | Default |
| TextCluster_prob3 | N | Default | Rejected | Default | Interval | Default | Default | Default |
| TextCluster_prob4 | N | Default | Rejected | Default | Interval | Default | Default | Default |
| TextCluster_prob5 | N | Default | Rejected | Default | Interval | Default | Default | Default |
| TextCluster_prob6 | N | Default | Rejected | Default | Interval | Default | Default | Default |
| TextCluster_SVD1 | N | Default | Input | Default | Interval | Default | Default | Default |
| TextCluster_SVD10 | N | Default | Input | Default | Interval | Default | Default | Default |
| TextCluster_SVD11 | N | Default | Input | Default | Interval | Default | Default | Default |
| TextCluster_SVD2 | N | Default | Input | Default | Interval | Default | Default | Default |
| TextCluster_SVD3 | N | Default | Input | Default | Interval | Default | Default | Default |
| TextCluster_SVD4 | N | Default | Input | Default | Interval | Default | Default | Default |
| TextCluster_SVD5 | N | Default | Input | Default | Interval | Default | Default | Default |
| TextCluster_SVD6 | N | Default | Input | Default | Interval | Default | Default | Default |
| TextCluster_SVD7 | N | Default | Input | Default | Interval | Default | Default | Default |
| TextCluster_SVD8 | N | Default | Input | Default | Interval | Default | Default | Default |
| TextCluster_SVD9 | N | Default | Input | Default | Interval | Default | Default | Default |
| TextTopic_1 | N | Default | Segment | Input | Binary | Default | Default | Default |
| TextTopic_10 | N | Default | Segment | Input | Binary | Default | Default | Default |
| TextTopic_11 | N | Default | Segment | Input | Binary | Default | Default | Default |
| TextTopic_12 | N | Default | Segment | Input | Binary | Default | Default | Default |
| TextTopic_13 | N | Default | Segment | Input | Binary | Default | Default | Default |
| TextTopic_14 | N | Default | Segment | Input | Binary | Default | Default | Default |
| TextTopic_15 | N | Default | Segment | Input | Binary | Default | Default | Default |
| TextTopic_2 | N | Default | Segment | Input | Binary | Default | Default | Default |
| TextTopic_3 | N | Default | Segment | Input | Binary | Default | Default | Default |
| TextTopic_4 | N | Default | Segment | Input | Binary | Default | Default | Default |
| TextTopic_5 | N | Default | Segment | Input | Binary | Default | Default | Default |
| TextTopic_6 | N | Default | Segment | Input | Binary | Default | Default | Default |

Essentially you have changed OVL_CMT to Rejected, TextCluster_cluster_ to Input (from Segment) and all TextTopic_1 to TextTopic_15 to Input (from Segment). TextTopic_1 through TextTopic_15 are binary variables with a default role of segment that needs to be changed to Input for use in the predictive models. The TextTopic_raw1 to TextTopic_raw10 contains the actual weights to derive the binary segment variables with a default weight of Input. It is up to you to either choose to use all of these variables or to use some variable selection techniques to select from these variables.

38. Make following changes in the properties panel of model nodes:
    a. Decision Tree : Assessment Measure to Average Square Error
    b. Regression : Selection Model to Stepwise, Selection Criterion to Validation Error
    c. Decision Tree connected to Neural Network (in this case the Tree is used to select variables): Number of Surrogate Rules to 1, Method to Largest
    d. Neural Network: Model Selection Criterion to Average Square Error
39. Attach a Model Comparison node (from Assess Tab) and connect all of the model nodes (Text Rule Builder, Decision Tree, Regression and Neural Network) to it as shown in Display C9.22.

**Display C9.22 Comparing Multiple Models via the Model Comparison Node**

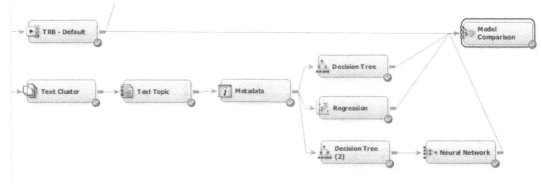

40. Change the Model Comparison Node properties panel Selection Statistic to Average Squared Error and Selection Table to Validation. The results from running the model comparison node are shown in Display C9.23.

**Display C9.23 Model Comparison Results**

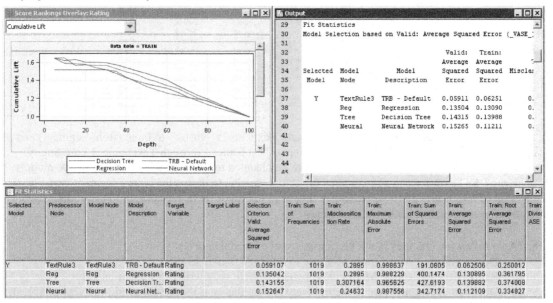

It seems that the text rule builder model has outperformed the other models using the chosen criteria for comparing models. Of course, if you change the selection criteria, the model selected might be different.

## Summary

Rules identified from the Text Rule builder revealed valuable insights on reasons for positive and negative experiences with a hospital visit. These rules are easily interpretable and can serve as a starting point for a sophisticated sentiment analysis exercise. The Boolean rules can also be directly used with SAS Content Categorization Studio for further analysis. From the results it is very clear that the hospital needs to improve on rest room maintenance, rescheduling process, and waiting times. Most of the patients seemed to be satisfied with the care and consultation they received from the physician.

Mistakes by experts in rating textual comments are expected (after all they are human!) and such mistakes may seriously impact model performance. In situations such as these the active-learning functionality in the rule-builder node can be used for improving model performance. In this case study, the text rule builder with active learning adjustments seems to outperform traditional data mining models (such as decision tree, regression, and neural network) built using output from text cluster and text topic nodes.

# Index

CPSIA information can be obtained
at www.ICGtesting.com
Printed in the USA
LVOW03s0134130216
474937LV00001B/2/P